PREVENTING AND SOLVING CONSTRUCTION CONTRACT DISPUTES

PREVENTING AND SOLVING CONSTRUCTION CONTRACT DISPUTES

H. MURRAY HOHNS

 VAN NOSTRAND REINHOLD COMPANY

NEW YORK   CINCINNATI   ATLANTA   DALLAS   SAN FRANCISCO
LONDON   TORONTO   MELBOURNE

Van Nostrand Reinhold Company Regional Offices:
New York   Cincinnati   Atlanta   Dallas   San Francisco

Van Nostrand Reinhold Company International Offices:
London   Toronto   Melbourne

Manufactured in the United States of America

Published by Van Nostrand Reinhold Company
135 West 50th Street, New York, N.Y. 10020

Published simultaneously in Canada by Van Nostrand Reinhold Ltd.

15 14 13 12 11 10 9 8 7 6 5 4 3 2 1

Library of Congress Cataloging in Publication Data

Hohns, H   Murray.
   Preventing and solving construction contract
disputes.

   1. Building—Contracts and specifications—United
States.   I. Title.
KF902.H64   343'.73'078   78-15788
IBSN 0-442-23481-3

Good concrete is made from
Portland Cement, potable water,
and fine and coarse aggregate.

Bad concrete is made from the
same materials.

Both have the same outward characteristics
and require testing by
tension and compression to determine their true value.

Sometimes good concrete must be tested to failure
before its quality is finally established.

PREFACE

Most texts on construction contracts and the problems and disputes that arise from such contracts are written from the legal point of view. Most are authored by attorneys whose orientation is obviously toward the law rather than the business aspect, human interest, or pragmatic elements of the problems experienced.

This text is intended to be different. Certainly it has been written with a curbstone knowledge of the law in mind. The emphasis, however, is not on the law or the lawyer, for in reality they are but tools created in theory for use by people to resolve disputes and problems. No legal references are cited: The emphasis of this effort is on the people involved and the best ways available to them for solutions and possible avoidance of problems they all experience sooner or later.

The thoughts which I share have been gathered over more than a quarter of a century in construction. Since 1965 I have been called in as an expert on hundreds of construction problems, most of which were in the United States, and collectively have involved billions of dollars, uncounted hurts, numerous broken promises, many lifetimes of work ended or temporarily in despair, some wonderful victories in court, a few losses as well, some recoveries of unexpected amounts of money, indeed results good and bad, some fair and unfair, some right, and some seemingly wrong.

My work has caused me to experience or learn the point of view of almost everyone who could be involved in a construction project from an injured unskilled laborer or his widow, to men who are recognized as leaders in the design profession, and presidents of some of the largest contracting firms in the world.

This book is intended to be impartial. I have not set out to favor the view or position of the designer, owner, contractor, subcontractor, surety, insurer, or investor. Furthermore, firms from each of these categories and others have been clients for many years, and individuals in each discipline have become good friends.

I have learned over the years that people are the cause and solution to virtually every construction problem. The man who is the student of people and who has learned to judge and motivate his peers by setting an example, invariably ends up ahead in disputes as well as in other aspects of life.

This book would not be complete without acknowledging the understanding and love of my wife Lili, who has spent many days alone while I traveled the country doing my job; my late Dad who taught me the four skills of communication which have been so valuable; and my partner Blair Wagner who has been everything the word partner infers. Others who have been important to me are Gerard N. Verrier, Robert D. Wallick, William P. Thorn, and John M. Poole.

H. MURRAY HOHNS
Mt. Holly, New Jersey

CONTENTS

1
THE ORIGINS OF CONSTRUCTION DISPUTES

BASIC RULES TO CONSTRUCTION CONTRACTS

The field of construction contracting has a number of significant character-istics, two of which bear mention at the start of this text. Perhaps the most important characteristic is that construction is an art, not a science. It is of all businesses, a people or service business. Second and just as important is that the construction industry is one of the largest dollar-generating seg-ments of the American economy—indeed the world economy.

Thus, individual contracts or projects can and often involve huge sums of money to be expended in near or distant places over long periods of time. Such major undertakings seem romantic, filled with challenge and potential for reward and success. Certainly many find both the challenge and the reward, but in reality the construction contractors and the professionals who design that which is to be built, usually work for less profit and indi-vidual personal compensation than any other profit-based industry.

The successful, serious student of the process of construction contracting and its whys and why-nots must have a grasp of the total concept of what is at stake. This total concept which is so necessary to the profitable func-tioning and understanding of the workings in the construction world, can be more readily focused upon if viewed in the context of a number of important points. These points or rules will greatly aid in the decision-making of the individual confronted with the need to purchase construction services, administer a construction contract, or be responsible for the per-formance of those construction services.

The first and foremost rule is that nothing is absolute in a construction contract. No amount of careful study, choice of words in the contract, architectural treatment, engineering design and drafting, legal review, con-cern, or money will assure absolute execution of the buyer's wishes or the seller's intentions. The specification book may say in great detail what is wanted, the contract documents may define the work unequivocally, everyone may understand all that is to be understood, the money may be ready to be spent, but no guarantee exists or can be obtained to assure that the results desired by anyone can or will be achieved.

Thus, the successful user of the construction industry is the one who can motivate, measure, and utilize or accept the efforts of others for the user's purposes. Furthermore, this person must be able to maintain a realistic judgment of what really is fair on an almost continuous basis in order to accomplish everyone's exit from the project with an acceptable degree of satisfaction. Since nothing is absolute an attitude permitting great flexibility is necessary to make the value judgments needed.

The reality of the first rule that nothing is absolute leads to the second rule or premise in the process of construction contracting: Nothing can be truly established as a fixed relationship or rule. If something worked once, there is no assurance it will work again. If one made and lost an argument last week, the same argument in a different setting may prevail this week. Those who are knowledgeable in arguing contract claims and disputes appreciate there are no hard and fast rules; instead there are principles established in the law and in the custom of the industry, both of which are constantly changing. Indeed, nothing in this text, including the rules in this chapter, can be taken as a hard and fast law. Every assumption made or fact believed in a construction contracting program must be properly regarded in the minds of all concerned as a possible variable before the job is done. Nothing really is certain, and nothing is sure except that where something goes amiss communications will slow down or break down somewhere, making the problem worse.

The rules of the industry go much deeper. Contracts are typically drawn by architects or engineers hired by the purchaser or owner. These design professionals rely on standard preprinted clauses and forms, most of which are found in or are the product of much study by committees appointed from leading members of professional associations, government representatives, specification writers, attorneys; indeed, a whole host of sources contribute to the documents that comprise a construction contract including, and far from least important, the last edition of the standard under study. Much of the effort goes into trying to correct the failures or problems in the earlier edition.

Unfortunately the ultimate interpreter of the contract is not the one who studied, wrote, or read it, but the court or board whose interpretation today tends to follow that which seems reasonable. Thus, in a dispute today, the special contract clause so carefully written to fill all the wishes of the owner by assigning all risk to the contractor often in the ultimate test will fail to do so if the judge, or a stranger, a latecomer to the whole dispute scene, finds the clause unreasonable. Thus, rule 3 states that even though the contract clearly says and requires something highly specific, the clause may very well be unenforceable, particularly if its enforcement is unreasonable, a term whose definition is itself one of arguable imprecision.

Rule 4 flows from rule 3 and gives the answer to the questions which arise from the latter. Why are contract clauses that are risk assigners or responsibility disclaimers attacked? Why can they not be enforced? Why do contractors not take the losses the contracts spell out as their possible risks? The answer is rule 4: One cannot bypass or lay off its own risk to others without being challenged as to the basic question of reasonability. If contested, the one who benefits from avoiding a risk will generally end up paying the costs involved if the risk was not avoided. The owner who has bought a poor piece of ground usually will not be able to pass the risks of the use of that ground to the foundation excavator or builder; it is simply not fair or reasonable. But reasonable or not, it certainly is worth the try by the owner or its designer to pass the risk to the contractor since contract disclaimers always work if they are not challenged and sometimes work even when challenged.

Rule 5 could well be rule 1 for it too is very basic to all construction projects; simply stated it is that there is never enough money to do what is desired or required. This basic lack of money is almost an absolute. Even when everyone involved has enough in his budget, be sure that somewhere in the program something will happen and someone will run out of money and cause all the other rules being cited to come into play.

Rule 6 is that costs will always increase. No one has yet been born who has been able to figure out how to eliminate cost increases. No matter what is started on earth its costs will increase. The only way to eliminate cost increases is to build less, which is a far worse solution than the problem of cost increase. This very complicated and sensitive situation is even more complicated by rule 7, which states that any cost increase is cause for criticism because of an inherent human cry, particularly in America and in its media, that everything acquired must be acquired for less. Everyone wants the best but only a few can truly afford the cost. Since we are led to believe all have the same rights, those in charge are urged by the public voice that the way to make the best available to all and protect the public first is to force prices down.

Rule 8 follows rule 7, the only answer to the urgent need to cut costs is solved by advancing existing technology to the razor's edge. Every major project seems to be seized upon by those in charge as a chance to increase the existing state of the art dramatically and save money by so doing, thus justifying the whole idea of the undertaking. The bigger and more complex the project, the larger the step forward in technology needed, and although unstated by the design profession until its recent inundation of claims for error and omission, the bigger the risk and uncertainties involved. Only one certainty is really safe and sure: The next innovation, interchange, method, material, or control will probably cost more, even if no one admits it and everything spoken or written about the new improved whatever says it saved money.

Rule 9 is that the complex and demanding projects of today require competent experienced people to manage and execute the work, but competent, experienced people are always in short supply. Wisdom is a trait possessed by a limited number of people, most of whom cost more than the salary budgeted. Usually too, the wiser a man gets, the older he gets, until he is old enough to be wise enough not to have to care. Competent, experienced people are not readily available. They are almost never waiting to start a new project and they are never available to finish an assignment totally before they are needed elsewhere to start a new, bigger, and more important project. Needs and challenges at times create competent people, but incompetent people always create needs and the consequence of challenges met but not conquered.

Rule 10 relates to the seemingly constant problem that there is never enough time to do the job. Even if the contract period is far longer than needed and an early finish is foreseen as easy, pressure to get it done sooner will be shortly exerted from all directions: contractor, owner, manager, designer, and supplier. This pressure to get it done sooner goes hand in hand with the usual experience that surplus time is always diminished by frittering away the excess days available until the job cannot be done on time. Then the job gets the right amount of attention.

Rule 11 is almost part of rule 10. It is that no matter how late one gets there is always an excuse which will ultimately get absolution for the delinquent one. This excuse traditionally is never spoken until the blasé optimistic attitude that "we will get it done on time" is so obviously patently absurd it cannot be uttered even by the most supreme ego. At that moment in history the cavalier "leave it to me" becomes (and without a change of expression) "What did you expect? You know I was delayed."

Unfortunately, even with good excuses or bad ones with absolution, time lost is simply gone. It almost always cannot be recovered, and acceleration, double shifting, overtime, more men, and the like are never available from owner, contractor, or designer as a voluntary means to make up lost time. Acceleration always costs more money and never really seems to help as much as it costs.

Rule 12 is very important for it is basic to the United States and is published in Western society as a foundation stone to all standards of fair play. This rule is that the free enterprise system of government and business conduct demands competition. The pragmatism of this demand is translated into open competition by the low bid to a prestated circumstance as supposedly the only fair or the best way for public bodies to purchase their needs. Often ignored initially and sadly is the second part of this rule that competition on a "pure" basis will always result in underbid jobs which coupled with the already existing short supply of money (and the fact that inescapable inflation is here to stay) simply leaves someone in the industry in an untenable situation. Too often at the end of the building project it is the buyer.

Rule 13 says that the industry or its users will never eliminate changes or modifications to contracts. No matter what is said, done, planned, decreed, or demanded, changes will be made. Someone will learn or re-member something that needs correcting of that already done or at least an adjustment to that to be done. Someone will want to make a change, and as soon as word is passed that no change can be made, everyone will find change necessary.

The last basic rule to be noted in this context is that no solution is known to any man dead or alive which can eliminate the problems of a society which is unable to estimate the cost or length of time to construct major projects.

These rules are very real world foundations of the contracting industry. The best professional knows that while the typical construction contract is hopefully written to overcome all the problems that emanate from these 14 basic rules, words on paper just do not always work no matter how well they are written.

The dilemma found by those subject to these rules is compounded by the element of profit. Virtually all who are engaged in the industry are profit-oriented. Invariably, this orientation stems not so much from in-herent avariciousness but from the basic need for survival. Each firm must make a profit to survive, and all the individuals involved in the quest for profit are eager to prove their particular self-worth so that advancement to a higher station in life can be achieved, a very definite essential for one and all to survive as individuals.

Problems in the profit element and related individual ambitions have to occur since the rules say there is not enough money for everyone to make a profit, particularly when there are not enough competent people or time. Whether more money is involved, or more technology in the design or process, or the work is more complex and larger, the difficulty of rendering the intent of the owner on plans, specifications, and contract documents has become so difficult that disputes must arise, and they do. Disputes in today's construction marketplace involve hundreds of millions of dollars.

The volume of disputed dollars creates a substantial amount of litigation and related activities each year. This litigation or due process takes place before the Federal Courts, Procurement Agency Contract Appeal Boards, State Courts, Owner Courts, Arbitration Panels, Special Boards, and the like, and behind all these formal hearings are countless problems, major losses and delays, failures of every description, letters of claim, complaint, and protest, denials, owner and designer review, face-to-face confrontation, negotiation, some settlements, and some surrenders. This entire panorama is a daily occurrence in every major construction job. The volume of disputes and their ultimate consequence has engendered long, complex regulations and rules for their solution, particularly in the federal sector. Indeed, some of the remedies for solving a dispute are as complicated or even more so than the administration of the original contract.

Specialists in contract administration, problem resolution, and dispute or risk analysis have appeared in the past two decades. There is now a National Contract Management Association and a Project Management Institute. Both these organizations have worked to establish the contract manager as a professional, one whose status may someday be recognized along with that of the doctor, lawyer, engineer, and architect. More than 3000 people have been professionally certified by their peers as contract managers. However, the basic purpose of contracting is to construct something, not to provide a professional with an opportunity to manage. Too often the project turns into building a paper file instead of a building. These organizations work in part as clearinghouses to share problems and solutions both real and imagined by their memberships which are daily involved in the world of contracts.

The volume and complexity of contract changes and their ramifications have led to a number of men and women becoming specialists in unraveling contract disputes. Certain attorneys, engineers, and business administrators have earned wide respect as experts in solving disputes.

Nearly everyone in the construction industry is involved in a dispute today. The owner has disputes with contractors, construction managers, and designers. The general or prime contractor has disputes with architects and engineers, clients, subcontractors, insurance companies, suppliers, unions, state laws, federal laws, tax laws—indeed, every area of its business is fraught with disputes. The architect or designer also has disputes with its subconsultants, its clients, the contractors, the manufacturers on whom design concepts are based, and so on. The subcontractor has the same disputes. Woven through all the disputes are the responsibilities of the guarantors: i.e., the products liability insurance carrier; the professional liability insurance carrier who insures the designer for certain types of errors and omissions; the contract surety firm who underwrites the performance, payment, and maintenance bonds required; the workman's compensation and liability insurance carriers; state and federal agencies responsible for safety. All these and more combine as potential disputants over major dollars.

Oddly enough, there are few central sources of data or texts which deal with the field of contract disputes in a comprehensive way. Perhaps this is explained by the fact that a comprehensive text on disputes would have to be Britannical in length because the subject is so broad in scope and myriad in detail that it cannot be easily captured. Yet, although the detail is vast, the individual problems great and diverse, and the procedures for solution varied, most construction disputes are essentially very similar. Their existence and resolution have origins and solutions with much in common.

This text deals with the problems, the procedures, and the practical aspects of the construction dispute.

THE GENESIS OF CONSTRUCTION DISPUTES

Everyone knows that construction disputes or problems originate from a variety of causes, of which perhaps hundreds or thousands of specific ones can be named. However, a survey of more than 300 relatively recent major disputes leads to the conclusion that their causes can be largely traced to five sources.

First, errors, defects, or omissions can exist in the contract documents. This can be a simple or a catastrophic error in plans or specifications on the part of the designer, or can stem from an error in reading or understanding the intention or scope of the contract documents by the person who is the ultimate doer of the work and who must first divine and then meet the intent of the designer. Contract document errors are major sources of construction disputes.

The second origin of construction disputes is the failure of someone to count the costs of the undertaking in its beginnings. It could be that the owner did not properly understand the magnitude of the project's cost and then finds itself in a position in which it must force the situation in order to meet its economic limitations. Or perhaps the designer's fee was not large enough to accomplish the studies, investigations, or whatever is needed to deliver a product timely and satisfactorily to its client. The designer decides it must take short cuts in order to avoid financial loss.

The contractor often will not estimate a job properly: It could fail to provide for means, methods, or materials, and thus become involved in a contract based on a difficult bid. Often a subcontractor will not include everything the general contractor insists or believes was written into the subcontract proposal and agreement, or perhaps the subcontractor made an error. The initial failure to count the cost in a construction program accurately by any and all of the participants is fertile ground in which to grow construction disputes. Builders, subcontractors, suppliers, and designers need to make a profit, as do people in all endeavors. Poor estimates often reduce or eliminate the profit margin, make the party with slim or negative profit projections supersensitive, and lead to a search for every opportunity to create income to recover estimate errors.

Third common cause of dispute is the changed condition. The soil borings or investigations indicate clay, but upon excavation the soil turns out to be ledge rock. The hole is wet, but it was supposed to be dry. Another common cause of changed conditions in the fast track days now in vogue is the failure of the preceding contractor to be done on time. The conditions under which the job was envisioned or bid have seemingly changed. Very often, too, the real problem is that no one really knows who envisioned what except that it was different than the uneconomical situation now being experienced. Visions and prophecy are hard to recall.

Changed condition disputes often begin with the problem of deciding if the condition discovered or experienced is in fact a changed condition. A change of condition can also have an impact on an entire program and cause what are known as ripple damages. More often than not the scope and quantification of ripple damages comprise a dispute of major proportion.

In recent years, consumer reaction has become a fourth major origin of disputes. The rise of tenant and condominium owner associations, and a sense or a reality of public awareness have made the ultimate user of a facility a source of dispute. The rise in the last 20 years in the courtroom and the media of the recognition of individual rights is a great contrast to the earlier doctrines in our land of "buyer beware" or the "sovereignty of the king." The recognition and enforcement of implied duties of the builder have created many concerns not only for those who build and design but also for those who finance and act as insurers and guarantors in the construction industry.

The idea of total liability arising from the right of users or bystanders to rely on products or services which fall short of acceptable levels of performances is a serious origin of construction disputes. Unhappy buyers today know how to seek relief and many are pursuing relief effectively or at least noisily across the country.

The last prime cause of construction disputes is the people involved. The construction business is a people business. It is not a science; it has little repetition or assembly line type of work in the execution of its processes; it is a field in which people new to each other are forced to work together for short periods of interdependent intense activity where costs accumulate rapidly. In the construction industry, people design the buildings, bridges, dams, roads, or whatever is to be built; people draw the plans, write the specifications, read the plans, bid the jobs, estimate the prices, and make certain assumptions, and in so doing put their individual and collective reputations and fortunes on the line; those same people make plans and set all sorts of personal goals. People are the industry's real stock in trade. The practices of the people of the industry, their personal habits and idiosyncrasies, their interpersonal relationships, the positions and power, the hierarchies and tradition that encumber management in all disciplines, the unions, the open shops, the equipment manufacturers, the suppliers, the very basic fact that the construction business is one of reaction to circumstances rather than planning, and the necessity for people to maintain their own security are problems that are all fertile ground for disputes. Finally, the customers of the industry are people as well.

Some people in the industry have learned how to make money from changes and disputes. There is no question that certain management techniques can make delay-oriented disputes an extra source of funds; in some cases, construction claims can and have produced fee type income which is higher than the profit margins contemplated when the project was bid or the profit actually attained during construction. However, undue claims too quickly or too often asserted can lead to dismissal of legitimate claims by a battered owner as more of the same. Contractors are well advised to put forth only claims that are real. Owners are similarly advised to recognize hardships and adjustment which their actions or lack thereof have caused.

2

IN-DEPTH DISCUSSION OF DISPUTE ORIGINS

CONTRACT DOCUMENTS

The Basic Relationships and Standard Forms

There are five sets of contractual relationships which are generally found in a construction program. These include the relationship of the owner to the designer, the designer to another design specialty consultant(s), the owner to the prime contractor(s), the prime contractor to its subcontractor(s), and the prime contractor(s) or subcontractor(s) to supplier(s).

These five basic relationships have been studied over the years by interested individuals as well as professional committees of varied membership from all corners and segments of the industry, along with private and public attorneys, all in all a potpourri of concerned people. The result of these studies has been the publication and wide usage of standard form contract documents, the most popular of which are those distributed by the American Institute of Architects.

The AIA now publishes standard forms as follows:

(Copies of the following Documents, current edition, are readily available. See the latest AIA Catalog and Order Form for a complete listing of AIA Documents and their latest edition.)

Document Number	Title

A-SERIES—
OWNER–CONTRACTOR DOCUMENTS

A101	Owner–Contractor Agreement	Form—Stipulated Sum
A107	Owner–Contractor Agreement (Small Construction Contracts)	Form—Stipulated Sum

A111   Owner-Contractor Agreement   Form—Cost Plus a Fee
A201   General Conditions of the Contract for Construction
A201/SC   General Conditions of the Contract for Construction—Federal Edition
A305   Contractor's Qualification Statement
A310   Bid Bond
A311   Performance Bond and Labor and Material Payment Bond
A401   Contractor-Subcontractor Agreement Form
A501   Bidding Procedures & Contract Awards
A511   Guide for Supplementary Conditions
A701   Instructions to Bidders

B-SERIES—
OWNER–ARCHITECT DOCUMENTS

B141   Owner-Architect Agreement Form
B352   Duties, Responsibilities, and Limitations of Authority of Full-Time Project
              Representative
B431   Questionnaire for the Selection of Architects for Educational Facilities
B551   Statement of the Architect's Services
B707   Owner–Architect Agreement—
            Interior Design Services
B727   Owner–Architect Agreement—
            Special Services
B801   Owner–Construction Manager Agreement

C-SERIES—
ARCHITECT–CONSULTANT DOCUMENTS

C141   Architect-Engineer Agreement Form
C431   Architect–Consultant Agreement Form
C801   Joint Venture Agreement

D-SERIES—
ARCHITECT–INDUSTRY DOCUMENTS

D101   Architectural Area and Volume of Buildings
D200   Project Checklist

E-SERIES—
ARCHITECT–PRODUCER DOCUMENTS

E101   Technical Literature for the Construction Industry

G-SERIES—
ARCHITECT'S OFFICE AND PROJECT DOCUMENTS

G601   Land Survey Requisition
G602   Subsurface Soil Investigation Requisition
G610   Owner's Instructions for Bonds and Insurance
G701   Change Order
G702   Application and Certificate for Payment

G702A   Continuation Sheet for G702
G704   Certificate of Substantial Completion
G705   Certificate of Insurance
G706   Contractor's Affidavit of Payment of Debts and Claims
G706A   Contractor's Affidavit of Release of Liens
G707   Consent of Surety Company to Final Payment
G707A   Consent of Surety to Reduction in or Partial Release of Retainage
G708   Architect's Field Order
G709   Proposal Request
G711   Architect's Field Report
G712   Shop Drawing and Sample Record
G801   Application for Employment
G802   Invoice for Architectural Service
G804   Bid Document Register
G805   List of Subcontractors
G807   Project Directory
G809   Project Data
G810   Transmittal Letter
G811   Employment Record
G813   Temporary Placement of Employees

J-SERIES–
STANDARDS OF PRACTICE

J330   Standards of Ethical Practice
J331   Code for Architectural Design Competitions

All of these are available from the AIA headquarters, Washington, D.C. For reference, AIA forms A101, A107, A111, A201, A201/SC, A401, B141, B801, and C141 are printed in Appendix A. These AIA contract forms are under constant study and revision. New editions appear every few years to answer, include, or exclude various pressures or problems which the review committees believe are important. The AIA contract studies and revisions are concomitant with efforts of other groups. The National Society of Professional Engineers, the American Society of Civil Engineers, The Associated General Contractors of America, the Mechanical Contractors Association, and others all have standard contract forms very similar to those of the AIA but in each instance the specific source has tailored the form to the flavor or finer subtleties important to the particular sponsoring group. These too are also revised from time to time. Appendix B contains a number of these other forms. The serious student of contracts will benefit from studying the differences and developing his own understanding of the thinking or prejudice from which such differences or preferences originated.

The final contract standard forms to be mentioned are those used by the higher levels of government. The federal government buys construction through a number of agencies, for example General Services Administration,

Army Corps of Engineers, Air Force, and Navy, and each agency almost always has its own contract form which to some extent is different from all its sister agencies. Each of the states has its own contract form and more than one contracting agency, and each agency, if of any size or self-esteem, has its own standard form. The same applies to various other government levels as they wind down to smaller forums and groups where the AIA or NSPE contract form is typically used.

In spite of the origins, prejudices, and ambitions of their authors, all these standard contracts essentially set forth very similar rules and have the same kind of structural format. Most construction contracts are short in themselves, but they incorporate by reference many documents which comprise the total agreement of construction.

The Contract Conditions

The documents referenced as part of the contract typically include a front end or "boiler plate" which sets out the general conditions or rules which are to be followed and the roles of those who will be governed by the rules. The front end can and usually will include an invitation to bidders, which is a multipaged statement describing very briefly the project and the salient terms and conditions a prospective bidder must (but occasionally does not) fulfill in order to file a responsive bid. The front end will include a bid form for all bidders to use, thus standardizing the responses of all offerers so that evaluations or comparisons can be rapidly and hopefully unerringly made. Strangely enough the bid form frequently does not provide enough room for the bidder to fill in the information required. Often, the IFB or RFP (federalese for invitation for bid or request for proposal) will require certain information to be listed on the bid form, but no place for the information is provided on the form. Careful review of the bid requirements and the bid form proposed is the first step an owner and its designer should take to avoid construction contract disputes.

Not only does the bid form require owner review, but the entire front end proposed by the designer should be examined by the owner. It is not sufficient to assume that a front end is acceptable because it is in common use on hundreds of millions of dollars in work. The author recently was retained to review the boiler plate typically reprinted over and over on all types of projects by one of the top 50 design firms. It was found to contain inherent contradictions, poor English, antiquated provisions, and all types of ambiguities and confusion which can boggle the very best technical and legal minds. Those who are purchasing the construction must read the boiler plate the designer has prepared; good business would indicate the contract conditions be ready 90 days before the project goes to bid. This will allow time for review and alteration. If there is uncertainty in the reviewer's mind regarding what the rules or general conditions should best be, then a premiere construction (not architect or engineer) consultant, or the largest local contractor, or the nearest Associated General Contractors office should be asked to review the documents.

The most common set of general conditions or standard contract clauses between owner and contractor is that published by the AIA as form A201. The 1976 edition is found in Appendix A.

The next commonly found section in contract documents is the supplemental general conditions, again a section dealing with generic relationships between the parties, but now directed somewhat more to the specifics of the project or the buying agency. Supplemental general conditions also follow a pattern and are often preprinted by all the organizations who have developed their own front end.

After the supplemental or special general conditions, a number of categories of essentially nontechnical provisions are typically found. One such category is often called temporary facilities. Then there are almost always revisions or amendments to the general conditions or supplemental general conditions by which the preprinted forms are adjusted to the specific problems or requirements of the project at hand.

These will or can include specific instructions on phasing, time, or scheduling requirements, allowance items, other contracts, and similar information which the contractor must know in order to bid and perform the job to meet the particular concern of the owner. The temporary facilities could be included in this portion as well.

In projects which are federally funded and to some degree when funded at state or other public levels, the front end also contains what can be voluminous sections dealing with the social concerns and endeavors of the buyer or the source of its funds to set minimum wages, rights of minorities, labor standards, safety standards, and all sorts of social reform, such as training or apprentice programs and other related ideas. Most of these social programs, however, are of little interest or concern to the average bidder. It is the rare contractor who will price or include the costs of these social services beyond assuming that its project management will somehow for the same cost perform the social features of the contract as well as the construction features. And so, in practice, almost all the attention goes to the construction features and not the social concerns. It is appropriate to note that the author has yet to see any government agency carry any cost or time allowance in its own estimates and schedules of a project to cover implementation of the new social programs. Agencies and contractors both seem to ignore the reality of the cost and time to involve or force new social order or concepts as a part of a construction project.

The result most often is that the social concerns, which are perhaps the least interesting aspect of the project to contractor and designer, are omitted, allowed to lag, or given only lip service. The media, both trade and nontrade, the concerned citizen, and the politician who invariably must approve the project funds, however, place great emphasis on the social concerns. The result has been (to the sad experience of some) that the offhand treatment of the social implications of the contract was the Pandora's box to technical or economic dilemmas, most of which would never have become issues except for the spotlight on the project from "unimportant" contract clauses.

The Technical Sections

Once the relationships, responsibilities, interactions of the parties, and methods of relief and change have been defined, the contract documents typically turn to the technical aspects of the work. Again many interested groups have set forth systems, methods, or formats to list typical information needed to construct the project. The most common technical format is that of the Construction Specifications Institute (CSI); if its format is followed, Divisions 2–16 of the specification book deal with the various technical categories of the work. Section 2 deals with earthwork, Section 3 with concrete, Section 9 with finishes, and so on. While other agencies may have their own particular technical format, the technical sections of construction contract documents whether in California, Maine, Florida, federal government, state government, private, religious, commercial, and so on, are all very similar and familiar to those in the industry.

Origins of Dispute in the Documents

Specifications are typically written by people who have become specialists in that endeavor; some are entrepreneurs and offer free lance specification writing for different design firms. Most specification writers have little field experience, almost none have management experience; their expertise is in writing specifications, not necessarily applying specifications.

Over the years, technical specifications have usually been produced by copying what was done last time, cutting up old specifications with scissors and pasting the good parts together, along with some new handwritten notes and material. It is no wonder specifications are so often out of date, in conflict, or speak to situations not found on the project. The introduction of word processing equipment, which came on the scene in the 1960s and 1970s, has led to specifications by computer, memory typewriters, and so on. The problems, however, are the same. Too often, insufficient attention is given to the writing of specifications and the special ability to read the "intent" of the specification is not possessed by the low bidder soon enough. The designer who forces performance by the contractor beyond what the specification obviously says is a key cause of construction disputes. Sometimes the disregard of a high-handed designer in knowingly forcing performance beyond specification level is but pulling the trigger for a later, more difficult confrontation which might otherwise never have happened.

Owners should review the technical specification effort during its process and the product as well. Specifications always seem to be done too late and often by people who have little connection with the day-to-day thinking and efforts of those responsible for the design. Owners should set safeguards and police the designer to prevent the specification from loss in quality.

The specification is often used to ensure a preference, give favor to a certain supplier, or to effect a specific result desired by the owner or designer which is counter to the pricing economics of the competitive market of suppliers to the general contractor at bid time.

When the preference is intentionally worded so that the owner gets both competitive pricing from an unsuspecting bidder and later claims the wording only allowed its preference (which might not have been the competitive price), disputes will arise. Somewhere the contractor will seek to recover its unforeseen exposure and loss. Sometimes the unspoken but desired preference is skillfully woven by the designer into the written word so that the competitive aspects of the laws governing public buying are subtly avoided to give those in charge of the design the ability to circumvent the choices provided or required by the law.

Disputes and problems will constantly come to the fore until those who write the specifications and contracts become loyal to the precept that society is best served by professionals whose loyalty is first to the policies and laws enacted by the collective wisdom of the legislators. For example, in Massachusetts nothing may be specified for a public job unless it is manufactured equally by three firms. The design professional who is willing to do its work properly can use this legislated restraint to the public's advantage. The designer who refuses and tries to subvert the law because somehow it believes its choice is superior to the law will inevitably end up in a confrontation that will be painful to many. Designers who try to inveigle competitive pricing and then insist on a certain product are a real cause of today's high error and omission insurance premium.

The construction industry and its users are constantly plagued with problems and disputes because specifications are not written in the office to conform to the way the game is played in the field. The biggest offense of this type could well be the typical caveat in virtually all specifications that no extra work can be done unless so ordered and agreed upon in writing, but every day on every project, extra work is authorized orally. This violation sooner or later leads to trouble. Often one of the items then at issue is the authority of the one who was giving oral direction and modification. Most contracts make the field men simple automatons who can only say no. They, according to the documents, have no responsibility or authority except to reject. Yet sophisticated, educated, experienced people daily give and take orders for extra work in direct variance to that which is written in the contract and was known and was agreed upon by all the parties and their leaders before work began.

It is far better to wait and get the change in writing. The wisdom of those who pondered over the contract clauses and perhaps spent days to decide whether or not it was better to write or speak changes into existence and agreement is assuredly greater than the reaction of hurried men concerned with immediate problems and profits. Experience indicates that contractors are the losers in hasty change order situations. Contractors make too many mistakes on change orders. For all the pricing methods used in changes, experience shows that most contractors and owners simply do not understand what is involved.

Good contracting means more than people who are capable of building a project well. It means that certainly, but it also means people who are willing to stand by their agreements and to do that which must be done to make the contract meaningful.

Those who choose not to wait for written modification agreements run the risk of being unable to convert what everyone thought had been said into dollars. For example, 250 verbal changes which are later followed in writing without a hitch, do not guarantee that the same hitchless procedure will occur on verbal change number 251. Again, experience shows that the inconvenience of waiting for written instructions is well worth the delay.

Owners who write the contract, however, must be prepared to pay the price for the delay and the costs it causes the contractor by the methods it has chosen to require to make modifications. Too often the owner or the designer tries deliberately to beat the contractor out of a fully due and proper adjustment for ripple or delay cost due to change.

Disputes over contracts arise because people will not live by the principles of the contract. Designers particularly try to force more than the bargain really provides. This improper attitude causes a defensive reaction in the contractor and arguments arise. The trend in the thinking of those who hear the facts in construction contract disputes today is to go to the intent of the parties at the time the bargain was struck. Thus, in the dispute process the hearer of the facts will seek to find the initial spirit or the thinking of each signator to the agreement. Then these questions follow: Who had the risk? Was it reasonable and fair? Was the contract or specification writer's intent objectively stated in its words? The writer's subjective intent, no matter how strong, must take a backseat to a reader's average understanding of that written if the intent is not plain to all. More often than not contractor losses from unfair or questionable contract interpretations by the designer or owner who wrote and then initially interpreted the words are recovered.

Unfortunately, disputes often arise in the field because the inspector has neither intelligence, training, choice, desire, nor authority but to believe and enforce the written word. Most inspectors are convinced every contract clause is enforceable because it is "in the book." "You bid it, you must do it." Thus a minor problem can quickly become blown out of proportion. Even worse, too many designers who know better take refuge in the written word and lead the owner to believe it will prevail. Consider the compounding of this entire scope of problems by realizing that the average specification writer or design professional genuinely believes that the words he puts to the page are law and are binding on the contractor, when in fact the words are ultimately binding only if some unknown hearer of facts so decides. Contracts are not law, they are agreements to be enforced under the law and due legal process.

Disputes arise because low-level personnel are given great authority to reject but little or no training or power to resolve the problem. In a controversy of any scope, lines quickly get drawn, personalities and jobs go on the line, and a dispute becomes alive.

Other Contract Documents

In addition to the printed specification book, other contract documents exist. Foremost here are the drawings which are part and parcel of the specifications, and both are intended to interact.

The question often arises as to whether the drawings or the specifications take preference in the case in which they conflict. There is no real fixed hard rule unless it is so stated in the particular contract document or something is obvious, such as larger scale details are easily recognized as more thought out than smaller scale details. The party who leans on the word or detail obviously contrary to the stronger understanding or intent of the total contract will eventually be out of the money in the dispute. Here again the hearer of the facts goes to the ideas or thoughts of the parties way back when. If one uses conflicts in the document unfairly to its advantage, sooner or later it will be challenged and it will lose the challenge.

During the bidding process before the proposal period is closed, other contract documents known as addenda or modifications to both specification book and drawings are almost always issued. This is information found needed after the initial package has gone to the printer to be reproduced for distribution to the bidding public. The lack of time and costs of reissuing totally corrected documents are so high that the addenda route is followed. Printing costs can run $50,000–$200,000 for major projects (just for plans and specifications for bidding).

Other important documents to the parties are those which are incorporated into the contract by reference in the specifications or somewhere prominently or obscurely on the drawings. For example, a document known as ACI 301-72 is frequently mentioned in Section 3, the concrete portion of the technical specifications. ACI 301-72 mentions ACI 347-68, another lengthy standard on formwork practice. By such reference, many requirements are assumed and counted on by the owner to be understood and many risks are assumed by the contractor. Some documents incorporated by reference are further qualified to be offered for the bidder's information only with all sorts of fancy disclaimers as to how the information may be used or the degree upon which it can be relied.

The Lack of Perfection in the Contract Documents

Owners, contractors, designers, and everyone involved in construction readily recognize and are quick to admit publicly the very obvious fact that a perfect set of contract documents simply does not exist. All somewhere have mechanical drafting errors or lack a needed dimension or detail. Many have errors which stem from the human nature of the designer and draftsman. Not only are there human errors, but changes always occur as projects undergo the design and construction processes. Certain selected equipment carefully specified becomes obsolete—something better has come along, but it is bigger or needs more power. There are changes in space usage to accommodate revised owner needs; something unforeseen occurs, and the documents and work scopes must be adjusted. The more complex the project, the more ramifications a change has. The shorter the period allowed for design, the more addendas that are required, and the more opportunity for error. No one man may know or remember every place a certain detail was shown. The larger the project, the more the people, the drawings, the thoughts, and ideas—consequently, the larger the project, the more errors there are. Contract documents are one major origin of disputes. Document errors become the fault of the owner when they cost the contractor unbid or unforeseeable dollars. Document errors become the fault of the designer when in the judgment of its peers and the custom of the industry the errors are gross and inexcusable. Document errors become liabilities when someone who has a right to rely on the professional is severely hurt or damaged. Punitive damages are starting to be considered as collectible against a professional when the hearer of the facts finds that the professional's refusal to come to grips with its duties are offensive to any reasonable standard of behavior.

FAILURE TO COUNT THE COST

Disputes continually arise because someone failed to count the cost at the beginning when the cost should have been defined. Few contractors bring claims on projects which come in near or under the construction budget; few owners seek liquidated damages when projects are done on time or close to it. If designs are waterproof and the products the designer specified fulfill the sales representative's claims, disputes are few and far between.

Contrary to the opinion of most owners, few contractors are deliberately claim's conscious. Most supervisory project personnel who work for the parties on the project have little real knowledge of disputes or what is involved in litigation or arbitration. Most of those involved in getting a job done have solved complex problems on a daily basis of face-to-face confrontation for such a long period of time that they come to believe they know it all.

Thus, they prefer to argue among themselves and write what they believe are clever letters to establish a record, and most distrust and resent the lawyer. Contractors who have made money on a job usually do not invent claims or pursue spurious claims. Most often, a contractor who is clearly entitled to a valid contract adjustment via a claim will ignore the situation if the job has come out well enough to live with. Contractors like to get the job done and over with. They fancy themselves builders; claims take too long to hold their interest.

Disputes arise when jobs do not come out well, and too often the reason for this is the failure initially to figure the cost accurately. This failure to count the cost initially is not confined to just the contractor, it applies to the owner who sets out unrealistically to build a factory, as well as the designer who sets out to design it for less than it will really cost either in design or construction. In construction, major dollars and work scopes are calculated and committed in very short periods of time. It is not uncommon that someone fails to count something, and ends up with a price that is too low. What is worse is that most of those in the industry simply do not have the money to pay for their errors. The man with the best intentions and $5.00 simply cannot pay for his $2000 error. Ironically, too, it would seem to some observers that those with the money to pay for their errors lack the degree of intention needed to dig deep enough to square the account totally.

In the competitive contracting market, contractors work at extremely low markup levels. Construction is perhaps the zenith in high-volume, low-markup business. Usually general contractors bid in the public market with a fee of 3 percent over its estimated field costs. Subcontractors of significant importance on the project run a 5 to 8 percent fee. Less major subcontractors will mark up labor by 40 to 60 percent and carry material at cost. All sorts of pricing systems exist, but all have one common trait: They are too low! This fee must carry the home office overhead before net profit can be counted. The hardest part is that the dollars have to be collected after all the delays and difficulties involved in retainage, back-charges, punch lists, and the like are resolved. Thus, to the contractor boxed in with retainage and other cash flow problems there is no room to absorb cost overruns.

Construction pricing methods frequently are not able to take into account the erection process that will be ultimately required in sufficient detail. The modern designer does not want to tell how a job should be done or pre-scribe or reveal any sequential restrictions not strength-related. Thus, millions of dollars of work are priced under severe time pressure using established unit prices calculated from the estimators' experience and which to some extent may have been proven in ongoing or recent projects. This policy often ends up with insufficient dollars on hand when an abstruse procedure for execution well known in design does not surface until the detailed means and methods of construction are worked out by the contractor. Hidden time requirements such as curing periods between adjacent concrete pours, or shoring procedures referenced in a distant ASTM or ACI manual, can make the failure to count initial costs very acute to the participants in the process. Designers who do this sort of recondite specifying, if pursued for relief, usually have to pay the costs.

The failure of a contractor to understand and/or correctly bid or price the work initially is a major reason for disputes. It is compounded by the ever present confident overbearing optimism inherent in all contractors that they somehow are charismatic and can overcome the dilemma of an obvious low bid. Experience is such that the contractor, regardless of its optimism or charisma, generally does not have the ability to make up money left on the table when bids are first opened.

The contractor in such a situation should instantly seek to identify where it went wrong. If a low price can be traced to inefficient labor or productivity factors, a reasonable chance of problem solution can be expected. One can sometimes engineer or dream up a more economical way of doing a task. Cost controls, bonuses, piecework deals, contests, and other similar ideas can overcome low labor unit prices.

If, however, the low price is traced to omitted quantities, the contract should not be pursued. Omitted quantities of materials are almost impossible to make up. Better to get out of the bargain even if a penalty is involved. Instant losses are better than long-term erosion of capital. The instant loss leaves one free to pursue gain elsewhere. The pluck to solve a low price by acceptance and performance of a cheap job, while admirable, often results in the demoralizing process of working hard for a year or two or more and knowing money is being lost every day.

For some reason heroes in the construction industry are not those who cut losses but rather those who make profits. The old adage of "misery invites company" is particularly true. Losing jobs generally become more sour and everyone in management wants to avoid involvement with the bad deal which only makes the loss worse.

Prices that have been offered and quickly are recognized as obviously too low should trigger intense efforts to negotiate, buy, or beg out of a bad situation. Forgetting the penalty or bid deposit aspect, the owner or purchaser also benefits greatly from this. While everyone knows it is nice to buy at low prices, in construction the price is set before one sees the product, and bargains or obviously too low prices at the bid letting or contract negotiations often disappear on the way to the certificate of occupancy.

The Psychology of People in Construction

It was noted earlier that construction is not a science, it is an art. Construction is really people, and the successful contract administrator, or disputant to a contract interpretation or unfortunate occurrence on a project, is well served to know a little about people. If traced down to its real roots, a construction dispute has just two people involved. They may be very visible to each other and all concerned, or they may not be known to each other and insulated from it all by many layers of other people. The successful disputant or claimant is often the one who knows the people in construction and how to meet them and their psychological needs to their immediate benefit and the disputant's ultimate benefit.

What are construction people? A question which provokes different images to everyone. Tough, hard-driving, demanding, strong, ignorant, shrewd, calculating, honest, dishonest, straightshooter are all typical adjectives applied at times to individual members of the industry, and even though the adjectives are in some cases antonyms, still the people they describe have some commonality.

Construction people have certain recognizable traits or makeup to their personalities. They are first a gregarious people. All want to belong to the crowd. The herding instinct is very strong in the industry's people. All seek and need that sense of acceptance or approval. They have a need to emulate the leaders or their concept of the leaders of the profession. Words like belonging, imitation, loyalty, morality, recognition, superiority, status are descriptive of the human elements of gregariousness. Try to make the other party feel as if he belongs to the pack. Find out the group the other party feels is important. Show him how resolution of the dispute will help him achieve or strengthen his membership in the group.

The next trait to bear in mind is found when a person feels accepted into the group. Almost always he will turn pugnacious. Now that he is in the group, he wants to fight for its needs and to seek personal recognition from his peer group. Words like aggressive, anger, defiance to rules, competitive, curiousity, hate, hostility describe the pugnacity in people. Give the other party something to fight for or with; perhaps a third person can be blamed for some of the dilemma. Use the trait of the other fellow's pugnacity to advantage.

More important perhaps than pugnacity is that everyone has the problem of ego survival. It is one thing to lose money on a contract problem, but it is a lot harder to lose face. All people have an idea of themselves which they feel must be defended. The construction industry is one in which the public and peer knowledge of what everyone is doing is vast. There are few secrets in the industry. Everyone's prices and problems are known; word travels fast. Disputes can often be more easily resolved when all the egos involved can survive.

Not only are people typically quick to protect their self-image, they all want to extend the position they currently hold or claim is theirs. Thus, any message couched in terms of new acquisition, promotion, saving money, or being protected will be heard and very often receive action. Everyone wants more space, a better promise of a better future, and the chance to increase the recognition of one's self-worth. Appeals to ambitions, goal realization, and increase of power help resolve disputes. The clever disputant deals with its adversary with these ideas considered and brought out.

The survival and extension of one's ego include as well the pleasing of a person. Some people like to enjoy themselves, some like to suffer, others want to argue, some like detail, some the big picture. The successful disputant looks for opportunities to help the people on the other side find pleasure in the dispute proceedings.

Construction disputes and confrontations arise because the people involved have needs. From the contractor's side the needs are usually dollar- or profit-related. The designer has his ideas, his building or design which might be his monument to himself, his reputation, his artistic temperament, his dollars, his insurance premium, and similar needs. The owners have needs as well: political careers, corporate careers, the need to have the space for a certain day. When something unanticipated or not properly recognized interferes with the fulfillment process, goals and security are jeopardized, communications become strained, and strains seem always to be followed by demands, refusals, other more intense strains, hard, then harder positions, and dollar losses. Unfortunately or perhaps otherwise, it is not in most people to recognize an error, particularly their own, and say "I'm sorry" and seek to make amends. In construction most are unable to pay for their mistakes, it is simply too expensive; and, alas, those who can afford to pay for the mistakes generally remember lots of other errors by the other party which even if already forgiven somehow must now be reconsidered.

Once confronted with a problem too expensive or complicated for ready resolution, the claim or dispute process begins. At this juncture several more typical human traits should be borne in mind in addition to those already considered. People are creatures who like order. People do not like chaos. Random thoughts, unconnected conditions and ideas, jumping from this to that, tend to confuse and obfuscate the paths necessary for problem solving. Human beings react best to order. Strangely, maybe sadly, people prefer well-ordered lies to chaotic truth.

Not only is order important, but personal habits are as well. There is a need in everyone to follow their habits and traditions. Appeals to keeping the law, following moral codes, being a good sport, being an early riser, or obeying the government are often rewarding. People have a need to obey.

If a construction problem exists, chances are that sufficient research and realism will find people as the root cause. In the quest of profits or career improvement, construction people have been known to be greedy, never satisfied, resentful, quick to cover themselves, quick to improve themselves, legalists one moment, rationalizers the next. They are often over their heads, lazy, not inclined to do good, incompetent, yet protected by the needs of others and the system, indifferent, discouraged, surprised, sick, or about to get sick. They can be self-righteous, alcoholics, lovers, power hungry, unfair, braggarts, tired, bored, and may have already been there once before. Some are young, full of ideals, new ideas, and a desire to improve the world. Some are great, servants of all, planners, men with visions, but each is a unique individual who is faced all day long with the problem of understanding the other guy and being understood while trying to fulfill one's goals when there is never enough money, time, or people.

People are a prime cause of construction disputes, and the only solution to these disputes as well. The rise of society's present attitude that everyone has rights has led to much of the activity in disputes. The vast amount of contention and society's attitude toward one's rights must be viewed from the more mature point that each day is one of opportunity, not guarantee, a foundational position which seems to be lacking in current thought.

3

THE COMMON CATEGORIES OF CONSTRUCTION PROBLEMS

The preceding chapter described the most common primary origins of construction disputes or problems. These problems from which claims arise come to the job in three main categories: design deficiencies, the construction process, and consumer reaction.

THE DESIGN DEFICIENCY

The design problem or deficiency which leads to a major dispute is generally beyond an error of omission. To be significant the design error usually must alter the means, method, environment, duration, or the conditions of the construction process. Any number of factors can influence this. The most common places in which design errors are made are in the foundations, in the construction of the frame and enclosure, in the utilization of spaces such as ceiling sandwiches and pipe chases, in contract documents where both the method and materials and the required end result are specified, in project duration, and in connection with related performance by others on which the project in question must at some point rely. Some of these problems bear discussion.

The Underground or Subsurface Problems

The cause of underground problems can generally be traced to the handling, display, and interpretation of subsurface investigations. The location, depth, number, and type of borings or subsurface investigations are established by the engineer who needs the information for the foundation design.

The borings will also locate the level of groundwater, at times a key economic consideration in the building's foundation and its construction. Borings are typically done by a drilling contractor usually hired directly by the owner, often on a competitive basis using specifications prepared by the architect, the structural engineer, or a soils engineer.

The owner obviously wants to get the most out of the borings and subsurface investigations. These investigations are first used for the design of the foundation; then they are typically offered to the contractor for its use in bidding and constructing the job, and it is in this latter use that the disputes begin.

It is well known in the industry that variations in the earth's composition constantly occur underground. It is also well known that each bidding contractor cannot adequately investigate the site to define the soil characteristics in the time period available for bidding. The owner thus has a problem. It wants to put the risk of the excavation and methods of foundation construction on the contractor; it also wants to buy the project for the best price. And so the dilemma: How can the owner pass its soil boring and subsurface knowledge to the contractors to help keep the bids down by sharing the best soil information available and still pass the risk to the contractor if the work is more difficult or expensive than the borings indicate?

The route followed to move this risk over to the contractor is to write a clever specification. The following are random sample quotations from four different jobs which are typical specification writer ideas of how to make soil characteristics available in order to get the most accurate price and simultaneously put all the risk on the contractor.

The first is from a school building in New England with local designers

Subsurface Conditions:

1. The Owner has explored subsurface and foundation conditions by having authorized the making of borings on the site.
2. The Owner makes no representations regarding character or extent of soil, rock or other subsurface conditions and/or utilities to be encountered during work. Subsurface formations, including water levels, included in reports have been interpolated from completed borings, correctness of which is not guaranteed.
3. The results of the subsurface investigation are available from the Architect. The Bidder must make his own deductions of subsurface conditions which may affect methods or cost of construction of the work. Each Bidder, before submitting his Bid, must visit the site and examine all conditions that may affect his work, and make his own deductions as to subsurface conditions. No claims for damages or other compensation will be considered should the scope or progress of the work be different from the ones anticipated by the Contractor.

The second is from a major hospital in the Midwest which was designed by an eminent engineer and had a construction manager as well.

All Contractors bidding on this work are cautioned to visit the site and determine all local conditions, including subsurface conditions that may in any way affect their work. The Owner has commissioned Soils Engineering Consultants, to perform a subsurface investigation which has been used as the basis of the foundation design for new construction. Reports of this study are available through the Construction Manager to interested parties for information only. Neither the Owner nor the Architect assume any responsibility for the results of the subsurface investigation other than that it reflected the best information available to them at the time of the design. Except as provided in the paragraph below, the subsurface conditions at the construction site shall be deemed to have been known to each Contractor prior to the submission of the Bid, and shall not be the basis for a claim or excuse based upon failure of a pre-supposed condition.

The Contract shall be based on the assumption that there is no bedrock on the site. In the event that bedrock is encountered, the Contractor shall immediately notify the Construction Manager and shall not proceed further until measurements are made for the purpose of establishing the volume of bedrock excavation and approval is given for the methods of removal. Adjustment will be made in the Contract Sum based on the required amount of bedrock excavation and Unit Prices established by the Architect and Construction Manager.

This sample is somewhat peculiar since the borings showed bedrock or boulders and thus the specification almost has to lead to a claim for time and general condition dollars in addition to the unit price for bedrock excavation.

The third example is from an upstate New York sewage treatment plant designed by one of the top 20 designers:

Unexpected Underground Structures

Should the Contractor encounter underground structures at the site materially differing from those shown on the Plans or indicated in the Specifications, he shall immediately give notice to the Engineer of such conditions before they are disturbed. The Engineer will thereupon promptly investigate the conditions, and if in his opinion they materially differ from those shown on the Plans, or indicated in the Specifications, he may make such changes in the Plans and/or Specifications as he may find necessary. If, in the opinion of the Engineer, an increase in cost results from such changes, the Contractor's bid price will be adjusted as provided for in Article 15 of the General Conditions.

Information on the Plans and any statements in the Contract Documents referring to the condition under which the work is to be performed or the existence of utilities or other underground structures are not guaranteed to be correct or to be complete representation of all existing data with reference to conditions affecting the work. Every effort has been made, however, to make this information complete and accurate on the basis of all data and information which could be procured by the Engineer. The Contractor shall make his own examination and shall draw his own conclusions as to the underground facilities which will be encountered, and he shall have no claim for damages of any kind on account of any errors, inaccuracies or omission that may be found. If, in the opinion of the Engineer, permanent location of a utility not otherwise provided for, is required, he shall direct the Contractor, in writing, to perform the work. Work, so ordered, shall be paid for under Article 15 of the General Conditions. Protection and temporary removal and replacement of existing utilities and structures shall be part of this Contract and all cost in connection therewith shall be included in the prices established in the Proposal.

The final sample is from a private Dow–Jones index listed firm building a $25 million plant in California using a top industrial architect engineer.

Site Examination

Before submitting his Proposal, each Bidder shall inspect the site of the proposed work to arrive at a clear understanding of the conditions under which the work is to be done and to be aware of the actual elevations, obstructions, and conditions apparent by an inspection of the site and evaluation of the Subsurface Geological Investigation. No allowance or extra consideration on behalf of the Contractor will subsequently be made by reason of his failure to be aware of such conditions. These were intended for design purposes only and are made available to bidders only that they may have access to identical information available to Owner. The boring logs are for Contractor's information only and are not intended as a substitute for personal investigation, interpretations, or judgment.

Existing utilities and underground structures shown on drawings are for the convenience of the Contractor and exact locations are not guaranteed. Verification of locations of and protection of these and other utilities and structures not shown, if any, is the responsibility of the Contractor.

All 4 of these statements have common goals and all the authors seem to have had the same teacher. The data regarding the subsurface characteristics are first defined as information only, not part of the contract documents, even though the information often is printed on the contract drawings or bound into the specification book. It is here that the specification writer goes astray, for those who write and enforce the specifications (most often the owner or designer) forget or fail to perceive that while they may be the author and the first interpreter of the specifications, the ultimate interpreter is a judge or panel with possibly little acquaintance or sympathy with fancy word schemes for risk avoidance. There is no question that the ultimate interpreter in recent subsurface disputes has been interpreting the soil disclaimer on the basis of reasonability.

The questions asked to determine if the disclaimer is reasonable ultimately become twofold. The first is whether the bidder did anything special to investigate the site, or did it rely only on the information printed or made available with the contract documents. If the bidder dug a test pit, drilled a hole, investigated the subsurface site characteristics in any way, or had previously worked on adjacent sites, and used this special knowledge to price or plan its bid, then regardless of the degree of such separate knowledge, it has become a special bidder. It can be argued that the bidder no longer has the right to claim reliance on the owner's information, indeed it has not so relied. The bidder who makes any major independent analysis of the subsurface essentially picks up the risk of the excavation.

The other method employed to win the argument for extra dollars to remedy the discovery of site work surprises is to establish early one's total reliance on the owner's information and that the information was sufficiently indubitable to permit such reliance. Sophisticated contractors will establish that they relied on the owner's information after they are announced as low bidder but before the contract is awarded. Congratulatory remarks to the owner and designer by the contractor's estimator and project manager on the good job in the documents on soil definition carefully recorded in preaward postbid meetings can often lead to a record which establishes the risk as the owner's, and the contract finalized with both parties aware of contractor reliance on the owner's information only.

Both parties to the contract will do well to establish in the preaward postbid period the degree of reliance on the subsurface information. The experienced buyer or owner can often lead the contractor, who is eager to see its low bid be followed by a contract, into asserting it had made all sorts of inexorable investigations to see what the soil really was.

This same principle applies to existing underground utilities, oil tanks, and old foundations, the location of which may not be precisely known. Often slight inaccuracies or error in location of underground obstructions can lead to considerable expense and delay. The ultimate responsibility for these costs could be established by reliance. Was it reasonable to rely on the contract documents? Did the contractor rely on the documents? Did it have recondite knowledge beyond the document?

The courts and review boards across the country have generally been ruling in favor of the contractor on underground claims of added cost. These decisions have evolved into common law, where the consensus of the decisions leans strongly toward contractor recovery. Consensus, however, does not guarantee a winner in a horse race any more than it does in a construction dispute. The essential element in this type of claim may well be reasonability and reliance but this does not negate the possibility of the superior ability of either or both parties to persuade each other or the hearer of facts as to who is really right. People often win arguments when they are wrong simply because they know how best to argue. Many times the skill of counsel will overcome the letter of the law.

Risks

The contractor should assess and display its reaction to risks to its advantage in any project it prices. The present practice of the typical contractor is sadly deficient in risk defense in comparison to the owners and designers.

Usually the contractor makes an estimate on a job consisting of three elements: 1) general conditions or field overhead, 2) the labor and material for work the contractor could or may want to do with its own forces, and 3) the subcontractor work.

Exposure to assuming responsibility for risk is largely found in the estimate techniques and summaries which make up the element of general conditions. Sensible estimator techniques in this area must be stressed since the estimate often records and binds the thinking of the contractor forever. For example, the usual job specification will set out that all work must be done between 8:00 A.M. and 4:30 P.M. on a five-day work week. Thus very obviously no overtime is contemplated, seen necessary, or wanted in the mind of the writer of the contract. The contractor, however, almost always includes an item of overtime in its general conditions. Perhaps it is only the overtime for the concrete finishers which could very well be within the perspective of the writer of the contract, but unfortunately, almost without exception only the word overtime is used in the estimate. Figure 3.1 presents a very typical general condition summary with overtime noted boldly and in a catchall manner.

Five years later in the construction dispute process the original estimate is produced as part of discovery. The contractor pressing a delay claim suddenly is confronted with the undeniable fact it recognized during the bidding stage that the job could not be timely done without overtime. The contractor has admitted it foresaw the potential of delay without overtime, and the urgency or ease of claim collection can be jeopardized. How much? In one experience in Florida, a $920,000 loss stuck because a contractor had listed large dollars for overtime on a nonovertime specification and also showed an additional item called contingency of $10,000 on its estimate. The judge said, "You knew it couldn't be done, you bought the risk."

In the present business world theory of risk minimization there is no place for a word like contingencies in estimates. All the parties involved, bidder, owner, or engineer, must think out its program and not have contingencies. Contingencies are unidentified but obviously recognized risks. Delay could well be in that definition of contingency. It may never have been in the mind of the person who wrote and quantified the contingency, but the persuadability of others five years later may make it difficult to overcome the connotation of the word contingency, particularly to the ultimate hearer of the facts. Modern contractors who are claim conscious should forget the word contingency.

The owner and designer should be just as leery of the word contingency. The 5 percent override on the base contract price carried by the typical owner in all its data processing or other displays on the construction budget should not be characterized, discussed, or written as a contingency. Let it be a reserve for upgrades in scope. Avoid the use of words which could mean more than intended. The adversary could well sharpshoot a good claim because of the words used in the planning process.

The place to put the dollars for contingency or overtime in nonovertime specifications is in a larger fee. Fee is a common word to define markup on the estimate of all field cost to cover the home office overhead and provide a profit. The fee or markup commonly used at bid time is in the range of 2 to 4 percent for major general contractors. This means that a general contractor with a $1,000,000 overhead must do more than $30,000,000 per year at 3 percent to break even. Studies of a number of contractors show that the home office costs usually run $4\frac{1}{2}$ percent of volume, and yet the bid markups of these contractors are often for less. There may be no end of good reasons for the difference at bid time but five years later in a major law suit for loss recovery those reasons are generally very hard to recall, and at best the low fee originally charged becomes just another item of persuasion or argument one does not need.

Those responsible for estimating, marketing, and buying have to be taught risk analysis and protection. If one adds a contingency fee to the estimate, what does it mean? Management must be aware of risk avoidance maneuvers and instruct its people what to do. Insurance companies are criticized because they go to great pains to teach the professional designer or manufacturer they have insured what to do or not to do. Good business practices are certainly good business. This includes training and coaching. The better one is trained and the more he practices, the better the player becomes and the better the score, particularly for the well-trained team. Whether the score is kept in runs, points, dollars, heads, or hides, the more practiced player will have more of what really counts.

The contractor, owner, or engineer sophisticated enough to increase its fee rather than particularize on paper controversial elements of cost is no more profit-minded or guilty of unfair business than if it had based its price on new methods of construction cheaper than those in common use and then kept all the savings as added profit for itself.

Defective Plans

A major source of disputes in design deficiencies is that categorized as defective plans. What are defective plans? Most people involved with plans have a working idea of the definition of this phrase, but in reality no standard exists

PROJECT OVERHEAD SUMMARY

PROJECT   *FACTORY BUILDING*   SHEET NO.

LOCATION *PALO ALTO, CA.*   ESTIMATE NO.

ARCHITECT   OWNER   DATE

QUANTITIES BY   PRICES BY   EXTENSIONS BY   CHECKED BY

DESCRIPTION	QUANTITY	UNIT	MATERIAL		LABOR		TOTAL COST	
			UNIT	TOTAL	UNIT CODE	TOTAL	UNIT	TOT
JOB ORGANIZATION:								
Superintendent	50	WKS	20	1000	500	25000		2600
Acctg. & Bookkeeping	50	WKS	10	500	300	15000		155
Timekeeper & Material Clerk	—							
Clerical	—							
Shop	—							
Safety, Watchman & First Aid	50	WKS	15	750	250	12500		13 2
ENGINEERING: - ENGINEERING INS.								
Layout	50 DAYS @ 300			—		15000		150
Quantities								
Inspection	Allow					5000		50
Shop Drawings	Allow					7500		75
Drafting & Extra Prints	Allow					2500		25
TESTING:								
Soil								
Materials								
Structural								
SUPPLIES:								
Office				1500				15
Shop				500				5
UTILITIES:								
Light & Power				7800				78
Water				1500				15
Heating				5000				50
EQUIPMENT:								
Rental				15000				150
Light Trucks				3000				30
Trucking				2500				25
Loading, Unloading, Erecting, Etc.				1500		3500		50
Maintenance				2000		1500		35
TRAVEL EXPENSES:								
Main Office Personnel				1500				15
FREIGHT & EXPRESS:								
Demurrage								
Hauling, Misc.								
ADVERTISING:								
SIGNS & BARRICADES:				750				2
Temporary Fences				8000				80
Temporary Stairs, Ladders & Floors				2500				25
PHOTOS:				250				2

Figure 3.1 Project overhead summary.

DESCRIPTION	QUANTITY	UNIT	MATERIAL		LABOR		TOTAL COST	
			UNIT	TOTAL	CODE	TOTAL	UNIT	TOTAL
Total Brought Forward								
GAL:								
DICAL & HOSPITALIZATION:								
ELD OFFICES:								
Office Furniture & Equipment				1500				1500
Telephones				2000				2000
Heat & Light				2500				2500
Temporary Toilets				700				700
Storage Areas & Sheds				1500				1500
RMITS:				5000				5000
Building								
Misc.								
SURANCE:				15000				15000
NDS:				38000				38000
TEREST:				8500				8500
XES:	N/A			4000				4000
TTING & PATCHING:				1500		1500		3000
Punch List				1500		7500		9000
NTER PROTECTION:								
Temporary Heat				12000		4500		16500
Snow Plowing				5000		2500		7500
Thawing Materials				1500		750		2250
MPORARY ROADS: (Site Conditions)				1500				1500
PAIRS TO ADJACENT PROPERTY:				2500				2500
MPING:								
AFFOLDING								
ALL TOOLS				4500				4500
EAN UP				10000		65000		75000
Final				5000		15000		20000
NTINGENCIES								40000
IN OFFICE EXPENSE	In Fee							
PECIAL ITEMS								
ESCALATION						20000		20000
TRAVEL						15000		15000
OVERTIME						50000		50000
				218750		269250		488000
	Overhead @ 10%					53850		53850
	Sales Tax 5%			10938				10938
								552788
TAL: Transfer to Meansco Form 110 or 115								

Figure 3.1   *(Continued)*

locally or nationally that precisely describes how to measure the plans for defects.

Indeed, everyone who has worked with plans knows that no set of drawings is complete or without error. Somewhere dimensions are missing, elevations or grades are in error, a detail is missing, or a detail is shown but not needed. Not only are these types of errors common, but all who work with plans know that drawings can always be refined and upgraded. Plans can always be made better, they can always be improved.

Thus all plans are to some extent defective and everyone involved in building uses defective plans every day. The question in plan deficiency disputes is when do the plans become defective to the point at which undue costs are generated from their use. The usual legal definition is that plans are to be prepared with the normal standard of care found in the profession, but no precise standard exists.

The designer has the advantage of its subjective knowledge of the intent of the plans. In some cases pressures from the client will be exerted for degrees of performance in excess of the objective intent of the plans. This, plus poorly drawn plans, poorly drawn details, poorly prepared notes on drawings, and poor specifications may reach a point where in the opinion of one's peers, a level of acceptable performance has not been achieved.

In the case of errors of omission from a set of plans, the decision of adequacy on the part of the professional is much easier to make than those which bear on methods or performance levels to be met upon completion. If an engineer omits the exit lights totally, are the plans defective? Does the owner have the right to rely on the boiler plate clauses requiring the contractor to meet all codes to overcome the lack of exit lights? Probably not, but then there is no fixed rule to answer these questions. The solution generally comes from people genuinely willing to confront such situations daily and work out the answer. This nice sounding method, however, is a hoped-for method of solution at best. It does not work all the time and is complicated by the lack of precise measurement.

Big disputes can arise from defective plans. Experience has shown that major dollars can be lost by one party to a contract because of a drafting error. Although this type of error can be very expensive it may not be uncommon to the profession. In this case, the size of the loss, while impressive, is substantively immaterial to the issue, for the question is not the extent of damage but the degree of professionalism used or lacking.

Engineers and architects are cautioned not to judge themselves in disputes over plans they have prepared which are called or seemingly are defective. The mature designer will acknowledge the existence of an error and do its utmost to mitigate the resultant costs of correction. The ultimate decision as to whether the designer is responsible for the costs is for someone else to make. The designer must remember, in spite of the noisy and menacing emotions of the moment, that it is neither the contractor nor the owner who is the judge of the designer. Their opinions may be temporarily unpleasant but they are opinions, not judgments.

The best reaction for a designer in an errant position is to move forward immediately, calmly, sensibly, and openly to measure and solve the problem and simultaneously to create a record of interested, fair, genuine professionalism with concern for all involved in any obvious overruns. The sophisticated designer will never admit responsibility for the error. It is a long road from discovery of an error in the design to proving it is the designer's liability, and many people lose their interest during the journey.

When a dispute over plans arises, the disputant will be forced to establish by the opinion of other plan producers and users that performance which constitutes the standard of care. Those who seek to prove that opinion will often find difficulty in locating peer professionals of sufficient stature willing to offer testimony of a critical nature to a fellow professional. Then, too, the degree of persuasive skills from member to member on the rosters of licensed professionals is greatly variable. Highly credentialed peer critics may be unable to put forth persuasively the arguments and conclusions or withstand cross-examination regarding the deficiencies needed to establish the errors and their maker guilty of gross error.

The best tactic to prove negligence or gross error is to seek legal precedence in similar cases previously tried in the courts. The second best tactic is reliance on the opinion of experienced professionals whose integrity has been long established and who are recommended and well regarded on the local scene.

The problem of omission is more easily solved since it most often can be quantified in a straightforward manner. The missing exit light costs so many dollars for the fixture, the conduit, the box, the wire, and the panel breaker. Perhaps by being purchased as a change or modification, the fixture will be more expensive than it would have been under the initial bid, but at worst this increase or penalty is relatively minor.

The traditional solution to the omission dispute is for the owner to pay for the omitted work. Although it has every right so to do, the owner infrequently pursues the designer for the difference between the so-called bid-price and the change order cost. However, the other types of design error most often lead to dollars of a magnitude that are pursued by the initial loser of the funds. These are the problems which add major costs due to increased duration of performance or due to a change in the anticipated method of construction. The change in method problem also comes in the form of directives requiring removal and redoing of work in place.

The designer who requires a contractor to wait or take longer in its planned performance period because of defective plans, inaccurate design calculations, changing technology, lack of space, poor fit, failure to coordinate, understand, or display interface requirements of components is often challenged.

The measurement of liability is still the same collective opinion of the peer groups and precedence found in earlier legal joinings on the same or similar issues. The liability, however, can be far in excess of the omission or error. The ancillary costs of a construction problem almost always exceed the direct costs. The owner and the contractor have the right to expect the designer to produce a set of plans which will allow the project to be built. The law says the owner warrants to its contractor that the plans, if followed, will produce the desired result and that the project is constructible. Thus, if an error by the designer prevents the contractor from reaching its ends, the question of liability and assessment of consequential costs exists.

Methods or Means and Specified Performance

The last major category of deficient design is that case in which the designer has specified both method or means of construction and the result required and then refuses to accept one of these two requirements. Examples of this problem are legion but one that occurs very often is when the designer specifies that a certain waterproof coating be applied in two coats on some exterior surface under stringent conditions. The contractor applies the specified coating exactly the way it should, careful inspection is made by the designer as work proceeds, and then to the dismay of all the waterproof coating leaks contrary to the specifications. Almost without exception, the work is rejected and the contractor told to do it over. The means to accomplish the end result has been specified and followed but the end result does not meet the specification.

The problem, while very simple and obvious in the foregoing illustration, can quickly get very complicated and significant when detailed structural or mechanical sequences are set forth. They are carefully followed, inspected before, during, and after construction, all approved, and, lo and behold, big cracks appear in the walls and floors or the fancy heating and cooling system does not heat or cool. The principle is still simple: If the designer tells the contractor how to do something, the designer cannot impose a result beyond the product of the specified means. This does not mean that the designer will not try nor does it mean that the contractor can view its responsibility for workmanship lightly.

Very often, the architect or engineer in its role as interpreter of the contract will find itself in the position of imposing the consequence on a contractor of an error for which the designer is at fault and will ultimately be found responsible. The imposition of the consequence of this error on the contractor can mean penalties beyond all fair play such as financial failure, loss of personal health, even suicide.

Can the contractor recover or pursue punitive damages for such a capricious and arbitrary act by a designer? Again, there is no hard rule, but it would appear that the common law evolvement process is now setting the stage for such punitive recoveries. Some recent liberal states who set the initial trends in law at lower court levels have found that failure to act in good faith while fully aware of the consequences even in a contract relationship is really a tort action and punitive damages are assessable. This is still to a great or perhaps total extent curbstone law and wishful thinking, but all who use the courts know that today's jurists are zealous to enforce the tenet that people who pretend to bargain and act toward each other in contract situations in good faith but in fact are acting with fraud, duress, and skulduggery should be brought to task.

THE CONSTRUCTION PROCESS

General

The next spawning ground for construction disputes is found in the consequences of the construction process. Many times it can be difficult to differentiate between a design error and the consequence of the process.

Some items that give rise to dispute in the construction process are failure to plan and schedule adequately, failure to follow a plan and schedule, disagreements over what material is really specified, disputes over what is really an equal, failure to supply adequate manpower, responsibility for lack of adequate subcontractor manpower, equipment changes that cause delay, changes, or modifications of scope that increase consequential costs beyond specified markup, failure by the owner to fulfill its responsibilities on time, failure by the engineer to approve contractor submittals on time, overpayment for work in place, underpayment for work in place, failure to provide temporary heat, hoisting, or elevators, lack of performance ability in one of the subcontractors, etc. certainly the list is almost endless, but by trade practice and custom essentially the items thereon relate to someone's failure to do that which was required by its contract, thus changing the basis expected and/or implied to be available to the extent that the means, environment, conditions, or duration of a project are significantly affected for any member of the process who feels sufficiently aggrieved to file a claim.

The construction process dispute is recently being pursued under both contract and tort concepts of law. If a designer who does not have a contract with the contractor does not do its work in the time and manner set forth in the owner–contractor contract or in the time and manner on which the contractor has a right to rely, i.e., duties well established by custom and the standard of care, the designer incurs a very real risk of being sued by the contractor in a tort action. Tort actions bring the added threat of punitive damages. This exposure to the design professional has never really been pursued until the last few years.

The reasons for this previous protection to the designer were numerous and included a reverence toward the designer from the contractor who was typically afraid of the consequences of angering someone who might be on the next job and have a great deal to say as to long-term contractor success. However, in the last 30 years contractors have become very mobile, and the need to depend on architect or engineer relationships for market goodwill has diminished. In those same years, government has become the biggest buyer of construction, and the influence of the designer on the contractor in the government situation has diminished as well. The final reason for the erosion of the designer's position of unreachable holiness is that the complexity of the process of construction and design is such that designer errors involving or altering the construction process cannot now be absorbed. They are simply too expensive. Contractors often cannot afford to pay for their own mistakes let alone someone else's, particularly when someone else is insured against the consequences.

The construction process dispute and almost all disputes involving large amounts of money are time-related. Much focus is put on contract time; the contract documents usually set forth a definite number of calendar days or a fixed final date for completion. There are at times more than one date in the documents when jobs are to be built in phases or specific milestone portions of completion are necessary. The second focus on time-related consequences is the juggling act an owner goes through when, for example, it develops a $100 million program which for reasons of management is spread over say ten or more separate contracts, all or many of which depend on each other for start, midjob milestones, and interrelated finishes. In this type of situation, the pace of everyone involved can and usually does break down and causes the need for adjustments for delay damages.

The No Damages for Delay Contract Clause

The number of claims for delay damages and the dollars involved in delay are such that the specification writers, lawyers, professional societies, government agencies, and others have and are now expending great effort to write no damage for delay clauses.

Typical no damage for delay clauses are as follows:

No Damage for Certain Delays

The Engineer may delay the commencement of the work, or any part thereof, if the Engineer shall deem it for the best interest of the Owner so to do. The Contractor shall have no claim for damages on account of such delay, or on account of any delay on the part of the Owner in performing any work or furnishing any materials to the Contractor, or due to extra work required, but he shall be entitled to so much additional time in which to complete the whole or any portion of the work required under this Contract as the Engineer shall certify in writing to be just. The Engineer shall forthwith notify the Contractor in writing of such determination. Such additional time shall be the sole and exclusive remedy for any delay caused by the Owner. The Contractor shall have no claim for damages on account of any delay on the part of another contractor.

Extensions of Time:

a. Should a Contractor be delayed in the prosecution or completion of the Work by act, neglect or default of the Owner, the Manager of Contracts, or the Architect, or the respective Consulting Engineers, or of any other Contractor employed by the Owner upon the Work, or by any damage caused by fire or other casualty for which the Contractor is not responsible, or by insurrections or civil disturbances, or by order of the court or any other public authority, or by combined action of workmen in no way caused by or resulting from default or collusion on the part of the Contractor, then the time herein fixed for the completion of the Work shall be extended for a period equivalent to the time lost by reason of any or all the causes aforesaid, which extended period shall be fairly determined and fixed by the Manager of Contracts; but no such allowances shall be made unless a claim is presented in writing to the Manager of Contracts *within forty-eight (48) hours* of the occurrence of such a delay.

b. In the event that any or all extensions of time granted pursuant to Par. G4-D.6, a, extend cumulatively for more than one hundred eighty (180) calendar days in length, the Owner may, in order to expedite completion of the Project, unilaterally declare the affected Contract to be terminated, and then proceed to complete the Contract in any way it sees fit; the Owner will make an equitable settlement of the remaining portion of the Contract, under the foregoing circumstances.

c. Apart from extension of time, no payment or allowance of any kind shall be made to any Contractor as compensation for damages on account of hindrance or delay from any cause in the progress of the Work, whether such delay be avoidable or unavoidable, and there shall be no liability upon the Owner or its authorized representatives or assistants, either personally or as officials of the Owner, it being understood that in such matters they act as agents or representatives of the Owner. THE PARAGRAPH IS A PRINCIPAL CONDITION OF THIS CONTRACT, AND EXECUTION OF THE CONTRACT SIGNIFIES ACCEPTANCE THEREOF BY THE CONTRACTOR.

d. The Contractor agrees to insert these provisions in any and all subcontracts entered into by him relating to this Project.

The courts typically have held that no damage for delay clauses are enforceable. Owner attorneys are quick to point out that the parties to the contract obviously contemplated the possibility of delay. There are, however, certain exceptions to the enforcement of no damage for delay clauses and the contractor must understand the game rules of the jurisdiction in which they are working. In one state which is severe in fighting any claim for delay damages, the landmark case to win such damages awarded a contractor entitlement for lack of access to the job, a cause which one can argue is really only another name for delay, but seemingly this was very different to the court. Smart contractors use the right buzz words. Every contract needs a glossary.

The contractor and its counsel should realize and be able to use the argument that there are other basic contract provisions, some of which may not even be written in the specification book but are strongly implied and part of all contracts. These are the duties the owner has, and they may have more weight in the totality of a contract than in an isolated no damage for delay clause.

The courts generally hold that owners have a duty to cooperate with a contractor. Failure to provide access, to deliver owner-provided equipment, return submittals, make decisions, make payments on time, or other similar problems simultaneous with some sort of less than constructive attitude toward such a duty have often been made into persuasive arguments which can and have overcome the best written no damage for delay clause.

The duty to disclose fully all available knowledge is important. If an owner holds back information even inadvertently, the no damage for delay clause can go out the window. Further, the duty to bargain in good faith is an implied portion of every contract. Owners who do not realize they are active parties to a contract or do the work on their part of the contract poorly run the risk of being challenged.

When someone ignores the unwritten elements of a contract and attempts to put undue, unforeseen, and unwarranted costs on the other party by the exculpatory disclaimer under no damage for delay there is strong exposure to the court ultimately ignoring the disclaimer. Further, the court has upheld that extraordinary delay is not included in this type of disclaimer nor is active interference resulting in delay allowed to pass without adjustment.

One typical problem on time-related disputes is that the initial determination of whether an adjustment is due is made at a low level in the inspection division. Usually, the time extension request is denied or time is given but dollars are not. As the dispute escalates, the engineer or architect retreats into the temporary safety it finds in the contract language, the owner follows its designer, nothing seemingly happens except to the contractor's finances, and a claim is born. The result is that is too late when the owner must finally come to grips with the situation. The best time for an owner to work out the real effect of a delay on the contractor's time and money is at the very time the delay is occurring. It will be easiest then to measure the impact. Typically, the longer an owner waits, the more it will cost.

Not only does the owner with an enforceable no damage for delay clause find itself exposed to an argument over duty, reasonability, and intent, it must also bear in mind the federal contract notes and agrees that the contractor is due an increase for ripple damages from changes.   The typical federal changes clause reads as follows:

3. Changes

(a) The Contracting Officer may, at any time, without notice to the sureties, by written order designated or indicated to be a change order, make any change in the work within the general scope of the contract, including but not limited to changes:

(i) In the specifications (including drawings and designs);

(ii) In the method or manner of performance of the work;

(iii) In the Government-furnished facilities, equipment, materials, services, or site; or

(iv) Directing acceleration in the performance of the work.

(b) Any other written order or an oral order (which terms as used in this paragraph (b) shall include direction, instruction, interpretation, or determination) from the Contracting Officer, which causes any such change, shall be treated as a change order under this clause, provided that the Contractor gives the Contracting Officer written notice stating the date, circumstances, and source of the order and that the Contractor regards the order as a change order.

(c) Except as herein provided, no order, statement, or conduct of the Contracting Officer shall be treated as a change under this clause or entitle the Contractor to an equitable adjustment hereunder.

(d) If any change under this clause causes an increase or decrease in the Contractor's cost of, or the time required for, the performance of any part of the work under this contract, whether or not changed by any order, an equitable adjustment shall be made and the contract modified in writing accordingly: Provided, however, that except for claims based on defective specifications, no claim for any change under (b) above shall be allowed for any costs incurred more than 20 days before the Contractor gives written notice as therein required: And provided further, That in the case of defective specifications for which the Government is responsible, the equitable adjustment shall include any increased cost reasonably incurred by the Contractor in attempting to comply with such defective specifications.

(e) If the Contractor intends to assert a claim for an equitable adjustment under this clause, he must, within 30 days after receipt of a written change order under (a) above or the furnishing of a written notice under (b) above, submit to the Contracting Officer a written statement setting forth the general nature and monetary extent of such claim, unless this period is extended by the Government.   The statement of claim hereunder may be included in the notice under (b) above.

(f) No claim by the Contractor for an equitable adjustment hereunder shall be allowed if asserted after final payment under this contract.

Thus the owner faces much precedent and pressure. Exposure to ultimate persuasive arguments by skilled experienced contractor counsel makes the owner's decision to rely on the disclaimer language in a contract dispute one of judgment and risk, certainly not one safe from alarm.

Other Time-Related Problems

Other aspects of time are also very important. Some of the more common of these aspects are the following:

a. The contract time is too short.
b. The contract time is too long.
c. Who is responsible for use of the contract time?
d. What does a schedule mean?
e. What about time extensions?

The Contract Time is Too Short.

This is very often the case and almost the rule in large contracts. The owner or designer has set a contract time which is too short, and short by more than say 10 percent. What does this mean?

A number of considerations come into the picture. The contractor has a right to rely on the time stated in the contract to figure its bid. Everyone knows that the tradition of the industry is that contractors always take the stated time and price the general conditions or time-related aspects of the work accordingly. If the contractor prices the general conditions on a period longer than the stated number of days, then it has taken upon itself the risk of time.

Who judges if the time is too short? To the greatest extent it is one's peers, or the opinion of those involved and those in the industry qualified and willing to make such a judgment. Can owner reliance on the basis that the contractor bid the work with full knowledge of the contract period and thus is responsible to complete on time or pay for the consequences lead to an owner position on time-related matters that will prevail? Perhaps, but it really depends on the contractor's experience and persuasive ability.

Unfortunately for the owner in this aspect of a construction argument most contract durations are set very informally. People who are experienced to varying degrees simply guess at the job length. Sometimes studies are made but mostly someone guesses or gives experienced "guestimates," or other owner needs to meet certain deadlines for completion which are totally unrelated to the construction process are the deciding factor for contract duration.

Owners should require the designer to calculate the construction period and so make a study of how long the job should take. The study does not have to be disclosed to the contractor unless it shows something unusual or unique. It is just another routine design calculation in the file. The worst situation for the owner is to be asked or challenged and have nothing in the file. On a recent project involving a $20 million monumental building with 650 days completion specified, about two weeks after the contract was signed the successful general contractor casually and politely asked the owner to help it with the schedule. "How did you calculate 650 days?" was the query. The owner did not calculate it nor did the designer. It turned out to be a guess and a bad guess by 250 days. Who paid? The owner who had set the time without any thought to reasonability paid, the contractor unquestionably had relied on 650 days. The adjustment was just over $1.6 million. Everyone was upset from the job's start and the entire job was filled with problems as everyone scrambled daily for clefts in the rocks in which to hide.

The best risk management technique for the owner is to let the contractor set the time of performance. In many circumstances both time and money can be used to evaluate the low bidder.

Life cycle cost estimates, systems building, and the like have all made inroads on the idea that only the low bid must be honored. The venturesome contracting officer is becoming more and more evident. The example below is an excellent specification for this technique. In its invitation to bid, the documents state:

IB-5 Completion Time

A. Time is of the essence of the Contract.
B. It is desired that all Work necessary to make the power station concrete floors between rows 12 and 18 available for Purchaser's use as soon as possible under a normal 40-hour-per-week schedule and the building be enclosed to permit heating in the areas bounded by column rows 12, 18 A, and M when required to maintain suitable working conditions.
C. It is desired that the intake structure be constructed after the floors are completed so that the intake structure and circulating water system will be completed within 18 months after award of

Contract. Work on the dock area must be coordinated with installation by others of heavy turbine generator components on turbine foundation constructed under Contract A. It is desired that the portal house and dust control system be installed as soon as possible to allow Purchaser full use of the coal handling equipment installed by Contract B.

D. The Bidder shall indicate on the Bid Form, the completion schedule for the work upon which has bid is based. The completion schedule set forth in the Bid Form shall take precedence over the desired schedule set forth above.

E. The provisions for changes in the contract time set forth in Article GC-10C. shall apply to the dates set forth in the Bid Form.

F. Completion times and schedule for submission of Compliance Submittals set forth in the Bid Form will be considered in the evaluation of the Bids.

Compliance Submittals

All Compliance Submittals which require Engineer's acceptance prior to the manufacture or fabrication of the materials and equipment to be furnished shall be submitted within *90* days following the Date of Contract, and in accordance with the following schedule: (Assume Contract Date is 28 days after bid date.)

Days After Date of Contract	Type of Drawing
_____	Reinforcing steel—initial submittal building floors.
_____	Reinforcing steel—final submittal building floors and walls.
_____	Metal panel walls.
_____	Roof deck.
_____	All other.

Submittals requiring revision shall be resubmitted within _____ days after receiving Engineer's review action copies.

Field Work Schedule:

This Bid is based on using _____ labor for all work on the project site and the undersigned will commence work upon Notice of Award and move onto the site _____ days after the Date of Contract (see Article GC-2).

The undersigned agrees that time is of the essence in the performance of the work included in this contract and that unless the contract time is changed in accordance with Article GC-10, that liquidated damages set forth on page A-2 of the Agreement shall be assessed for each calendar day the work is incomplete. Contract time is set forth below. If the Date of Contract is not within 28 days after the Bid Date, the Contract Time will be extended the number of days the Contract Date is deferred. Notice of Intent to award the contract is normally issued in 7 to 14 days.

All work necessary to make the following major phases of the work or items of work ready and available for the Purchaser's use shall be completed within the following contract times: (Assume Contract Date is 28 days after bid date.) (See Article IB-5, C. page IB-5).

Obviously, the bidder, knowing what the owner desires for a time of performance, is going to offer something reasonably close. If not, one stands to lose the job to any competitor who cannot be trusted to be "square." Thus the bid form just quoted put the risk of time on the bidder but still subtly (or maybe not) did an effective job of covering the owner's needs.

Owners who set times that are simply irresponsible face challenges by contractors for breaches of implied duties. Contract breach leads to total cost damage calculations and collection. Good contractors faced with impossible times of performance will do well to retain a scheduling consultant early. The contractor should give notice of possible claim for the cost of extended performance and for acceleration. Early contractor statements on impossibility of performance and efforts to overcome will potentially set the stage for successful recovery. A good general rule is that the one who sets the contract time warrants that the work can so be done in accordance with all the other conditions of the contract. If detailed studies show that the curing time between adjacent pours make the job impossible to do timely, then the error and liability are on the one who initally made the error, not the estimator who did not take the time to discover the designer had made an error.

The Contract Time Is Too Long

What does one do when the time of performance is too long and the time available is simply not needed? Most contract language does not speak to this directly and since the case stated is a very unusual situation, the lack of attention is not terribly distressing.

In Chapter 1, the rules set forth noted that this particular situation is almost always altered by one of the people involved so that the excess time soon disappears. The contract period if set too long can have one immediate effect—a higher price which occurs when the bidding contractors base their estimates for general conditions, escalation, and the like on a longer period of time.

The owner is then at an immediate disadvantage, for the soft costs, which are anywhere from 5 to 18 percent of the price of its facility, are based on a longer contract period. This can get into some numbers which by themselves are substantial, but they are not easily identifiable or significant to the owner in the overall picture. For example, on a $20 million job, with general conditions through all the trades running some $1.6 million the overstated length will add approximately $25,000 per month in time-related soft costs. Six months too long would be $150,000 more buried in the bid, not too significant when one compares $20,150,000 or $20 million, but to the contractor's side of the bargain the dollars can be highly significant to its profit picture.

What should a contractor do when it sees the job getting done soon? Obviously it has the opportunity to finish the job faster or slower, whichever is cheapest. If it completes ahead of schedule, it then has the problem of getting an unready or unprepared owner to take over. Some lower levels of government such as a county sometimes cannot take a job without approved funds for maintenance and staff. Early completion can make that seeming plum turn into upset for a contractor. My experience has yet to see a contractor recover for costs it claimed because the owner would not accept the project sooner than the contract indicated. However, my experience is limited to just two cases that pressed forward to litigation. In both of these the contractor bid the project using the total number of days for its general conditions and issued a schedule for the full time at job start only to discover later than a chance for additional profit existed and in fact took place but to a somewhat less extent.

Contractors are most often in the position of having to prove damages on a non-total-cost basis and defend owner claims or arguments against a total cost assessment. As an example in a recent case, a contractor who had bid a job with a $970,000 fee brought a $5 million claim including $4 million for subcontractors and $1 million for its own account, proving its portion of the claim by time studies. The owner's defense in part was to display to the court through discovery of the contractor's records that the claimant had already made $1,270,000, some $300,000 more than hoped for at bid time, and now wanted another million. The outcry of indignation against the contractor was fierce. The judge seemed to be similarly impressed. (The owner shortly thereafter settled the $5 million case for $250,000.) Hence, the contractor is in a more difficult position to increase profits in a claim as compared to recovering losses.

The sensible contractor who sees an opportunity to do a job in less time than that allowed should estimate costs on the shorter period it believes is attainable. The estimate should contain some sort of scheduling considerations to reinforce the propriety of the conclusions. Obviously, the fee or markup should reflect by its size the contractor's assessment of risk.

The contractor's first schedule should display its plans according to the bid. If the owner is the federal government, it will typically send the schedule back rejected and request the time be fully used. In this case, the contractor is best served by writing a friendly letter explaining how and why the early schedule is necessary and meaningful and then resubmit the schedule by filling in the float to the job end date by an activity as "postcompletion cleanup if required." Sometimes this works, sometimes not. If the government tells one to reschedule the work for larger periods, it runs the risk of being accused of interference to operations.

The contractor is best served by straightforward comments on construction time. Experience indicates that the contractor will do best to rely on the stated time for its bid; if the job gets done sooner, fine, perhaps there are some extra profit dollars. If the time is much too long and the competition is sharp, it is probably better to call the owner and explain the situation before the bid. Then submit the bid as seen and be fully armed for winning an argument if the owner drags its feet during construction.

Who Is Responsible for the Use of Contract Time?

Some people in the industry take the position that the contractor may do as it pleases as long as it delivers the project in the degree of completion required in the time stated. The position of these proponents is as follows: The contractor is totally within its right to do the first 10 percent of the job in the first 50 percent of the time if the contractor so chooses providing of course the contractor then completes the remaining 90 percent of the work in the last 50 percent of the time. In other words, a contractor theoretically cannot be considered as behind schedule unless the completion date after any due adjustment has passed.

This same line of thinking also leads to the conclusion that the contractor does not have any need or real obligation to give the owner or its designer a schedule. All such information is totally the contractor's business; it will deliver on time if left alone to pursue its work.

Although many find merit in these ideas, the trend in construction is away from such positions. Owners have found they cannot be placid parties in a construction contract. They must comment on performance constantly or run the risk of implying acceptance of performance which may have been totally unacceptable, but with the defects thereto so patent that failure to speak could forfeit the right to ultimate later challenge.

Today contractors are invited to bid on jobs in which specified detailed scheduling, planning, and progress or status monitoring methods are set forth to the benefit of all. Contractors that are quick to point to the duties of cooperation and disclosure of the owner and designer are thus required to make the same frank and open disclosures of their own thinking process. Owners who permit general contractors to refuse to have full attendance of subcontractors at job meetings are being bamboozled.

The planning of the utilization of the time available for construction is best done openly with the input of all who have concerns taken into account. The contractor who falls behind schedule early is often helped by the public peer pressure of being behind. Those responsible for the slippage can often be pinpointed and helped when a base for measurement is available.

The advent of the critical path method (CPM) of scheduling and its display to some degree of most every participant's obligations to each other make the secretive shroud of contractor operation and planning no longer a viable or defensible situation.

Contractors, by means of certain types of contract conditions, must now openly share how they plan to use the contract time. Anyone's failure to accomplish the planned use of time along the way may affect the economics of the general contractor, subcontractor, the owner, and so forth, but the basic premise of how the time could or should have been used is common knowledge. All the participants in the project have the right to measure their ultimate experience on the job against a common document which shows the experience implied and planned during bidding or in the immediate days right after the Notice to Proceed.

Effective use of contract time subject to the constraint of weather and other noncompensive risks is a right common to all who share in the contract. Abrogation of these rights to just a few of the participants will generally not be tolerated. Government or other owner penalties of reduced payment to those behind schedule have the problem of being potentially self-defeating, an argument that is heard daily. Experience would nonetheless indicate that the owner is best served by enforcing the entire contract even if it momentarily appears to be self-defeating in the specific case.

What Does a Schedule Mean?

The industry and the law are undergoing major considerations as to the real meaning of scheduling. The previously noted recent advent of critical path techniques has started the formation of a not yet totally resolved industry custom or policy and common law on schedules.

The broad aspects of scheduling are, however, becoming clear. Essentially the functions of scheduling and coordination have the greatest sphere of influence over any participant in the program. The possession of greatest influence in this aspect over the economics of the various participants can vary from general contractor, to owner in the case of a multiple prime state law, or to the construction manager. These broad aspects indicate that those who participate in the construction process have a right to rely on the best use of the contract provisions for the project as a whole in accordance with the tradition of the industry as well as the specified scheduling and planning process.

Those who have followed the industry on a total basis will fully agree that the CPM schedule has aided all involved on a job. The general contractor who inadequately performs the specified scheduling work runs the risk of challenge by the subcontractor and owner if they thus have a bad experience on the project with regard to time.

The minimum risk situation is that in which the one responsible for scheduling openly offers its best ideas, exchanges them as far as reasonable, and openly publishes all scheduling information. To fail to do so is to invite criticism and attack from almost every participant or firm interested in the program. Good quality, openly displayed sensible scheduling, even if ultimately unachieved, serves everyone well.

What about Time Extensions?

The usual method in the contract process is to recognize two types of time extensions: compensible and noncompensible. The compensible time extension comes about when the period of performance is modified by an action or lack of action by the owner. It can include constructive changes and total or partial suspensions or slowdown of work. The compensible time extension is generally expensive and is usually initially resisted by most owners from both the time and dollar views put forth by the contractor. The noncompensible category includes but is not limited to weather, strikes, earthquakes, war, acts of God, and other occurrences recognized as a risk to both parties for which no compensation except time is due or expected.

Coupled with these two types of time extensions is the intrinsic practice of most contractor management personnel to claim an extension of time for any occurrence to which it might be somehow entitled. Somewhere back in Adam's first day after leaving the Garden of Eden, word must have been passed that it was smart to ask for extensions of time. Whether or not the reason for the extension was truthful or really delayed the job is immaterial; smart people in charge always ask for time. Thus contractors always seek time extensions for all and every reason.

The owner should grant every request for noncompensible time extensions. If the contractor asks for ten days for bad weather, the sophisticated owner will immediately grant the request. As difficult and unpleasant as it may be to admit, in practice job completion and contract completion dates have little to do with each other. The contractor will get the job done when it gets it done. The contract completion date generally has little real meaning in this process. Only at the end when everyone tallies up the figures is the time issue faced or related to reality. Thus every noncompensible time extension requested by the contractor should be granted by the owner. Such an action makes the owner a good fellow, certainly reasonable, understanding, and acquiescent. Perhaps it means the loss of liquidated damages, but experience is that liquidated damages are always too low to be of substantive value anyhow. Indeed, it is only in recent years that anyone has really paid attention to liquidated damages.

Many still believe that liquidated damages cannot be collected unless a bonus is offered. This is not true although the damages, if so high that they are obviously a penalty, can turn out to be uncollectible. If the liquidated damages are reasonable or too low, they are held not to be a penalty, but are readily accepted as a preagreed measure of damages. The typical contractor in a litigated dispute over time extension often ends up trying to preserve the stated liquidated damage amount since it is typically far lower than actual damages and almost never can be increased since the figure was set by the owner unilaterally and then agreed to by the contractor.

The owner is best served by issuing every noncompensible time extension. The contractor is best served by only notifying the owner that such a noncompensible situation has occurred and that it may be necessary some time in the future to require an extension of time. There may be a better time extension which has dollars attached yet unknown. Just a polite letter of notice is all the contractor needs in the file for most of the job.

The Consumer Reaction

The final category of dispute origin to be covered in this chapter is the reaction of the consumer. The consumer can be any of a variety of people. It can be the owner, private or public, one person, or a corporate body. The corporate owner in a dispute situation always boils down somewhere to one person who at times is hard to find. The word "management" is the protective umbrella used to shield some individual.

The consumer can be a tenant or it can be a purchaser of the building such as a condominium buyer. It can be a user who pays money to come into the building or comes in to spend money or go to work. The aggrieved consumer can have no relationship with the contractor, designer, or owner and still have a reaction to their endeavor.

Good buildings can sell images. Bad buildings can provide much press and publicity. In some cases there is no such thing as bad free advertising—here and there a problem building or structure can be a blessing. Everyone in Boston knows the Hancock Building.

The condominium owners association. The condominium owners association is a relatively new group found on the construction litigation scene. It has become prominent in the past five years in certain sections of the country as a force finding fault and seeking relief from the construction industry.

There is a great deal of strength in the condominium owners association. The average condominium has between 200 and 300 owners or units which collectively can finance and press hard in pursuit of the contract when the residents feel the builder has done them in. The difficulty from the contractor's point of view is that just a few vocal residents can alarm and incense those who might be more placid and willing to solve whatever difficulties they may have with their new home by themselves. Quite often the condominium owners association is made up of retired people who are pleased to have the excitement of a court battle or are titillated with the idea that a substantial sum of money might be recovered or substantial free improvements might be made to their home.

The attorney for the condominium owners association can generally finance its efforts with modest contributions from each owner. A $50 contribution or assessment per unit can be $10,000–$15,000, enough for hours of legal service. When the builder or the developer is solvent, the builder or developer or both bonded, or an insured architect or bonded subcontractors are involved, the condominium owners association and its counsel can start and press litigation which will have a collectible dollar at the end of the rainbow. There are times, of course, when no assets can be collected. The builder who has failed and depleted its assets is pragmatically beyond judgment. Buyers who have dealt with a marginal developer or contractor run the risk of having damages that cannot be recovered.

The condominium owners association has a hole card since its dispute is generally heard in front of a jury of people who are sympathetic to the little guy. If the developer is a defendant and it has hired a general contractor, a major problem for the developer often exists with an AIA A201 compulsory arbitration agreement between developer and contractor and a lawsuit between the condominium owners association and developer and no consolidation procedure provided. The developer can find itself in two separate proceedings and the real possibility of a third proceeding which can be a separate arbitration with the designer group as well. The more contracts entered into by the developer can just mean more arbitrations. The condominium owners association is a serious consumer group for potential litigation which is complicated by common contract forms, not drawn for the developer's interests. Experience indicates that serious condominium owner associations will be served well by bringing demands for correction and adjustment against the owner.

New policies and regulations now exist for insurers underwriting surety bonds for condominium construction contracts in specific parts of the country. Sometimes performance and payment bonds on condominium projects are very hard for general contractors and developers to obtain.

The success of the condominium owners association in the legal process has led to some spurious suits being brought by such groups. A condominium owner's lawsuit needs merit just like every other lawsuit to be worth pursuing. Occasionally, parties in lawsuits win or lose undeserved dollars. The prospect of winning undeserved dollars is poor reason for a condominium owners association to start a lawsuit and those that do usually end up with assessments against all the owners for unproductive but high legal fees and new officers on the condominium owners association.

The public owner. The typical public owner, such as a school board, division of hospitals, or municipal corporation, is not particularly set up or organized to display or assert its reaction to its new building or alteration project if it does not do what it was supposed to do. The adage "out of sight, out of mind" applies for most of the problems experienced as this type of owner takes over and uses a newly constructed facility. The maintenance and occupancy of the building are generally the responsibility of a set of people other than those who were responsible for its design, creation, and construction. No one knows who did not do what, or had agreed to do something else, or whatever. It is only infrequently that the designer passes out manuals which say that the building was designed to support such and such or that so many kilowatts of electricity were to be consumed, and so on. Instead the use of the building is something that is learned by chance by the people who take over and determine for themselves how they think the building should best be used. It is the rare building which is put into use as programmed by the engineer.

It has not been until recently that any emphasis has been placed on getting the contractor or the designer to pay for its mistakes. The prime move toward this idea has been the staggering cost to repair, replace, or alter a system that does not function acceptably let alone as it was purported to. The complexity of today's installations as far as keeping indoor atmosphere cool, dry, hot, quiet, etc., or treating one's sewage makes it necessary for an owner who has not been well served by its contractor or designer to pursue one or both for relief.

Until some years ago there was a tendency among the design profession to be unwilling to find fault with other members of its profession. More recently, responsible professionals in design are now willing and forced to criticize those among them who have not done the type of design work expected. The same has occurred with contractors. As demands have been made upon the construction industry to produce more and more in a shorter period of time, workmanship has decreased. The tendency of the employees to favor the union hall to his employer has led to a lack of skill, interest, durability, and workmanship in many, many cases. The owner now is unwilling and unable to absorb the results for design and poor workmanship.

Warranties.   The contract will typically spell out the obligation of the contractor regarding guarantee. The AIA contract form is as follows:

> The Contractor warrants to the Owner and the Architect that all materials and equipment furnished under this Contract will be new unless otherwise specified, and that all Work will be of good quality, free from faults and defects and in conformance with the Contract Documents. All work not conforming to these requirements, including substitutions not properly approved and authorized, may be considered defective. If required by the Architect, the Contractor shall furnish satisfactory evidence as to the kind and quality of materials and equipment. This warranty is not limited by the provisions of Paragraph 13.2.

13.2.2   If, within one year after the Date of Substantial Completion or within such longer period of time as may be prescribed by law or by the terms of any applicable special guarantee required by the Contract Documents, any of the Work is found to be defective or not in accordance with the Contract Documents, the Contractor shall correct it promptly after receipt of a written notice from the Owner to do so unless the Owner has previously given the Contractor a written acceptance of such condition. The Owner shall give such notice promptly after discovery of the condition.

The courts typically hold that the warranty means what it says and will enforce the contractor's obligation to meet the requirements.

Many problems in warranty come about when the contractor is unable to duplicate the agreement it made with the owner, its suppliers, and its subcontractors.   The typical national manufacturer of major equipment will offer a warranty for one year from startup, a period that does not coincide with the agreement between owner and contractor.

Obviously, the owner feels the contractor knew this and should have priced its bid to offer the warranty stated regardless of startup caveats from the manufacturer, but most contractors do not do so and instead a squabble over warranty periods occurs. The owner has no reason to give in on warranties and if it does so, it has been outplayed. The 1976 AIA A201 edition attempts to deal (see Appendix A) with this problem to the owner's discomfort.

Owners who accept buildings with problems that are known at the time of acceptance, accept the problems as well, unless the problem is noted as an exception. These problems are called patent defects. An unknown defect is called a latent defect and is something wrong that no one could observe or properly realize. In the case of discovery of a latent defect, the owner must give the contractor the opportunity to repair or correct the defect before it can go off and seek relief in some preferable way.

Aggressive contractors do well to press continually for substantial completion certificates, to urge partial occupancy, and constantly fill the atmosphere with a stated willingness to make good all defects in workmanship. The aggressive contractor will simultaneously be careful to vigorously argue verbally that it is responsible for only those defects arising from failure to follow plans and specifications, and request proof of departure each time a defect is raised for correction. Aggressive owners will do well to hold substantial sums of the contractor's money. Reduced retainage at occupancy generally means increased grief after occupancy.

The high degree of sophistication needed to understand how to run one of today's projects puts another burden on the owner. Typically owners do not maintain the new building and its components as required by the operator's manual. Often, too, no one on the owner's staff understands the theory of the boiler plant, fan room, or controls. An owner with but a small amount of missed but required maintenance may have needlessly given the aggressive contractor enough muddy water to make the owner attempt to have the contractor perform the correction process of equipment defects so frustrating and lacking in joy that the owner inevitably decides to face the problem itself, and all parties go their separate way.

The implied warranty or right of reliance on buildings or parts thereof to do certain things has received a great deal of press and publicity. Certainly the owner or user has a right to pursue recovery of damages if the project does not deliver that which all have a right, stated or not, to expect. The product's liability litigation as it relates to construction again seems to follow the doctrine of reasonability. The subject of product's liability litigation is outside the scope of this text.

4

THE KEY TO CONSTRUCTION DISPUTES – TIME

Those involved in a construction problem quickly learn that it is not the hard (or the nuts and bolts) dollars which are important, rather it is in the time-related costs that the huge damages arise to all concerned. The owner with the late facility has all sorts of extra costs or losses with which to contend in a delay posture; the contractors suffer the added costs of stretch-out, escalation of wages and other costs, and the loss from their inability to focus management on new spheres of work.

Time is the key and the use of time is planned by scheduling. Every construction job requires a schedule. Some require simple bar chart schedules, others require more detailed studies, and some require complex computer-produced critical path method or construction manager control system programs. The latter techniques are very popular, and becoming more so to the extent that sophisticated scheduling techniques are now required for certain types of work. The bar chart is the usual standard when CPM scheduling is not used.

The collective published opinion of the construction industry, which matches the author's experience, is that the advent of CPM techniques in scheduling has measurably shortened construction times; furthermore, these techniques can and do provide meaningful information on which key economic decisions can be made. The advent of the CPM schedule in no way negates any of the rules set forth in Chapter 1; it does, however, provide an information device which can be simply prepared, refined to any degree, and used with success.

The same limitations or expectations can be found in a bar chart. It, too, is an information device and as a scheduling technique can be detailed or refined to almost any level and then followed similarly to CPM schedules with some degree of success.

Strangely, the majority of scheduling done in the construction industry is of far less quality or interest than one would expect. Until about 1960 the question of how one should use the construction period traditionally was claimed to be the exclusive right of the contractor. Further, the general contractor then and now has traditionally protected its use of the job's duration from any subcontractor's right as well. For instance, a standard contract clause in any general to subcontractor contract reading as follows can be very successful in protecting the general contractor on the time aspects of subcontractor performance:

> It is understood and agreed that work is to commence and all materials and equipment needed for installation of the work be furnished as soon as construction is ready to receive same. Work shall be completed with utmost dispatch, time being of the essence of this agreement.
>
> It is understood that all shop drawings, details, colors, samples, etc., be approved by the Architect prior to starting any actual work.
>
> The product shall be delivered in accordance with these mutually agreed shipping dates:
>
> First Floor—April 10, 1979
> Second Floor—May 15, 1979
>
> It is understood and agreed that time is of the essence and if for any reason this subcontractor fails to furnish and install its materials and equipment in order to maintain the progress of the job and complete its installation on time, the subcontractor represents it full well understands the consequence, and that it shall be held responsible for all costs resulting therefrom to the owner, contractor, and other subcontractors due to time lost in construction.

Thus the schedule which the subcontractor agreed to follow was that which at that time best suited the immediate or perhaps future needs of the general contractor. The schedule could be written, graphed, verbalized, or even dreamed. The following or finish trade subcontractors were for all practical purposes contractually obligated to make up for the failure of other earlier subcontractors to be completed on time.

A general contractor's statement has been put forth and successfully argued at many a progress meeting to the effect that a job is not late until the adjusted contract period has actually been exceeded. In addition, some owners believe that somehow, in spite of every indication to the contrary, the contractor who has done everything too slowly for 600 days is suddenly going to repent and complete the entire work schedule in the 100 days remaining, a belief that is always in vain.

The complexity of recent projects, the failure of so many jobs to be done on time, the advent of the computer and the CPM schedule, and the appeal for wisdom have led to today's schedule specification.

THE BAR CHART

A typical bar chart specification is as follows: AIA201, 1976 states:

> The Contractor, immediately after being awarded the Contract, shall prepare and submit for the Owner's and Architect's information an estimated progress schedule for the Work. The progress schedule shall be related to the entire Project to the extent required by the Contract Documents. This schedule shall indicate the dates for the starting and completion of the various stages of construction and shall be revised as various stages of construction and shall be revised as required by the conditions of the Work, subject to the Architect's approval.

A typical bar chart is shown in Fig. 4.1.

The bar chart is a simple linear chart. However, it states a number of important facts which must be borne in mind in dispute positions. First the bar chart can be analyzed for a critical path. For instance, the bar chart in Fig. 4.1 was for a certain correctional facility and displayed a number of facts. In Fig. 4.2 the critical path is shown in the matched portion of each bar. The jail equipment follows the enclosure of the building. There is every expectation that the building will be enclosed and heat will be available for installing the master bar, cell doors, and locking mechanism portions of the jail equipment.

The initial progress schedule through a bar chart displayed the critical path shown in Table 4.1.

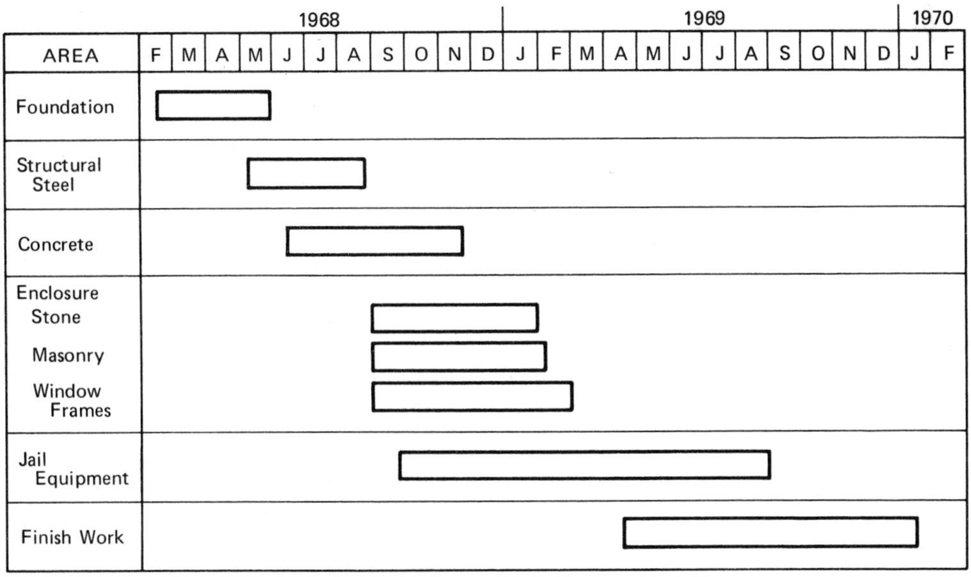

Figure 4.1 Jail initial progress schedule.

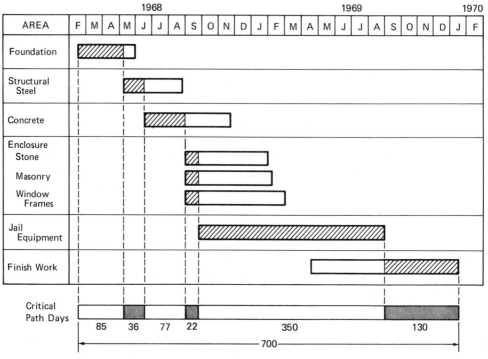

Figure 4.2   Jail initial progress schedule with critical path.

The bar chart can thus at times be used to develop and argue a critical path and possibly a sequence of work on which contractor or subcontractor can claim a right to rely. The contractor who ignores its scheduling obligations under a contract potentially faces claim by a subcontractor whose economics have been affected by poor coordination. The prudent general contractor best plans effectively and then is expected and obliged to work at

TABLE 4.1   Initial Progress Schedule: Critical Path—June 19, 1968

	Critical Start	No Longer Critical	Days Critical
Foundations	02/15/68	05/10/68	85
Structural steel	05/10/68	06/15/68	36
Concrete floors	06/15/68	08/31/68	77
Walls			
Stone	08/31/68	09/21/68	22
Perimeter masonry walls	08/31/68	09/21/68	22
Windows	08/31/68	09/21/68	22
Jail equipment	09/21/68	09/06/69	350
Plaster, paint, and finish	09/06/69	01/15/70	130
			700

the normal level of the industry to ensure that the contract work progresses reasonably on schedule.

The subcontractor has the right to rely on the cooperation of the general contractor and certainly to expect a minimum of interference with the subcontractor's work plan. The subcontractor also has the right to rely on the general contractor for job management with a normal degree of competence. If the general is negligent or indifferent in these aspects, both a contractual and a tort theory for recovery loom as possibilities to the subcontractor. Torts are the key to the potential for punitive damages; there is no question that punitive damages are almost impossible to collect, but who knows what will happen in a dispute? It is not easy for subcontractors to prove such negligence by the general contractor, nor is it easy to prove that damages flowed from the scheduling failures. However, the latter is an argument and a potential problem that all should bear in mind.

The general contractor also has the right to rely on the subcontractor to perform on schedule; this requirement is a universal contract provision in subcontracts. It is strange, however, that delay damages suffered from a delinquent subcontractor are not often pursued by the general contractor.

THE CPM SCHEDULE

The critical path method (CPM) is technically known as the I–J or activity on arrow method of network diagramming. This method had its genesis in the late 1950s in refinery turnarounds, Polaris missile programs, and several other significant endeavors. The technique was taught to the private construction industry by seminars given across the United States; it caught on rapidly with both large buyers and producers of construction.

Initially hailed as the ultimate solution to all job problems, the technique went through a number of reactions by the user. It is now 20 years of age, part of every major school's management and engineering curricula, and very much a part of life in the construction industry. The most recent years have seen more and more emphasis and attention given to the CPM scheduling work on each job. The use of CPM in claim evaluation and presentation makes it an ever increasing focal point for attention.

CPM will become even more important as time goes on; capable logicians have developed and are gaining respect in the industry. The logician or scheduler so important in wartime for fueling tanks and trucks and feeding the troops is just as important economically in the construction process. As the position of the scheduler gains recognition and a higher salary level, it will elevate the place of the schedule.

A detailed analysis of CPM is beyond the scope of this book. However, for the sake of the reader totally unfamiliar with this technique the following is noted: The CPM process simply means drawing the activities involved in the project in their sequential relationship to each other and showing each activity as an arrow.

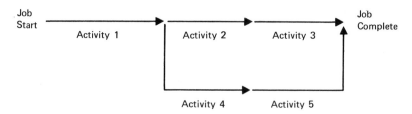

In the accompanying illustration, activity 1 is the first to be performed. When activity 1 is complete, activities 2 and 4 start. At the end of activity 2, one can do activity 3. Activity 5 follows activity 4. The job is complete when activity 3 and 5 are done.

In CPM each activity arrow is described by two numbers; one at the tail called the I number, one at the head called the J number. The activity numbers are shown in circles or nodes and are also called events.

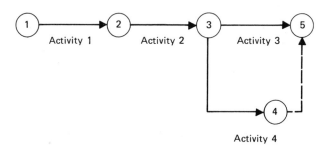

Activity 1 has an I–J number of 1–2, activity 3 has an I–J number of 3–5. The J number of one activity is commonly the I number of another activity. The activity can be described further:

10	Duration in days, weeks, etc.
$2000	Value of activity
920	The specification section
PTR	The painter or responsible subcontractor
10 men	The type of manpower

Graphically all the information can be shown as follows:

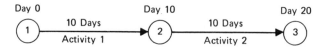

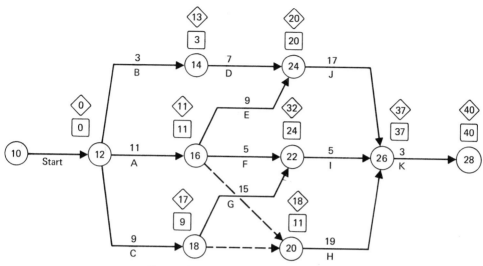

Figure 4.3   CPM network diagram.

CPM time calculations are based on the premise that subsequent activities cannot start until the preceding activity is complete. This is not necessarily entirely true in the field but is a limitation of the calculation process. In this sample, activity 1 starts on day zero, takes 10 days, and is completed on day 10. Activity 2 starts on day 10 and is completed on day 20.

The CPM diagram in practice will consist of many arrows; it can look overwhelming to the tyro but is really very simple. The ten-arrow diagram in Fig. 4.3 shows all that is involved in a 2000-arrow diagram. The dashed arrow is a dummy or network restraint to show activity relationships on the diagram when a number of activities have dependencies. The dummy has no duration.

TABLE 4.2   Schedule Generated by Diagram of Fig. 4.3

I-J	Activity	Early Start	Early Finish	Late Start	Late Finish	Float
10–12	Start	0	0	0	0	0
12–14	B	0	3	10	13	10
12–16	A	0	11	0	11	0
12–18	C	0	9	8	17	8
14–24	D	3	10	13	20	10
16–22	F	11	24	27	32	16
16–24	E	11	20	11	20	0
18–22	G	9	24	17	32	8
20–26	H	11	30	18	37	7
22–26	I	24	29	32	37	8
24–26	J	20	37	20	37	0
26–28	K	37	40	37	40	0

Figure 4.3 shows the early and late event times for each activity. The early times are in the box, the late in the hexagon. The early event times will be the early start of the activities flowing from the event or node; they are calculated by a forward pass from job start by adding the duration from each arrow to the largest event time of its I node. The late times at the nodes are the result of a backward pass or subtraction from the total duration calculated in the forward pass.

The ten-arrow diagram in Fig. 4.3 generates a schedule that can be displayed as shown in Table 4.2, where the float is the number of days an activity or chain or path of activities can be delayed without affecting the end date. Zero float activities are critical. The chain of critical activities is the critical path. Any delay or extension of work on the critical path increases the duration of the job.

THE PRECEDENCE DIAGRAM

The precedence method is very similar to the I–J technique. It is event-oriented rather than activity on arrow. The precedence diagram is typically drawn with boxes containing an activity number, description, duration, and all the other information found in a CPM diagram. A box precedence diagram, which is based on the same logic as the ten-arrow CPM diagram in Fig. 4.3, is illustrated in Fig. 4.4.

The precedence diagram can also be drawn without the boxes. This technique, sometimes called succeedence, is shown in Fig. 4.5. It too is the same

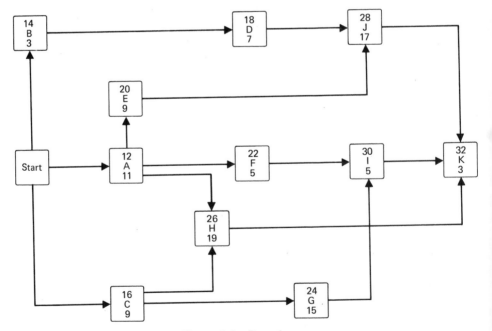

Figure 4.4   Precedence.

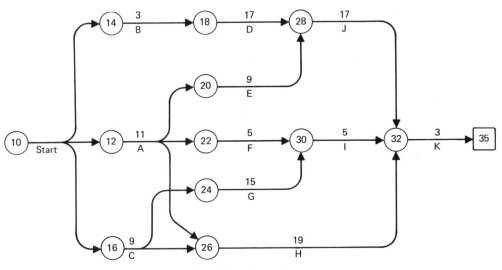

Figure 4.5   Precedence on arrow.

diagram or schedule as shown for the I–J techniques. Precedence techniques and the computer programs for them allow the use of overlaps. Overlap can be of great value in network planning on a complicated job of several thousand arrows. It can also be a terribly difficult problem to keep straight in revising and updating the several-thousand-arrow schedule.

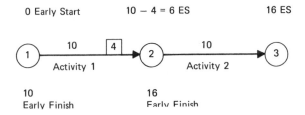

The same ten-activity schedule is also shown with its early start times a bar chart in Figs. 4.6.

COMMON CPM SCHEDULE SPECIFICATIONS

There is no end to the variety of specification requirements written for CPM schedules. Several which were used in the real world have been selected as examples for this text.

Contractor Prepared Network Analysis System:

1) The General Contractor shall prepare a CPM type work progress schedule with his own, "in house", personnel. It is intended that maximum input planning shall be provided by Contractor's supervisory personnel. In the event the Contractor does

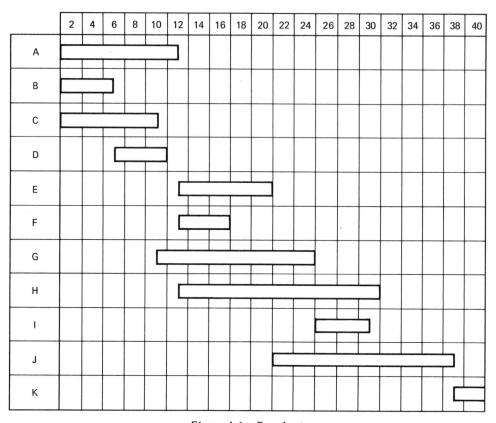

Figure 4.6   Bar chart.

not have computer capability, the use of an outside firm approved by Architect may be used for making mathematical computations and analyses.

2) The system shall consist of diagrams and accompanying mathematical analyses:

    a. Diagrams shall show order and interdependence of activities and sequence in which work is to be accomplished as planned by Contractor. The basic concept of a network analysis diagram shall be followed to show how the start of a given activity is dependent on the completion of preceding activities and its completion restricts the start of the following activities.

    b. All activities of Architect and Owner that effect progress shall be shown also.

    c. In addition to detailed network diagrams, provide a summary network consisting of a minimum of 50, and maximum 150 activities which is based on and supported by the detailed diagrams. Summary network shall be time scaled with weekends and holidays shown.

    d. The mathematical analysis of network diagrams shall include a tabulation of each activity showing:

        1.0. Activity code identification number.

        1.1. Preceding and following event numbers.

        1.2. Activity description.

        1.3. Estimated duration of activity (calendar days).

        1.4. Earliest start date (by calendar date).

        1.5. Earliest finish date (by calendar date).

       1.6. Latest start date (by calendar date).

       1.7. Latest finish date (by calendar date).

       1.8. Slack or float (calendar days).

       1.9. Monetary value of activity.

       1.10. Specification section in which activity is a part.

   e. Analysis shall list activities in sorts or groups as follows:

       1.1. By the preceeding event number from lowest to highest.

       1.2. By the amount of slack or float time.

       1.3. By responsibility in order of earliest start dates.

       1.4. In order of latest start dates.

3) Submission deadlines:

   a. A preliminary network defining operations during first 90 calendar days after notice to proceed shall be submitted prior to Contractor's first application for payment.

   b. The completed network analysis consisting of detailed mathematical analysis, network diagrams, and anticipated monthly payment schedule shall be submitted prior to Contractor's third application for payment.

4) Contractor shall participate in a review and evaluation of proposed network diagrams and analysis with Architect and Owner.

5) Contractor shall submit at monthly intervals a report of actual construction progress by updating network diagrams and analysis. Any change to progress shown on "critical path" shall require Contractor to revise network analysis system before next application for payment.

CRITICAL PATH METHOD SCHEDULING

1. GENERAL

   1.1 Contractor shall provide all the information necessary for the Critical Path Method (CPM) Consultant, employed by the Government, to develop a network plan demonstrating complete fulfillment of all construction contract requirements and for the CPM Consultant to keep the network plan up to date in accordance with the requirements of this section. The Contractor shall utilize the plan in planning, coordinating and performing the work under this Contract (including all activities of subcontractors, equipment vendors, and suppliers). The principles and definition of the terms used herein shall be as set forth in the Associated General Contractors of America (AGC) publication "CPM in Construction, a Manual for General Contractors," Copyright 1965; but the provisions of this section shall govern the planning, coordinating, and the performance of the work under this contract.

   1.2 Not Used.

   1.3 The Contractor agrees that the CPM Consultant's Project Network Diagram is the Contractor's committed plan to complete all work within the allotted time and assumes full responsibility for the prosecution of the work as shown.

2. INITIAL SUBMITTAL

   2.1 To the extent necessary for the CPM Consultant to reflect in a network arrow diagram the Contractor's plan for completion of this Contract, the Contractor shall be prepared to meet with and assist the CPM Consultant, and furnish information subsequent to the award of this Contract.

The CPM Consultant will submit for the Contracting Officer's review an arrow diagram describing the activities to be accomplished and their dependency relationships together with a computer-produced schedule showing starting and completion dates for each activity in terms of the number of days after receipt of notice to proceed. All completion dates shown shall be within the period specified for Contract completion.

The General Contractor shall be responsible to reflect, through the Consultant, all subcontractor work as well as his own work in the network diagram in logical work sequence and time estimates to show a coordinated plan of work for all contractors, thereby providing a common basis of acceptance, understanding, and communication.

In conformance with contract specifications General Provisions Clause 7 and General Conditions Clause 12, entitled "Payments to Contractor," the General Contractor shall furnish a breakdown of the total contract price by assigning dollar values (cost esimates) to each applicable network activity, which cumulatively equals the total contract amount. Upon acceptance by the Contracting Officer, the values will be used as a basis for determining progress payments. The General Contractor's overhead, profit, and cost of bonds shall be prorated through all activities. The CPM Consultant will furnish monthly, a cash flow curve indicating graphically the total percentage of activity dollar value scheduled to be in place based on both early and late finish dates.

2.2   The Arrow Diagram shall show the sequence and interdependence of activities required for complete performance. In preparing the Arrow Diagram, the Contractor shall assist the CPM Consultant by breaking up the work into activities of durations of approximately fifteen (15) working days each, except for nonconstruction activities such as procurement of materials, delivery of equipment, and concrete curing.

The diagram shall show not only the activities for actual construction work for each trade category of the project, but also such activities as the Contractor's work of submittal of shop drawings, equipment schedules, samples, coordination drawings, templates, fabrication, delivery and the like, the Government's or Architect-Engineer's review and approval of shop drawings, equipment schedules, samples and templates, and the delivery of Government-furnished equipment or partition drawings, or both. Activity duration (considering the scope and resources planned for the activity) shall be furnished on the diagram.

Failure by either the Contractor or the CPM Consultant to include any element of work required for the performance of this Contract shall not excuse the Contractor from completing all work required within the Contract completion date(s).

Seasonal weather conditions shall be considered in the planning of and scheduling of all work influenced by high or low ambient temperatures for the completion of all contract work within the allotted contract time.

2.3   The Contractor shall also furnish the CPM Consultant the following supporting data for inclusion in the Arrow Diagram:

2.3.1   Average number of trade workers per day, per work activity to suit the durations, in work days, shown on arrow diagram for normal 5-day, 35-hour work week or for a time period as otherwise specified.

2.3.2   The usage on the site of major construction equipment.

3. REVIEW AND APPROVAL

3.1   Within ten (10) calendar days after receipt of the initial arrow diagram and computer-produced schedule, the Contractor shall meet with the Contracting Officer or his Representative and the CPM Consultant for joint review, correction, or adjustment of the proposed plan and schedule. Within five (5) calendar days after the joint review, the CPM Consultant will revise the arrow diagram and the computer-produced schedule in accordance with agreements reached during the joint review and shall submit two (2) copies each of the revised arrow diagram and computer-produced schedule to the Contractor and Contracting Officer. The resubmission will be reviewed by the Contracting Officer and, if found to be as previously agreed upon, will be approved. An approved copy of each will be returned to the CPM Consultant. After the CPM Consultant has received the approved copy of the Arrow Diagram and Computer-produced schedule, he will substitute calendar dates on the computer-produced schedule, in lieu of the number of days from date of notice to proceed and will furnish three (3) copies each of the computer-produced schedule as thus revised and the arrow diagram to both the Contracting Officer and Contractor.

The Arrow Diagram and the computer-produced schedule generated therefrom, as approved by the Contracting Officer, shall constitute the Construction Contractor's project work schedule until subsequently revised in accordance with the requirements of this section. Revisions to the Official Contract completion date will be confirmed in writing by the Contracting Officer, per paragraph 7 below.

3.2   At an appropriate time in the above sequence of events, the Contractor shall sign his name in a signature box provided by the CPM Consultant on the reproducible original network drawing(s) to signify that the finalized initial arrow diagram(s) is in fact the Contractor's committed work plan and schedule. Subsequent to the Contractor's signature, the Contracting Officer or his duly authorized representative will sign in a space provided below the Contractor's signature to indicate recognition of the Contractor's schedule.

4. PROGRESS REPORTING AND CHANGES (AFTER COMPLETION OF REQUIREMENTS OF INITIAL REVIEW AND APPROVAL.) Reports current through 25th of month with input cut-off the 25th of month and submitted monthly by the sixth (6th) work day.

4.1   Once each month, the Contractor shall meet with the CPM Consultant and Contracting Officer's Representative(s) and provide the information necessary for the CPM Consultant to prepare and submit to the Contracting Officer by the sixth (6th) work day of the month a revised (updated) arrow diagram, as applicable, showing:

4.1.1   Changes in activity sequence logic so that no activity is shown started without all its preceders being completed.

4.1.2   Changes in activity durations and manpower of unstarted activities where agreed upon.

4.1.3   The effect to the network of any justified delays in activity progress. Before any delays are reflected on the network, the Contracting Officer must label the activity delay "Justified," in accordance with Contract General Conditions, Clause 5.

4.1.4   The effect of Contract Modifications (activity durations, manpower, logic and cost estimates) to the network.

When the Contracting Officer or his Representative and the Construction Contractor are unable to agree as to the amount of activity extension(s) or revision to be reflected on the Arrow Diagram, due to a modification, the CPM Consultant shall reflect that amount of activity time extension(s) or revision on the Arrow Diagram as the Contracting Officer or his authorized representative may approve, from the CPM Consultant's input, in his best judgment to be appropriate for such interim purpose. After final agreement has been made or a final decision rendered by the Contracting Officer as to any activity change or revision, the CPM Consultant shall revise the Arrow Diagram prepared thereafter in accordance with such decision.

4.1.5   Changing activity logic, where agreed upon, to reflect revision in Construction Contractor's work plan, i.e., changes in activity duration, cost estimates, and activity sequence for the purpose of regaining or improving progress.

4.1.6   Changing milestones, due dates, and the overall contract completion date which have been agreed upon by the Contracting Officer since the last revision of the Arrow Diagram (refer to Paragraph 7).

4.2   Once each month, usually at the same time when the network is updated, the CPM Consultant, the Contractor, and the Contracting Officer's Representative(s) shall jointly make entries on the preceding computer-produced calendar-dated schedule to show actual progress, to identify those activities started by date and those completed by date during the previous period, to show the estimated time required to complete each activity started but not yet completed, to show activity percent completed and/or dollars earned, and to reflect any changes in the Arrow Diagram approved in accordance with the preceding paragraph. After completion of the joint review and the Contracting Officer's recognition of all entries, the CPM Consultant will submit an updated computer-produced calendar-dated schedule (Monitor Report), to the Contractor and the Contracting Officer no later than the close of business on the sixth (6th) work day of the succeeding month.

The resultant monthly monitor CPM Computer Printout Data shall be recognized by the Contractor as solely his updated construction schedule to complete all remaining contract work, except that portion affected by interim Contracting Officer decisions.

4.3   In addition to the foregoing, once each month the Contractor will receive a narrative report prepared by the CPM Consultant. The narrative report will include a description of the amount of progress during the last month in terms of completed activities in the plan currently in effect; a description of problem areas, current and anticipated delaying factors and their estimated impact on performance of other activities and completion dates, and recommendations on corrective action for the Contractor's consideration. Within seven (7) days after receipt of this report, the Contractor shall submit to the Contracting Officer, in writing, an explanation of corrective action taken or proposed.

5. PAYMENTS TO CONTRACTOR

5.1   The monthly submission of the computer-produced calendar-dated schedule shall be an integral part and basic element of the estimate upon which progress payments shall be made pursuant to the provisions of the Clause "Payments to the Contractor" of the General Provisions and the General Conditions of this contract, and the Contractor shall be entitled to progress payment only upon approval of estimates as determined from the currently approved updated computer-produced calendar-dated schedule, as mentioned under Paragraph 2.1.

5.2   Payments to the Contractor shall be dependent upon the Contractor furnishing all of the information and data necessary to ascertain actual progress and all the information and data necessary to prepare any necessary revisions to the Computer-produced, calendar-dated schedule and/or the Network Arrow Diagram. The Contracting Officer's determination that the Contractor has failed or refused to furnish the required information and data shall constitute a basis for withholding any such payments until the required information and data is furnished and the schedule and/or diagram is prepared or revised on the basis of such information and data. Any disagreement with the Contracting Officer's determinations shall be a dispute concerning a question of fact within the meaning of the Clause of the Contract General Provisions entitled "Disputes."

6. RESPONSIBILITY FOR COMPLETION OF CONTRACT (UNLESS CONTRACT TIME IS ADJUSTED PURSUANT TO REQUIREMENTS OF PARAGRAPH 7 BELOW)

The Contractor shall, whenever it becomes apparent, from the current monthly computer-produced calendar-dated schedule that the contract completion date will not be met, take any or all of the following actions at no additional cost to the Government:

6.1   Increase construction manpower in such quantities and crafts as will substantially eliminate, in the judgement of the Contracting Officer, the backlog of work.

6.2   Increase the number of working hours per shift, shifts per working day, working days per week, or the amount of construction equipment, or any combination of the foregoing sufficiently to substantially eliminate, in the judgment of the Contracting Officer, the backlog of work.

6.3   Reschedule activities to achieve maximum practical concurrency of accomplishment of activities.

7. ADJUSTMENT OF OFFICIAL CONTRACT COMPLETION TIME

7.1   Activity time delays may and probably will occur during the life of a construction contract for one reason or another. However, this does not automatically mean that an extension of contract time is warranted or due the Contractor. It is possible that a strike or contract modification will not affect existing critical activities or cause non-critical activities to become critical, i.e., a strike or modification may result in only absorbing a part of the available total *float* that may exist within an activity chain of the network, thereby not causing any effect on the contract completion date or time.

The Contract completion time or times shall be adjusted by the Contracting Officer only when it can be clearly shown on a monthly updated arrow diagram and resultant computer-produced monitor printout that as a result of alteration to one or more network activities (with or without logic revisions), or additions of one or more network activities, the completion time or times were extended for reasons unforeseen and beyond the control and without the fault or negligence of the Contractor, including, but not restricted to, contract modifications, strikes, unusually severe weather, differing site conditions, suspension orders, acts of God, etc., as provided for in the Contract General Provisions.

7.2   In the event the Contractor requests an extension of any contract completion date, which is not automatically handled by the monthly monitor update meeting, he shall furnish such justification and supporting evidence as the Contracting Officer may deem necessary for a determination as to whether the Contractor is entitled to an extension of time under the provisions of this contract. The Contracting Officer shall, within a reasonable length of time after receipt of such justification and supporting evidence, make his findings of fact and his decision thereon and shall advise the Contractor in writing thereof. If the Contracting Officer finds that the Contractor is entitled to any extension of any contract completion date under the provisions of this contract, the Contracting Officer's determination as to the total number of days' extension shall be based upon the currently approved computer-produced calendar-dated schedule and on all data relevant to the extension. Such data will be included in the next monthly updating of the schedule. The Contractor acknowledges and agrees that if actual delays in activities which, according to the computer-produced calendar-dated schedule, do not affect any Contract completion date shown by the Critical Path in the Network, no extension of official contract time will be granted except as stated above. The requirements of the Contract General Provisions shall govern herein, as applicable.

8. CONTRACTOR USAGE OF CPM SCHEDULING

The General Contractor shall be required to maintain and use on a daily basis, as a working management tool, a current CPM network diagram and computer-produced printout schedule throughout the life of the Contract.

To accomplish this, the following activities shall be performed:

8.1 POSTING OF NETWORK

From the outset of actual contract field work, a current copy of the network diagram shall be posted in the job site field office by the Contractor for daily usage by him, his job superintendent, and/or any authorized field personnel.

8.2 DAILY MARKING OF NETWORK

Daily or as appropriate color mark-in actual progress on each applicable activity of the network to maintain a running account of job progress and to keep current true representation of the work areas for viewing. Also, mark on the network actual activity start dates and completion dates. Be prepared to alter the logic, if any activities are started without all its preceders being completed. The CPM Consultant is required to change the activity sequence, with the Contractor's input so that no activity is shown started without all its preceders being completed.

8.3 PRELIMINARY CPM MEETING

Approximately two (2) days before each scheduled monthly CPM monitor update meeting, the Contractor or his representative shall meet with the Government's field representative to discuss the previous month's progress, including activity sequence revisions, logic changes due to contract modifications, problem areas with corrective measures proposed or undertaken, and any real or anticipated delaying factors. This preliminary meeting will facilitate the making of a productive and meaningful CPM update meeting shortly thereafter.

8.4 EFFECT OF CHANGE ORDERS (MODIFICATIONS)

As requests for Contractor's proposals on GSA Form 1137, for added, changed, or deleted work, are received by the Contractor, he shall, in addition to the Contractual procedures for contract modifications, assess the overall effect on the existing CPM network diagram activities by establishing any and all revisions to the current network to satisfy the particular change request.

At the subsequent preliminary and monthly CPM meeting, the Contractor's proposed activity effect (activity logic, duration, manpower with or without dollar settlement), shall be discussed and if agreed upon, shall be incorporated in the network logic change for that month. If no agreement can be reached, unilateral logic and/or time revisions shall be made in accordance with paragraph 4.1.4.

8.5 RESOURCE LEVELING

With manpower and any major equipment noted for each noncritical activity on the network diagram, the Contractor has the opportunity to spread out work crew assignments within activity slack or float times, shown on the accompanying computer printout, in order to maintain a reasonable steady work force and avoid labor waste and inefficiency. Keep in mind that any float used for one (1) activity is subtracted from the float shown for succeeding activities within a particular activity chain, i.e., any other start time other than early start (ES) time will effect succeeding activities. It is recommended that manpower requirements be totaled for critical activities per day or week as well as for all early starts and late starts, then level out the labor peaks and lows. The logical sequence of activities as shown on the network diagram must not be arbitrarily altered to satisfy this leveling. The CPM Consultant is available for assistance in this area.

8.6 COMPUTER PRINTOUT INFORMATION

The computer-produced CPM printout sorts, furnished with every monthly updated network diagram, have certain information that the Contractor shall avail himself of daily, to closely monitor and maintain scheduled progress during each reporting period. Upon receipt of the first printout sorts and all subsequent monitor printout sorts, note the following for the duration of each reporting period:

8.6.1 All activities that could be started based on early starting times.

8.6.2 All activities where desired progress must be maintained based on utilization of total floats shown (resource leveling of noncritical activities).

8.6.3 All activities where progress must be increased per paragraph 6, "Responsibility for completion of contract," based on negative floats shown.

8.6.4 All activities where progress much be adhered to based on "critical activity" status, zero float.

8.6.5 All activities which must be started based on late start dates shown.

8.6.6 All activities which must be completed based on late finish dates shown.

8.6.7 All activities that are actually started, by dates.

8.6.8 All activities that are actually completed, by dates.

8.6.9 All activities which will have remaining durations carryover on to next month's CPM monitor report. At the monthly preliminary CPM meeting with the Government's representative, determine extent of all remaining durations.

8.7 WORK DEFECTS AND OMISSIONS (D's and O's)

Where it is recognized and agreed that for all practical purposes, a particular "preceder" activity has advanced in work sufficiently that only D's and O's and/or a relatively minor amount of work remains and the General Contractor has started or plans to immediately start succeeding related activities, the activity shall be reported as complete by recording "O" duration and inserting a corresponding completion date. However, sufficient dollars (w/percent complete) shall be reported as unpaid to reflect the actual outstanding work.

8.8   BENEFITS

The posted network diagram and computer printouts shall be the basis for a common understanding of past, current and future progress of the Contract work. Updating (regular review, analysis, evaluation, and recomputation) of the network diagram/computer printout, on a systematic basis, is an important factor in maintaining progress.

The CPM network is the Contractor's schedule to construct the work; and it is oriented to benefit the Contractor, i.e., to suit the Contractor's plan and schedule. To maintain this benefit, it behooves the Contractor to take an active positive daily interest in using this management tool for activity delay evaluations, progress forecasts, modifications reconciliation, progress payments and the like.

With the Contractor's dedication and commitment to the proper effective use of CPM discipline in the daily management of construction, the benefits in economy will be self-evident. Any avoidance of the discipline, that effective project (CPM) Scheduling requires, will result in utilization of this CPM tool to prove out such avoidance.

As is evident, certain specifications can provide some extremely onerous requirements for scheduling. The argument of too much specified scheduling is countered by the owner's experience or disappointment in the results of too little scheduling done on previous jobs.

The typical owner will generally accept either precedence or CPM logic techniques; precedence diagrams and schedules in the patois of the construction job are very often called CPMs. At times some owners can be very legalistic about the CPM technique. Certain government agencies place high emphasis on CPM techniques and quality. Experience indicates there are many delayed payments for late or faulty CPM work. As much as $7 million has been held back for what the government felt was faulty CPM work on one major federal job.

CPM schedules are too often criticized for their format and drafting techniques, only infrequently is the logic studied to see whether the plan set forth conforms to the contract or design. Too often the emphasis is on the wrong facet in CPM, although increasingly emphasis is placed on the entire scheduling process.

Problems with CPM schedules are frequent and comments are often made that CPM schedules do not work. Experience, however, is that CPM schedules do work and that it is the people in charge of the job that do not work. Indeed, those in charge do not want to make the effort to plan or schedule, or have their work be bound or measured by such a plan. Problems with schedules are really problems with people. Problems in construction are people problems.

Quality is available in CPM scheduling just as it is available in every other luxury or necessity one must purchase. Quality scheduling requires knowledge, time, interest, and money. Contractors who skip any ingredient will not get the quality specified or intended. Owners and contractors are becoming more demanding in the scheduling work.

WHO OWNS FLOAT

The bidding contractor should be aware of the consequence of scheduling requirements, which can be serious and very demanding. The second specification quoted earlier states that time extensions will be made only if the change is on critical path. Thus the float in the contract under its specification is claimed by the owner as its property, a claim generally and vigorously challenged by contractors. Federal board decisions exist favoring both positions.

It is the author's opinion that float in reality belongs to the project. Some use of a portion of the float by the owner will not affect the contractor nor will use of some float by the contractor affect its work for the owner. Owner use of the float, however, can cause definable and recognizable consequences on the finances of the contractor at a certain point.

When the contractor's economics are affected beyond the risk implied in the contract documents, an adjustment in dollars and time is due. It is very possible for changes by the owner to require the contractor to divert attention and added workers to noncritical activities to the extent that the overall job slips and an equitable adjustment is required.

Again, the float belongs to the project. After all, the project is in theory a mutual betterment to both parties and when exclusive use of the float by one party is found or determined to be at the undue expense of the other an adjustment is due. If the owner specifies the float as the owner's property per the sample specification, then the bidder can knowingly price the risk therein as long as the owner does not breach all the unwritten implied elements of the contract.

DOCUMENTING THE DELAY USING CPM DURING CONSTRUCTION

Most CPM specificiations call for the schedule to be updated on a monthly or other regular basis. The update process should include revisions to the logic to show changes in sequence and schedule, a new listing of the early and late times, and a narrative analysis describing the job status, revisions if any, problems and the reasons for their existence, and a record of progress during the updating period. This update process by itself will show very clearly to the scheduling engineer the existence and effect of any problems on the project.

The control of the CPM is also important. Many arguments are advanced as to whose schedule the CPM really is. If the owner provided a CPM consultant to the contractor, all is fine until the job goes behind, and then the contractor often will say that the schedule is the owner's and not the contractor's and thus is not binding. Sufficient protesting of this kind can and has resulted in the owner foolishly dropping the schedule as a waste of money. Contractors will often ignore the schedule, hold the entire idea in disdain, or use the process only for dollar calculations for payment. Experience indicates that such contractor ploys and indifference invariably bring major problems on perfecting claims' entitlement by the contractor if disputes arise.

The best device for recording the cause and effect of the construction delay is a reasonably well-done CPM updated monthly. The owner is best served when it buys the CPM from a responsible consultant, then insists that the contractor provide the necessary information and cooperation to produce a schedule showing all plans within 30 days of the notice to proceed, calls for updating meetings, attends them, and places great emphasis on the consultant's work, and then publishes and distributes the results of the scheduling work for everyone concerned.

Following this scheduling program will produce benefits for the owner and for the job. Typically the contractor will not like or push the updating process for it will demonstrate and point out contractor shortcomings as well as those of the owner and designer. The greater number of deficiencies in a construction project are found on the contractor's side of the table. Obviously the contractor has the hardest job of anyone in the process as well as the greatest exposure to problems.

The idea of using the CPM to show the effect of change orders during construction as described in the second sample specification quoted previously is not usually effective. Too often there is no resolution as to time or money until long after the change has occurred, and the contracting officer almost never directs adjustment of the schedule as specified. The effect of the change orders can be seen and shown during updates by skilled scheduling people, but this unbiased view is generally clouded and not wanted or dismissed by the prejudice of contractor or contracting officer.

Individual, everyday change orders usually have little effect on construction progress. They are, however, constantly seized upon as ways out for both parties to extend construction periods. Contractors often do not realize the pressure the owner's project manager is under from his superiors to get the job done. The change order can be the saving grace to a number of problems for the staffs of both the contractor and the owner. One fine point to note is that regular updating of the CPM will very likely minimize or depress the potential of superior persuasive skills of either party to expand a claim to overcome unrelated bad economics on a job.

The contractor is often best served by keeping no records. The fewer CPM updates, the easier one can set forth its recollection. Owners who cause delays in the pile driving should analyze the delay immediately and pay the claim due as soon as possible. The exposure will be less, the problem with the work and its effect clearly seen, and the ripples not totally though out or predictable. Instant resolution is cheaper and the contractor will be happy and will respect the owner. Aggressive contractors who experience delays driving piling are better off to give the required notice politely, along with vague concerned shrugs of the shoulder and pleasant letters stating that documentation of the cost is in process. Then eight months later when pouring the 12th floor concrete and having knowledge of how the job really stands the contractor is best served to announce the need to push the piling claim.

APPLICATION OF NETWORK ANALYSIS AFTER THE PROJECT IS COMPLETED

The network analysis technique is being increasingly used to demonstrate the effect of changes on a project. The type of analysis can be made whether or not CPM is used to plan the job. If CPM was not used, an as-planned network analysis is first done which portrays the approach to the work orginally envisioned by the contractor even if its approach was less than the best one. There is no fixed rule that one must have a contractor's CPM from the job with which to start this analysis. The network studies involved in impacting changes into an as-planned CPM can be relatively simple or extremely complicated, and are usually the latter. It is far from a simple matter to insert changes and problems into an as-planned CPM and let the computer crank out the effects. In many cases the degree of detail in the diagram does not provide that needed to display the impact of the change or claim. The complexities of relatively minor changes which involve a number of trades on rough-in and finish work cannot always be meaningfully shown in a normal CPM diagram. Sometimes the costs of the study are not worth the amount of recovery that can be proved.

The analysis of changes and claims by CPM analysis requires a scheduling engineer of experience and skill. Both judgment and knowledge of what really occurs on a construction job as well as the understanding of how logic insertions must be made are needed by the engineer to be able to evaluate the effect of the changes. The point at which countless small changes, meaningless in themselves on a one-by-one basis, become collectively significant is determined by a process of study and judgment that requires poignant acumen on the part of the expert.

The scheduling expert is also faced with the problem of reducing its massive CPM study down into a simple statement that can be rapidly assimilated and easily believed by the hearer of facts. It is rare to go through each and every point of calculation and study of a delay analysis in the courtroom, hearing presentation or deposition.

In delay analysis weeks and months of study must be done to permit the experts to find all the pertinent facts. These must then be effectively condensed and displayed in perhaps one simple graph of the conclusions which states the case convincingly. For example, exceedingly complex CPM studies were made to support a contractor's claim of increased costs in its work involving several million dollars worth of alterations in an existing city jail which housed some 1200 inmates. Detailed as-planned and as-built studies were made; they consisted of some 20 sheets of logic drawings and hundreds of pages of computer printouts to show first the planning and sequence which were dictated by the documents and envisioned at bid time, and then the as-built studies made to show the effect on the production costs when a maximum security system was imposed on the inmates and the bystanding contractor as well. The owner and contractor also found errors in design which prevented the original sequence of construction from working. The job finally took two years longer. The CPM analysis conclusively proved the exposure on a mathematical basis.

The chart in Fig. 4.7 was used to tell the expert's story to the judge and the jury. The backup documents were also in the courtroom; they were impressive in volume and gave credence to the obvious thorough knowledge of the expert. The backup documents had been examined in discovery. Only the conclusions mattered in the trial.

The CPM technique of evaluating effects of changes invariably displays the failing of both parties. If an as-built schedule is developed, it will automatically note the major problems of everyone concerned.

Figures 4.8 and 4.9 are detailed studies summarizing the foundation work on a 15-story major communication company building in a southern city. Each problem that occurred in the foundation work was carefully researched, described, and documented from the job records in a narrative report called Delay Notes. The effects of each problem are shown individually as a graph in Fig. 4.8. Figure 4.9 is a linear comparison of as-planned versus as-built studies and includes weather studies as well.

The study of construction delays by CPM techniques is expensive and time-consuming. It cannot be accelerated by adding more scheduling engineers. A single person or expert must develop a feel or understanding for the logic to address the history of the job properly and to be able to present the comments and corrections necessary. Expert CPM studies of good quality are typically done by a team of two men, one senior, one somewhat less senior. It is not work for young engineers, nor is it easy.

The total CPM work, including the delay note and damage reports on the communication building, required 1400 hours of consultant effort. To this must be added the input of the contractor's staff, the efforts of the lawyers to learn what is in the expert's findings, and the costs of the time of similar people on the defendant's side who must absorb and possibly counter the expert's findings. In this case several hundred thousand dollars were spent on CPM studies and their review. The claim exceeded $7 million.

Finally, poorly done CPM work in claim studies has a great potential for disaster. CPM quality has been compared to pregnancy: One is either pregnant or not pregnant—there is no in between. Errors in CPM logic simply make the CPM dates wrong. When CPMs are in error, they are of no value. Picking through a schedule generated by erroneous logic for dates that help

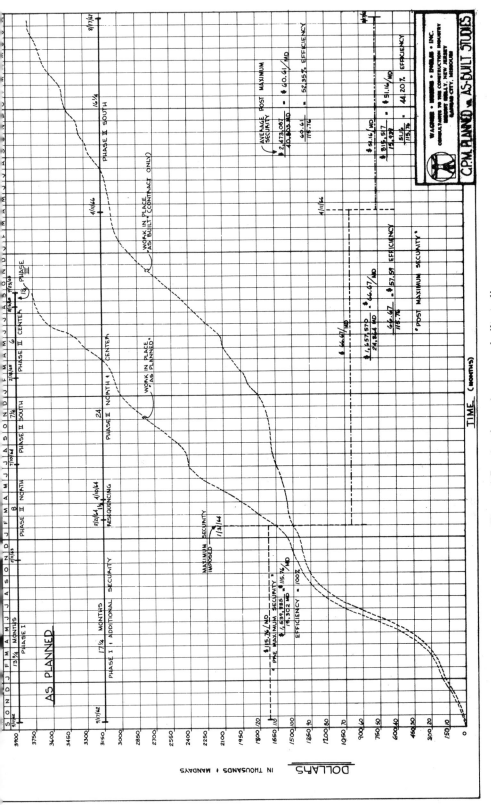

Figure 4.7 As-planned versus as-built studies.

77

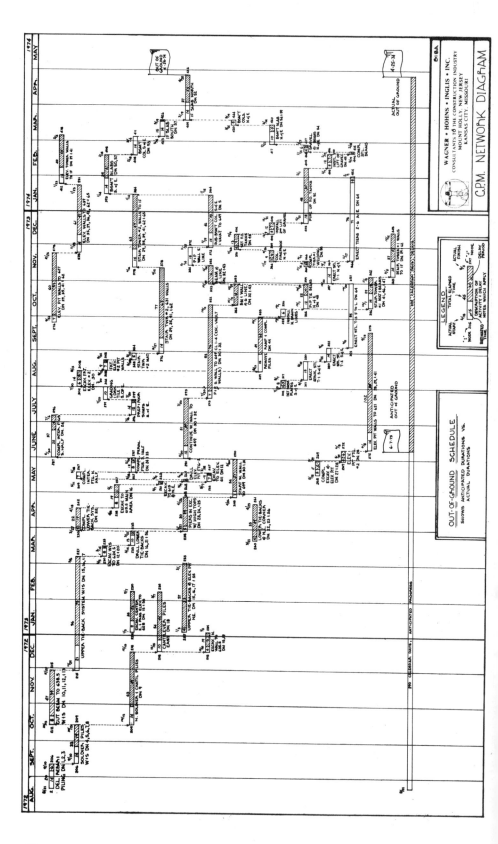

C.P.M. NETWORK DIAGRAM

WAGNER • HOHNS • INGLIS • INC.
CONSULTANTS TO THE CONSTRUCTION INDUSTRY
MOUNT HOLLY, NEW JERSEY
KANSAS CITY, MISSOURI

OUT-OF-GROUND SCHEDULE
SHOWS ANTICIPATED DURATIONS VS.
ACTUAL DURATIONS

LEGEND

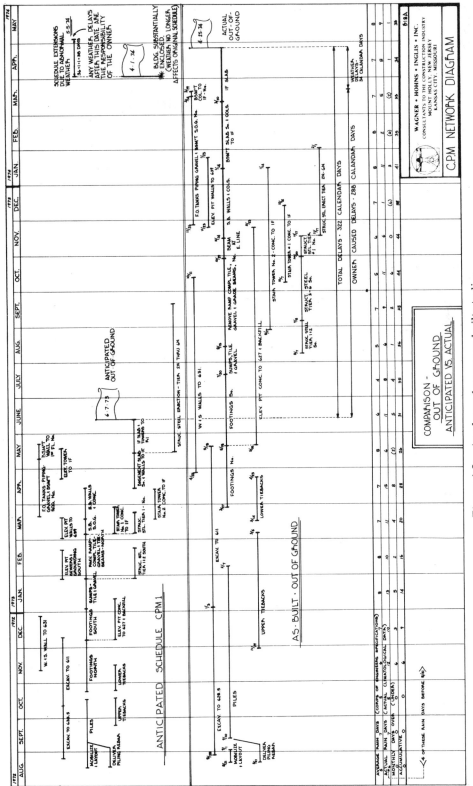

Figure 4.9　As-planned versus as-built studies.

79

one's argument is of no consequence nor definitive value. Do not confuse an error in logic with degree of detail. However, if a question exists as to the quality of the CPM, have it reviewed by an expert; the value of the CPM work can thus be discerned almost instantly.

A LAST WORD ON SCHEDULES

The schedule that best serves every participant in the project is one that reflects the currently known opinion and the planned activity of the parties involved. If errors are found or better ideas or knowledge evolves as the job proceeds, the schedule can be corrected or updated and adjustment to work plans made by all. Attempts to force a schedule, to set up claims, or to sandbag the other party usually do not work out. The schedule should indicate the risk, reward, and consequences of progress as planned; it should be used as a tool. A schedule is not an erection set diagram, but a representation of plans which needs a skilled person to read and make use of its information. The scheduling process must be held fluid by all and when it addresses itself to problems it should be recognized as binding by both parties even when it hurts.

5

WHAT DO YOU DO AT THE START OF THE PROJECT?

In this chapter the thinking of the parties as they enter into the construction contract is examined from the view of mechanics and risk. The owner should be aware of its risks and how best to frame them contractually to buy off the exposure; the contractor will be best equipped by recognizing its basic risks in the mechanics of the contract chosen by the owner and to have a policy to recognize and price the risks accordingly. The other participants in the process should be informed to the subleties of the risks involved as well. As this chapter is read, the reader should keep in mind its position. For example, where a point is made to the owner's advantage, the contractor should reflect on the best way to react for its own interest. Bear in mind that morality is not involved; contracting is an art and the artisans have all degrees and varieties of skill. The process is played out with rules, the score is kept with dollars, and sometimes referees are needed to adjust the score.

INCLUSION OF DISCLAIMERS IN THE CONSTRUCTION CONTRACT

Owners are well advised to write the contract with many well-thought-out all-inclusive disclaimers on risks even with the full knowledge of the reasonability doctrines and interpretations now the vogue in the courtrooms of the United States. The reasons for such a policy are numerous. First, the job may proceed without problems. Everyone knows open doors invite curious and often undesired and unwelcome guests. People, a category of life which includes contractor personnel, are reluctant to knock on or seek to open closed doors. Therefore, well-written contacts which clearly close all the doors and put the burden on the contractor are to the owner's advantage. Second, there is the very real possibility that the contractor will price the risk in its bid, or, intentionally or not, become a special bidder, or simply believe that the terms of its contract are enforceable and not pursue a claim. The contractor may not be claims' conscious or it may not choose to pursue its loss. Third, the judge, arbitration panel, or hearing board may well believe the contract terms in the specific case to be reasonable, and therefore enforce them. One should always remember that an agreement or dispute is won or lost on the persuasive ability of those involved. A poorly presented claim for dollars unquestionably due under the law is very often unsuccessful. Thus the better written the disclaimer, the better chances there are of prevailing and the more difficult is the burden on the contractor as plaintiff and its attorney.

TYPES OF DISCLAIMERS OR RISK TRANSFER CLAUSES FOR THE CONSTRUCTION CONTRACT

The owner should look to pass the risks of underground soil conditions and obstructions to the contractor. Included without limitation in underground risks are dewatering, shoring and bracing systems, disposal of waste, and location and quality of borrow.

The risks of a waterproof building or structure are to be passed onto the contractor. On a basement or waterholding tank, this can sometimes be accomplished in the specifications by adding the phrase: "The contractor shall at its own expense include any item, material, device, or process which in its opinion is necessary beyond that specified to guarantee a waterproof basement (or tank) for ten years. The contractor shall advise the architect engineer of this means, method, process, or material prior to ordering or incorporating same." This phrase should be added by the owner during the addenda period of the project. If the designer is aware of the owner's idea too soon it may diminish its zeal for designing a waterproof basement and rely on the contractor to do that which the designer must do. This same phrase will help in the roofing section of the contract as well. The technique will sometimes work in many places.

The owner should always bear in mind all previous occasions in which it has simultaneously heard the contractor blame the design and the designer blame workmanship as the cause. Whenever these two arguments are aired, the owner should subsequently include a design requirement similar to the foregoing for the contractor and a somewhat similar performance requirement for the designer.

Risk should be bought off on the contract time. The owner should never calculate or state the time for performance. If the owner must give a period of performance, then it should request a study on a time usage program from the designer. This should include studies on overtime requirements, acceleration cost premiums, winter exposure and costs, and similar features.

The owner will do better to have the bidding contractor state the time of completion and fill in a questionnaire or yes–no list as to whether the time needed includes overtime, double shifts, and the like. The typical contractor will say no to any extra cost item on a bid form, which is fine for the owner since the burden of this is then on the contractor.

Time extensions for noncompensible reasons should be set forth clearly to remove risk. An example of this is the AIA A201 paragraph 8.3.1:

> 8.3.1   If the Contractor is delayed at any time in the progress of the Work by any act or neglect of the Owner or the Architect, or by any employee of either, or by any separate contractor employed by the Owner, or by changes ordered in the Work, or by labor disputes, fire, unusual delay in transportation, adverse weather conditions not reasonably anticipatable, unavoidable casualties, or any causes beyond the Contractor's control, or by delay authorized by the Owner pending arbitration, or by any other cause which the Architect determines may justify the delay, then the Contract Time shall be extended by Change Order for such reasonable time as the Architect may determine.

The owner will note that the AIA A201 form does not provide a no damage for delay clause. Indeed Article 8.3.4. opens the door for a possible assertion of delay damages if necessary adjustments were properly assumed as forthcoming by the contractor.

One state writes the noncompensible time extension provisions of its contract as follows:

> The Contractor shall not be liable for loss or damage due to delay in completion, resulting from uncontrollable forces. The term uncontrollable forces being deemed for the purpose of this contract to mean any cause beyond the control of the Contractor including but not limited to failure of facilities, flood, earthquake, storm, lightning, fire, epidemic, war, riot, civil disturbance, sabotage and restraint by court or public authority, which by exercise of due diligence and foresight the Contractor could not reasonably have been expected to avoid. If the Contractor is rendered unable to fulfill any obligations by reason of uncontrollable forces, he shall exercise due diligence to remove such inability with all reasonable dispatch. Any delay resulting from such uncontrollable forces shall extend the completion dates correspondingly, provided that such extensions shall at all times be subject to the approval of the Owner.

Even more stringent is this example:

EXTENSIONS OF TIME

The time provided for completion of the Contract shall be extended (subject, however, to the provisions of this numbered clause) only if in the opinion of the Engineer the Contractor is necessarily delayed in completing such part by such time solely and directly by a cause which meets all the following conditions:

1. Such cause is beyond the Contractor's control and arises without his fault;
2. Such cause comes into existence after the opening of Proposals on this Contract and neither was nor could have been anticipated by investigation before such opening.

Variations in temperature and precipitation shall be conclusively deemed to have been anticipated before opening of such Proposals on this Contract except to the extent that the actual monthly average temperature varies from a temperature which is 10 per cent above or below the monthly normal temperature and except to the extent that the actual number equal to two plus the normal number of days of precipitation per month.

In any case, the variations in temperature and precipitation described in the immediately preceding sentence will be cause for an extension of time only if occurring between the actual time of commencement of the Work at the construction site and the time for completion stipulated in the clause hereof entitled "Time for Completion and Damages for Delay" (or such time as extended as provided for herein). In the case of portions of months the number of days will be pro-rated by the Engineer. Temperature and precipitation shall be as recorded by the U.S. Weather Bureau in its publications, including that entitled "Local Climatological Data with Comparative Data," which is applicable to the area in which the Work is to be performed, and in the case of precipitation, the normal number of days of precipitation (of 0.1 inch or more) per month as abstracted from the aforementioned publications area as follows:

Month	Normal number of days per month on which precipitation exceeds 0.1 inch
January	7
February	7
March	8
April	7
May	6
June	6
July	5
August	7
September	6
October	6
November	7
December	7

In any event, even though a cause of delay meets all the above conditions, an extension shall be granted only to the extent that (i) the performance of the Work is actually and necessarily delayed and (ii) the effect of such cause cannot be anticipated and avoided or mitigated by the exercise of all reasonable precautions, efforts and measures (including planning, scheduling and rescheduling), whether before or after the occurrence of the cause of delay, and an extension shall not be granted for a cause of delay which would not have affected the performance of the Contract were it not for the fault of the Contractor or for other delay for which the Contractor is not entitled to an extension of time.

Any reference herein to the Contractor shall be deemed to include subcontractors and materialmen, whether or not in privity of contract with the Contractor, and employees and others performing any part of the Contract and all the foregoing shall be considered as agents of the Contract.

The period of any extension of time shall be that necessary to make up the time actually lost, subject to the provisions of this numbered clause, and shall be only for the portion of the Contract actually delayed. The Engineer may defer all or part of his decision on an extension and any extension may be rescinded or shortened if it subsequently is found that the delays can be overcome or reduced by the exercise of reasonable precautions, efforts and measures.

As a *condition precedent* to an extension of time, the Contractor shall give *written notice to the Engineer within 48 hours after the time when he knows or should know of any cause which might under any circumstances result in delay* for which he claims or may claim an extension of time (including those causes which the Authority is responsible for or has knowledge of), specifically stating that an extension is or may be claimed, identifying such cause and describing, as fully as practicable at the time, the nature and expected duration of the delay and its effect on the various portions of the Contract. Since the possible necessity for an extension of time may materially alter the scheduling, plans and other actions of the Authority, and since, with sufficient opportunity, the Authority might if it so elects attempt to mitigate the effect of a delay for which an extension of time might be claimed, and since merely oral notice may cause disputes as to the existence or substance thereof, *the giving of written notice as above required shall be of the essence of the Contractor's obligations and failure of the Contractor to give written notice as above required shall be a conclusive waiver of an extension of time.*

It shall in all cases be presumed that no extension, or further extension, of time is due unless the Contractor shall affirmatively demonstrate to the satisfaction of the Engineer that it is. To this end the Contractor shall maintain adequate records supporting any claim for an extension of time, and in the absence of such records, the foregoing presumption shall be deemed conclusive.

A disclaimer for means and methods of construction is important. AIA A201 2.2.4 states the negative of this:

> 2.2.4   The Architect will not be responsible for and will not have control or charge of construction means, methods, techniques, sequences or procedures, or for safety precautions and programs in connection with the Work, and he will not be responsible for the Contractor's failure to carry out the Work in accordance with the Contract Documents. The Architect will not be responsible for or have control or charge over the acts or omissions of the Contractor, Subcontractors, or any of their agents or employees, or any other persons performing any of the Work.

The owner will be better served to write a disclaimer on this situation in a positive tone.

The risk of location or presence of unknown and live underground utilities should be passed on by disclaimer as should risks of ingress and egress to the site, temporary water, temporary power, temporary sanitation, and all other nonpermanent features of the work. Hoisting, heating, winter protection, and temporary facilities and operations should all be totally disclaimed by the owner, and included specifically in the general contractor's responsibility. The minimum unit's capacity or locations of temporary power and utilities to be supplied for construction should be specified by the designer in its work. There is a trend to make the contractor responsible for a trade provide the temporary aspects of the trade. This trend is particularly true in locales in which public law prescribes that four or five or more separate prime contractors be used. The question of which approach is best should be asked on each project, i.e., should the general prime or the electrical prime be responsible for temporary power?

The general conditions should always make the general contractor responsible for scheduling and coordinating all work regardless of the number of primes. The owner should always disclaim any responsibility for scheduling and coordination. The contract should contain language requiring individual primes who have been damaged by the actions or lack of action of another prime to seek first adjudicated relief against the delinquent prime before bringing an action against the owner.

Disclaimers on safety are very important, particularly since any significant injury to anyone on the site will trigger a lawsuit which will name everyone on the project as a codefendant. The "hold harmless" features of today's contract are well included by the owner seeking to put the total risk of the work on the contractor.

CHOICE OF CONTRACT SYSTEM

Today's owner, if not bound by strict government regulation, has a number of choices to accomplish its building program. It can hire an architect and upon completion of the plans, invite bids from a select or random number of contractors and award a single lump sum contract. The owner can often invoke the same principle of selection to other trades and allow the competing general contractor to use only firms on the owner's specified list for each specific trade. The owner (as many states do) can hire an architect and then let a number of prime contracts, one of which is a general contractor responsible in varying degrees by contract and state court decision for coordination.

Owners must bear in mind that they are a party to the construction contract and as such have a great deal of responsibility and work. The owner must react to every action or request of the contractor as well as to its lack of actions or requests. It is not easy to build a project. It is not easy to get contractors to do what one hired them to do. Quite often, too, the subtle underlying game of risk analysis and management of one's exposure to the other party makes erecting the building at times seemingly secondary to building a good job record.

The owner can request the general contractor in its single lump sum bid to name several or all of the proposed subcontractors and suppliers. This requirement helps to eliminate bid shopping and certainly aids in getting the job started unless excavation and other immediate start-up trades are not among those to be named. In Massachusetts and to some extent the other New England states this named subcontractor policy is carried even further by a state law that requires filed subcontractor bids.

The subcontractors file or deposit their bids together with a certified check in a legislated amount for bid security with the owner who publicly reads them aloud a week or two before the general contractor bids are taken. Once read aloud and checked for law compliance, the owner then forwards all the subcontractor bid opening information to each bidder. The subcontractors are allowed to restrict their bids from certain general contractors and the general contractors do not have to accept the low filed subcontractor. General contractors who wish to perform a filed subcontractor's trade themselves must bid the trade to the owner as a filed subcontractor, but can still go in with a cheaper subcontractor price in that trade, if offered. General contractors can require a bond on all or just a certain subcontractor; the filed subcontractors also have some rights superior to those of nonfiled subcontractors on progress payments and retainage. The filed bid system has both good and bad points. No conclusion as to preference is readily available.

Another choice for construction purchase is to use the construction manager concept. In this approach, in addition to the designer, a "professional" construction manager is added to the building team. The professional construction manager (CM) may be an alter ego of the designer, a professional CM firm, or a general contractor which has become a construction manager. Some owners have even set up an in-house CM division and do the CM work themselves.

The CM sometimes provides certain or all of the general condition items at the site and manages people, paper flow, schedules, and documents, and oversees and coordinates the construction as well. This type of responsibility is very similar to those found in the role of the general contractor who subcontracts all the work. However, in the CM case there is an implied and contractually stated shift of trust and concern from the usual general contractor's position of profit orientation to one in the owner's interest. In this posture of trust, the CM may contract itself with all of the individual trade contractors, in which case they still in most instances are subcontractors as far as the owner is concerned. Alternatively, the owner may contract with the trade contractors directly, in which case they all become individual primes who look to the owner and its agents for instruction, measurement, and payment.

At times, the CM is a pure professional, selling only people with all hard items provided by others. The question of whether to use a CM is the subject of countless trade media articles, seminars, professional association meetings, government research, and so on. There is no real answer to this.

A construction manager is not terribly expensive. To some extent it is argued that CM is just another service layer already implicit or inherent in the construction program. Some claim that CM advice during design is financially helpful in the selection of building systems, but many find that this is not true. The CM firm is most often a general contractor and the typical general contractor has little detailed knowledge of building trades or systems beyond excavation, carpentry, and concrete. Thus the CM has little staff resource from which to offer first-hand financial knowledge on the most important creative features of the building, such as heating, ventilating, air-conditioning, and electrical system.

A construction manager does, however, have advantages. It gives the owner who may not have experienced staff a sounding board of interested people who have a personal philosophy very different from that of the designer. Thus an automatic system of checks and balances is created which often has some nice bristles that are good for the owner. It gives the owner a professional agent who supposedly can be relied upon to do a professional job. Thus if the project is late, the owner can blame the CM, as can each subcontractor or separate prime contractor who also look to the CM.

The owner can seek a guaranteed maximum price from the CM. It would appear that the best way to use a CM with minimum risk is to have the CM contract with the trade contractors and do the general conditions itself. This makes the CM more of a contractor than a professional. The guaranteed maximum price CM is on the surface the best way to buy off risk, but always remember this text's first rule of construction: Nothing is absolute, there is no guarantee. The result achieved depends on the skill of those involved and on time and chance.*

The last contract method to be considered is the design–build concept where a single firm is retained by an owner to do the new project. This is a

---

*The risk and exposure taken by the CM far outweighs, to my thought, the fee available. Experience indicates Fast Track construction will be delayed and cost more than any other method.

very often used method in major industrial construction. Its users are very successful although several negative comments are made and much pressure against increasing the use of this system in the public sector comes from the professionals. Obviously the owner who uses design–build passes up the chance to hire one of the great architectural designers of the world. The claim is also made that the design–build concept restricts advancement of the state of the art of design and construction. Recently, however, the design professional is saying via the trade media that the recent marked shift in the industry and its users to hold the designer responsible for the cost of its errors will ultimately discourage and dissuade the private designer from trying to advance the state of the art.

The big drawback to design–build is that the biggest buyer of construction, government, does not want to buy this way. Therefore, the greatest number of firms in the industry are not geared to design–build methods. Hence any move to design–build is met by a large outcry from the majority of the industry. The student of contracts soon learns that contract methods and clauses and the courts which enforce them represent and bow to the pressure of the majority or to that vocal minority who seemingly are the majority.

Design–build concepts, if widely used, could encourage research and development by the builder instead of by the subcontractor or supplier. The only firms in the industry doing any real research are the large manufacturers; most of the other industry members are too busy trying to do what is specified or are too undercapitalized to do research. Interestingly, the leaders and biggest contractors are the design–builders.

Design–build minimizes the risk and inconvenience to the owner because it places the responsibility for it all on one source. The design–builder has both errors and omission insurance and surety guarantees available. The design–builder has an identifiable past performance record; anyone can view its earlier work and ask if the customer is satisfied. Design–build has many advantages.

KEY OWNER POINTS IN THE CONTRACTING SERVICES

The owner is best served when it retains locally available designers. Efforts should be made to use a lead designer who has a local office commitment and thus has the availability for spontaneous meetings or something as simple as the delivery of a set of prints on someone's way home. The same restriction applies to the prime designer's choice of consultants. The potential for problems and those actually experienced when, for example, a project in St. Louis is designed by an architect from New York, a structural engineer from Seattle, a mechanical engineer from Texas, an electrical engineer from New York, and a lighting consultant from Atlanta, generally outweigh the advantage of those special talents no matter how great they might be.

The same applies to the general contractor and its subcontractors and suppliers. Where possible, the owner seeking the best for itself aside from initial low bid price should deal locally. The owner will be best served over the long term by developing a competitive marketplace from which it can buy than to go to the marketplace that exists.

If the owner builds nationally and has the legal option or ability to select its bidders, it should try to design locally. The owners should then select several local bidders and balance the locals with a list of four to six national general contractors and three or four mechanical and electrical subcontractors. These latter should get enough work to follow the owner's market with interest and yet not all the work lest the locals become disinterested.

The owner must bear in mind that it alone or those in its role in the industry have the ability and the need to shape or change the industry. The consumer is the ultimate determinant of what and how items are brought into being. The owner who takes this broad philosophical approach as a foundational plank in its purchase of construction will serve itself best. Some owners, however, seem to believe that contractors are fair game and that any legal device to insulate themselves from risk is smart. For example, there are several public agencies which make the contractor execute nonreliance disclaimers as part of its bid. These particular agencies require the contractor to state that it has investigated the site to the degree necessary to bid the project and that no reliance was put on any information-only type of documents. One agency goes so far as to require the contractor to state that it did not rely on the plans and specifications except for physical quantity and final arrangement.

These and similar clauses are disturbing because they are contrary to what the owner customarily and actually intends. The competitive market is not polished or independent enough to withstand this type of play. Somewhere in attempting to pass risk, a degree of morality must be found. Although this is not a text on morality, it must be noted that in contracts the owner is the leader. Leadership is exercised by setting an example. Owners that write contracts fairly to cover all the known risks and deal openly and honestly with the contractors when problems arise end up ahead in the long run. Owners that do otherwise and cheat on the rules of the game are far worse than their contractor counterparts to a construction contract.

THE CONTRACT FORMS OR CLAUSES

Contracts are a lawyer's business and as such are beyond the scope of this text. However, a few pointers on contract policy are necessary to complete what an owner should do at the beginning of a project.

Owners should not blindly permit the designer to write the contract for design services or for construction. The owner can and will do well to start to develop its agreements with usual standard forms such as AIA A201, but this should be only the start. Appendix A contains AIA A201 in its 1970 and 1976 editions on a side-by-side comparison. The sensible owner will do well to review the changes and then make up its mind what its own contract will say.

The 1976 AIA contract requires arbitration of all disputes and then limits the owner's right to consolidated arbitrations. The owner should be sure that the relief or dispute procedure in all contracts with designer, CM, contractor, subcontractors, supplier, and whatever is the same or one could end up in separate arbitrations, federal court, and state court all at the same time.

Contract sections on insurance are to be examined in depth by the owner. A recent development is to have one insurer named to cover an entire project. This is claimed to result in meaningful savings and elimination of paperwork for all. It has been resisted by some contractors as confining.

RETAINING THE DESIGN PROFESSIONAL

The owner should approach hiring its design professional from a number of points of view. The first is that the designer almost always needs work. Only at rare and relatively short periods of history have designers been too busy to be attentive and dependent on the total market for new projects. The designer needs its client far more than the client needs its designer.

Next, it is not easy to retain a professional. Investigating and rating professionals require a great deal more effort than being a member of a committee which looks at renderings and slides of the designer's past projects. The successful selection of a professional is a lot of work and goes into areas far beyond the apparent loveliness of the designer's past efforts.

The first place to make a sounding on the designer is on the peer level. Good designers recognize good design and good design is a very important feature. The next place to rank or test the designer is with the contractors. Ask the contractors who might build the facility for suggestions of good designers. This time, the emphasis of the inquiry is on the pragmatic side of the project. Does the designer do its work on time? Does it equivocate? Can it run a meeting? Is it truthful? If a construction problem or dispute occurs, chances are that the designer's decision will initially trigger the action or reaction. Does the designer have any money? Most people have no idea of the designer's finances. Designers and staff are typically the lowest paid of all players on a project; quite often the design captain on a multimillion dollar project makes less money than any journeyman mechanic on the site. In addition to the economic problems, designers are everything else that people are and must be viewed totally when being considered for a new commission.

THE DESIGNER'S CONTRACT

The owner will very often rely on the architect to prepare a contract, and very typically the AIA form in Appendix A is used. However, comparison of the AIA form to the Navy form in Appendix C reveals great differences and demonstrates different degrees of detail and specific performance requirements which can be set forth in a design contract.

The owner should specify the minimum effort it will accept from its designers. Phrases such as "all necessary details including one-eighth-inch scale or larger floor plans of all areas" should be inserted throughout the contract; avoid use of one-sixteenth- or three-thirty-second-inch scale plans and elevations. Specific studies needed with requirements and recommendations should be included. Never let a designer pose a question to an owner without a written recommendation with reasons as to the best choice for solution.

A designer should tell you how much it will cost to heat, cool, monitor, operate, and repair the various building systems it is proposing and provide sufficient documentation to back up the cost which it sets out. These items should exist in the owner's file long before the job goes to market.

Designers must be required to explain the theory of the heating and cooling plants, or how the sewage treatment or process plant will operate. Too often owners end up with plants no one understands. The careful owner will include safeguards against this in its design contract.

CONSULTANTS

The owner is well advised to hire just one consultant as chief of design and let that chief designer hire everyone else. As soon as the owner hires the soil engineer, the boring contractor, or the surveyor, design responsibility is split and more parties must be part of a dispute process. It is better to have the designer hire all needed personnel, including the testing laboratory. The 5 or 10 percent fee charged as an override for subconsultant handling is often miniscule compared to the problems of retaining the subconsultant individually if something goes wrong.

There is a move today by some soil engineers to isolate themselves as a class of professionals that would like the owners to believe has a $50,000 limit of liability. A special insurance firm with policies and ideas helpful more to the engineer than the owner has been sponsored by some soil engineers. The owner has no need to be drawn into such situations and can avoid them by contracting with its designers.*

PROFESSIONAL LIABILITY INSURANCE

Professional liability insurance (PLI) or errors and omissions insurance is carried by most designers. If the designer does not have such coverage, there are serious questions as to whether it should be retained. The owner has a right to ask about the designer's insurance and upgrade or downgrade the coverage as it feels necessary. The owner must always review the designer's

---

*The limitation has no bearing on third party duties.

insurance. Some designers have more insurance than others, and sometimes there are large deductibles; a $20,000 deductible is not uncommon, and some large firms go to a $100,000 deductible. Premiums are very high because there have recently been many errors and claims, and many losses. One type of policy is called a wasting asset and provides the cost of defense of a claim as part of the face coverage.

Designers are learning to be more careful, and educational programs sponsored by the insurers and professional societies for all design disciplines are available and have become popular. The sensible owner will realize that it really pays the premium for insurance coverage. Once this is understood, the range, depth, and intent of the insurance can be handled to whatever degree is necessary.

There are some items on which a designer may not be covered by PLI. Some exclusions are the failure to complete drawings, specifications, or schedules of specifications on time. PLI can exclude designer failure to process shop drawings on time. In some cases the owner can get some protection in these excluded categories by requiring special insurance endorsements from the designers. PLI will generally not cover express warranties or guarantees. PLI will not cover the problems if cost estimates or planned construction costs are exceeded by marketplace influences. Endorsement and higher premium can also provide some coverages excluded in base policies.

Losses caused intentionally are not covered. Experience includes a hard fought contractor claim against an architect who had intentionally, deviously, and maliciously tried to default the contractor to protect itself. There was little question to the court about the architect's motives and the PLI attorney agreed that the architect was intentionally out to get the contractor and thus had no coverage. In this case the architect had no money either, nor did the contractor. PLI does not cover fraud, dishonesty, malicious or knowingly wrongful acts, errors, or omissions committed or at the direction of the insured.

The professional is very likely to be covered by the Continental Insurance Group (CNA) or Northbrook Insurance Company. These two firms write much of the errors and omissions insurance in the United States.

Policy period and territory are important. The policy should be checked to see if claims for errors made but unknown previous to the policy are covered. PLI policies can cover errors on a retroactive basis if the claim is made during the policy period. They also cover claims on errors during the policy period; problems can arise if designers switch insurers or indemnitors if any knowledge exists of possible claim.

Design joint ventures have the problem of being dissolved after the project is done and latent defects may not be covered by insurance for the claimant.

Territory is important at times. Some policies protect one almost worldwide but require the claim to be made in the United States or Canada. Others limit coverage to the United States and Canada. Firms working on non-domestic programs need to review the insurance situation in depth from the territory aspect.

The PLI policy also has certain particulars on settlement. The carrier will not settle a case without written consent of the insured but if the insured refuses to consent to a settlement, any loss above the offer of settlement is to the account of the insured. This hard fact is sometimes upsetting to the designer who feels that the principles involved are important and does not want to settle.

It should be remembered that insurance settlements are just settlements not judgments. The collective wisdom of the insurer, counsel, and experts, however, who view the entire problem from the dollar and cents perspective, puts the claim into the context of the insurance company's responsibility and forgets the principles involved.

Claims against designers take a great deal of uncompensated effort by the designer. The careful designer will report claims early and work closely with the insurer and the counsel chosen.

The owner should require certification of the insurance coverage, specific endorsement for the project in hand, and a 30-day cancellation notice on the policy. The owner should also have a copy of the policy.

OTHER CONSIDERATIONS IN HIRING A DESIGNER

The owner should require by contract and in practice enforce timely performance by the designer. If time-related damages are assessable and necessary against the contractor, they are due equally from the designer.

Too often design work uses and at times fritters away days which can never be made up in construction. If money is to be spent to accelerate a project, it is best spent in the design phase. It is far less costly to expedite design than construction.

The owner also cannot fritter away time in design reviews. The more design reviews, the more red marks there are on the drawings. Just as important, the less reviewers there are, the less red marks. No one who is asked to review a set of drawings can do the job without making red marks. The only proof of a good review is lots of red marks.

The designer can be rated on performance. The design contract can call for the design of waterproof walls or roofs, nonskid surfaces, or other obvious physical requirements. There is no reason why performance specifications cannot be included in the designer's contract.

The designer must be held responsible for all code compliance. The practice that specifications require such compliance by the construction contractor is a nice standard procedure. Sometimes it works, and the contractor will correct the design, but when details in the contract drawings do not meet the code, the designer generally ends up responsible. The owner should be sure to require the same compliance to all laws of the land from the designer as well as the contractor.

The contract for design should provide a cap for design fees if construction costs run over the budget. On the Louisiana Superdome, the bids were approximately $10 million over the budget which the designer had insisted was adequate. Shortly after the contract was awarded to the builder a bill in excess of $500,000 was presented by the designer who was retained on a percentage of cost. The $10 million budget overrun had created $500,000 in design fees.

The design contract should specify that the designer is to approve and certify the correctness of the contractor's price breakdown and monthly requisitions: to review and approve the progress schedule and cause it to be followed, monitored, and revised; to approve all change orders as to necessity and cost; and to comment and recommend action on each and every request made by the contractor.

The designer is to produce a set of bid documents by which the project can be built and will properly function. Adjustments or modifications caused by designer error or omissions are to be paid for by the designer. The initial interpreter of the design contract shall be the owner.

The owner must deal openly and fairly with the designer, as well as with the contractor. Cheap fees, short design durations, fund restrictions, and unreasonable reviews end up with performance levels below those needed. Fancily written contracts by owners do not necessarily overcome fees that are too low and it is the rare designer who can afford to pay for many mistakes at 100 cents on a dollar when its gross sales start at less than seven or eight cents on the same dollar.

Owners should not fall in love with their designers. Owners owe the designers reaction, supervision, criticism, and approval.

THE CONTRACTOR

At the start of the project or in its bidding, the contractor has a number of considerations to make. Some of the considerations described earlier include setting up the estimate form properly and reviewing the front end of the specifications in depth. When a contractor is low and is satisfied with its bid, it really should destroy its estimate. It has served its purpose and is no longer needed.

The contractor enters the typical project as a somewhat unwilling party to the contract, for it has no choice in the contract terms. The contractor is often chosen purely on price and its price calls for little more than estimating the cost of the workmanship on a set of documents. At times the contractor, in order to survive, must take work by contracts with terms to which it objects. The only saving grace is that the contractor does have the right to price risk into the job by putting down larger fees when the contract terms are too onerous to bear at normal markups.

This text has dealt in depth only with those contract safeguards and terms which are favorable to the owner. The contractor is well advised to look at the owner's conditions and price the work accordingly. Too many contractors bid work without reading the contract or, if read, they act as if they do not believe it means what it says. Experience requires emphasis that the contractor examine and copy all the information offered by the owner. A contractor who did not bother to go and look at soil samples and core borings in the engineer's office was unable to perfect a $400,000 claim from changed soil conditions. The hearer of facts felt it had failed to avail itself fully of the information available before it bid.

Contractors should particularly read the contract relief clauses before the job is bid. The contractor should rely only on the contract documents for all its estimate data. Brilliant ideas, new methods, experience with the site across the street, a contract time too short, and similar risk-related items are to be reflected in fee size.

Once low bidder, the contractor can request such changes as adjustments in notice times to make them workable; proof of the owner's ability to finance and pay for the project (note the ability is to be safeguarded by the contractor on a continuous basis of checks if needed): the possibility of waiving arbitration of disputes, or alternatively, agreement upon a single arbitrator; and all sorts of bookkeeping necessities and administrative details that can be worked out at the preaward meeting without voiding the public law concept that each and every bidder must have an equal opportunity.

During the bid period, the contractor should examine and get a copy of each and every piece of information offered by the owner. The contractor should carefully put away its bid set of documents and at least two other copies of the bid drawings. Original plans and specifications are always in short supply in construction disputes. Similar file copies in multiple copies or microfilm should be kept of everything supplied by the owner in the bidding process.

Before the contractor prepares the progress schedule on a different project, it should request the owner to furnish its scheduling. If the owner has provided a detailed schedule in the documents, as many now do, with error or impossibilities in performance, the problems therein should be brought up after the work has been bid.

The contractor should request a survey from the owner showing boundaries and a legal description of the site together with copies of all land use agreements. The contractor should protest low unit prices if set forth unilaterally by the owner in writing as soon as possible—prebid, preaward, or as soon as realized as too low. If unit prices are too low, they can be challenged if their use is sought for more than a reasonable amount of extra work.

The contractor should review the change clauses, the dispute clauses, the notice clauses, the time extension clauses, payment clauses, payment of material on site clauses, suspension of work clauses, and hold harmless clauses. The contractor should be sure that the subcontractor's contract forms used incorporate or flow down the same relief or dispute clauses from tier to tier.

Contractors should hire subcontractors on a when and as-needed basis with job progress the determinant and the contractor the initial interpreter of the subcontract. No subcontract for more than 5 percent of the project should be let without a bond for faithful performance, and subcontractors who are a key for job progress should also be bonded regardless of size. The subcontract and negotiations that lead to hiring the subcontractor should note that subcontractor delay could cause damages to the owner, general contractor, and other separate and subcontractors on the project. The general contractor should not pay subcontractors until funds for them have been received from the owner. General contractors must be careful in front loading estimates not to use later subcontractor dollars for front loading lest it be determined that the later subcontractor's dollars were collected early by the general contractor and not paid over. General contractors that supply items to the job for installation by a subcontractor should say how much the cap on the material quantity is as part of the contract.

The general contractor should note that the specified warranty or guarantee required by contract can be argued as no longer due as soon as any change order or other action or lack of action by the owner affects the end date. Contractors should seek to have liquidated damages stated in the documents rather than contracts which will allow the owner to pursue actual damage. Liquidated damages are generally too low. If too high, they can be attacked as being a penalty. Actual damages can be as long as telephone numbers before one knows it. Contractors should bear in mind that damages sought generally cannot exceed the amount normally foreseeable when the project was bid.

General contractors who agree to time extensions without agreement of subcontractors may be opening the door to subcontract claims. Careful language in the subcontract or in the negotiations leading up to the subcontract are necessary to overcome this very common exposure to subcontract claims.

The contractor will do well to try to establish in writing a friendly understanding among the owner, the designer, and itself on the following points as soon as possible:

1. The function of the drawings and specifications.
2. The designer's position with regard to contract interpretation.
3. All the points to be considered if a total or partial suspension of work by the owner occurs.
4. The purpose of cost allowances. How they are to be spent or incurred.
5. The responsibilities of the designer as to shop drawings and submittals. Make sure the designer meets its contract with the owner.
6. Review all modifications and hold harmless clauses.

7. Ask what happens if a subcontractor is unacceptable but its price was used and is too low to absorb.
8. Ask why subcontract provisions are in the general contract if no privity exists between owner and subcontractor.
9. Discuss disputes, arbitration, the arbitration panel, litigation, i.e., what is to be done, long before any dispute arises.
10. Define substantial completion.
11. Discuss the process leading to partial and final payments, and storage of materials and payment of stored materials. Define a bonded warehouse.
12. Discuss the architect's authority for changes and the change process. Set some sort of time limits on changes—no new changes in last four months of job.
13. Talk about guarantees, warranties.
14. Review early what is included in operating manuals and start-up and test requirements.
15. The "or equal" problem should be set forth in detail.
16. Quality control and performance testing methods and requirements should be defined.
17. Discuss retained percentage policies.
18. Discuss the possibility of future claims after final payment.

6

WHAT DO YOU DO WHEN CONSTRUCTION IS UNDERWAY ?

WHAT HAPPENS WHEN THE DISPUTE OCCURS?

Discovering the Dispute and Giving Notice

Very often nothing dramatic or noteworthy happens when an occurrence or event goes by or takes place that will ultimately lead to serious financial consequences or delay. There are major problems with problem recognition and internal reporting, let alone notice. For example, the error of a designer in a slab thickness or pump size may not be discovered until far too late. The discovery of a changed condition may not be realized by its finder.

On a construction job, disputes do not generally occur in straight out-and-out tones. Instead, situations happen or problems surface, changes are proposed or forced, their resolution becomes impossible to bear, and a dispute arises. Virtually every construction contract has provisions for giving the owner written notice that something has happened or been found which may increase the cost and time of performance for the work. The contract sets forth the contractor's duty to advise the owner within a set period of time for such an occurrence. The design professional is usually advised in the notice as well. Several typical notice provisions are as follows:

a. 1.03.2 Any other written order or an oral order (which terms as used in this subsection shall include direction, instruction, interpretation, or determination) from the Contracting Officer, which causes any such change, shall be treated as a change order under this section: Provided, That the Contractor gives the Contracting Officer written notice, by registered mail, within 10 days of receipt of such order, stating the date, circumstances, and source of the order and that the Contractor regards the order as a change order.

b. If the Contractor is of the opinion that any work required, necessitated, or ordered violates or increases its work under the terms and provisions of this Contract, he must promptly notify the Engineer and the Commissioner, in writing, of his contentions with respect thereto and request a final determination thereof. If the Commissioner determines that the work or that the order complained of is proper, he will direct the Contractor to proceed and the Contractor shall promptly comply. In order, however, to reserve his right to claim compensation for such work or damages resulting from such compliance, the Contractor must, within five days after receiving notice of the Commissioner's determination and direction, notify the Commissioner in writing that the work is being performed or that the determination and direction is being complied with under protest.

   In order to claim an adjustment for changes in the Contract Work, the Contractor shall also within 30 days of the date of a written order or the date of acknowledgement of an oral order notify the Commissioner in a written statement of the nature and amount of compensation claimed.

   Failure of the Contractor to so notify the Commissioner of his protest and of his claim for compensation shall be deemed as a waiver of claim for extra compensation of damages therefor.

The purpose of this notice provision is obvious. When followed, it gives the owner a chance to learn, evaluate, and respond to what is happening because it has become aware of its increased potential economic responsibilities to the contractor. Further, notice gives the owner the ability to pitch in, help solve the problems, to share them, or even to terminate the contract if the change is so great that the economics make cessation of the program the best way to go.

Contractors often view the notice requirements as a sneaky method of eliminating or avoiding contract claims. Experience indicates that this is not the underlying reason for the notice requirement. Many times in an owner's defense against a contractor's claims, great attention is placed on the lack of notice or the untimeliness of notice. This defense is particularly true in federal contracts where the letter of the contract is more frequently relied on rather than the romance of the situation, which the contractor might feel is more properly the issue to be addressed.

The sensible contractor searches out the notice provisions in its contract, makes sure that those on its staff who are administering the contract are well aware of the requirements, and if needed sends to the owner a polite letter such as the following:

Gentlemen:

   On November 15, 1976, we hit obstructions in our excavation which are not shown or referred to in the plans and specifications. We hereby give you notice that these obstructions are in our understanding far beyond the contract requirements and could well have a serious effect on contract performance including both time and cost.

We will certainly do everything we can to minimize these costs and are in constant touch with your field representatives regarding the problems. However, to meet the formal notice requirements of the contract, we are forwarding this letter for your files. Obviously, as you already know, we welcome any suggestions and help that you are able to offer in our mutual attempt to solve this problem.

For the sake of formality, we of course must reserve any and all rights the contract provides for adjustment to our price and time.

Very truly yours

A letter such as this—simple, friendly, nonthreatening, offering to help, trying to save money—is the type of notice letter that makes friends, and is also invaluable if a dispute arises.

Early recognition of activities or lack of activities by the owner which suspend all or part of the work of the contractor is very important since most contracts with a suspension of work clause call only for those extra costs to be due starting 20 days prior to the contractor's notice of the suspension. Failure to give notice of suspension of work will act as a waiver of those related costs even if the extra costs are as bold as can be.

Contractors are well advised to give notice as soon as possible. Notice does not mean war. One may give as many notices as it likes in the course of a job. Notice does not mean one will press further for contract adjustment. The best contract relationship is one of fair dealing and open concern for the other party. That tenor of concern in one party will invariably force the other party to follow the example set, and if not, the record will preclude unfair claims being perfected.

Disputes infrequently arise from changes in which physical aspects of the work are altered, increased, or decreased, equipment substituted, and the like. A certain amount of jockeying back and forth on the price adjustment of the physical change always goes on in the change order process since everyone involved has to prove his worth, but disputes of consequence do not usually arise. Feelings, if bent by being upstaged in a change order negotiation, can always be smoothed by the turnabout potential available on the next face-off.

Disputes over hard changes occasionally arise in government or private work where in the urgency of the situation, considerable work is authorized with the price to be determined later, and then when the time comes the price cannot be worked out to the satisfaction of both. It was noted earlier that disputes arise mostly in the intangible elements of construction changes where work is stopped or the quantities added are such that the length of time or the environment of performance is increased, or the sequence planned is changed or more or fewer men are needed, or general conditions, overhead, or productivity is changed. The quantification of this type of change and its refusal by the owner or the designer is where the disputes start.

Now that the contractor's notice has been discussed it should be noted that owners have notice duties as well. Owner claims have to be judged by the architect or engineer before arbitration can be demanded. Owners will do well to be sure they too have reviewed the process they must follow to register a complaint, lest they do so illegally and end up abrogating the rights of the contractor. Owners and designers are often too quick to forget or ignore the rights of a contractor. There may not be many rights left to the contractor but experience indicates that care by the owner not to violate the contract it wrote is good business. For example, a separate electrical prime contractor fouled up its own work and everyone else's on a lower level government body $12 million job. Its poor performance, chicanery on substitutions, plain cheating, and lack of workmen finally led to the termination of its electrical contract. Unfortunately the owner did not follow the procedure it had set forth for termination and when it sought the excess completion costs from the surety it was faced with a demand for illegal termination. The owner went to arbitration where it lost and had to pay $225,000 to the electrician. The owner was then exposed to claims from the other primes, all of whom ended up with some sizable adjustments. The $12 million job cost an additional $1,750,000 which should and could have been collected from the nonperformer's surety if the procedures set forth had been followed.

Preserving or Establishing the Record

After proper notice one of the most important rules in a dispute situation is to document properly the position one has suffered. If the contractor has suffered a delay that is continuing, its manager's daily log or report should state the delay and its effects. The men or work affected by the delay should have the time broken down on a daily basis, and the delay time should be assigned to a new cost category set up just for this purpose. Sometimes, this breakdown is very difficult and sometimes it is not precise at all, but if the breakdown is the best judgment of the man who is there and is responsible, it is hard to refute.

Two years later when that time sheet or report is produced stating that three out of eight labor hours were wasted or lost because of delay, and the person in charge had signed his name, that document becomes an extremely valuable piece of paper for proving damages. The good manager should point out in the daily report as much about the delay as it can. Good managers should not repeat the same statement regarding the delay every day. Men must be instructed to write a paragraph and verbalize what went on that day, where work was being affected, and what equipment was down or standing by.

With this type of documentation on hand, sooner or later the other side will fold. Without it, the burden of proof is on the plaintiff, and this burden is at times intolerable. An electrical subcontractor on a job in Oregon that lost its corporate existence also lost its claim in state court against the general contractor for delay and acceleration costs. It just had no records to prove anything more than a $1.1 million payroll overrun. The little bit of effort required to teach construction field personnel to document the effect of job problems properly is insignificant when compared to the difficulty of going into a courtroom five years later and begging the court to believe that the home office gross payroll accounting numbers are all that exist. Judges cannot award damages on that basis. There is no excuse for today's contractor to be without such documentation. The minicomputer data processing systems available are so easily and reasonably installed that no one needs to be information poor.

Another excellent vehicle for keeping the claim fresh and well documented is to update the notice letter continually by writing and telling the owner what has happened. The delay in the foundations caused the concrete work to move back and be done in the winter. Write to the owner and point out that the problems in the foundations have caused the winter work, but that the added costs are being minimized. Explain it in nice words. Again, if winter protection is extra because of the delay that occurred, the winter protection costs and extras should be flagged on the daily worksheets. Separate accounting codes should be set up. Field and office personnel should be instructed what to do when in a claim situation.

It is good to give copies of daily reports on a weekly or monthly basis to someone in the field on the owner's side of the project such as the clerk of the works or resident engineer, particularly if the reports are well put together and support the claimed position. This means that the owner has constant notice of the cost being incurred. The fact that the reports are regularly going to the owner will make the preparer more careful. By acceptance of reports the owner is tacitly giving assent to the necessity of one day coming to grips with the additional costs.

Simple, common courtesies to the owner are so important in the record—after all, in a claim, to a major extent the contractor is really spending the owner's money; certainly the contractor is trying to spend as little as possible and trying to help its customer. The personification of that helpful attitude in the construction dialogue or record that is preserved will be valuable beyond description in the courtroom or in negotiations. Constant courtesy and concern are extremely difficult for the other party to counter. It is easy to react negatively and strongly to someone who is obnoxious, cantankerous, obstreperous, overwhelming, arrogant, and demanding. It is awfully hard to argue with someone who is polite and sincere and is working every minute of the day to be helpful.

Experience shows that the contractor should and is able to write letter after letter documenting the claim with much more ease than the owner. Memos to the file and all sorts of job record-keeping systems are inevitably indigenous to the contractor manager.   In the litigation process, the effect of 50 well-written documents from the contractor to the owner is sometimes overwhelming regardless of their merit or self-serving content.   The author has seen one major government department of public works which did not write a letter over the three-year-life of a $13 million school job. It received many letters each week, however, and when the job was completed it was sued for $880,000.   It does not seem to have much defense. Contractors can outwrite owners and designers.

When a dispute occurs, the best source for a contractor to go to for help in developing a record is the attorney.   The next best place is a fine claim consultant.

HOW TO SELECT THE ATTORNEY OR COUNSEL

Qualification and Experience

Selection of special counsel to litigate or advise on construction matters should be based on a number of criteria which when really brought into focus are simply experience and performance.

One should bear in mind that attorneys are simply people with all the wonderful attributes and problems that all people have.   They are best judged by past performance and by their own salesmanship when you meet them.   Performance can often be judged by talking with the attorney's past clients.   Important questions to ask include the following: How well did counsel keep the client informed of the case's progress? How efficiently was the case handled? How much time did key people spend on the matter? Was the case successful?

Counsel's experience can best be determined through conversations with other attorneys.   Ask the lawyers you know to tell you the construction lawyers in the area.   Other obvious sources of legal expertise exist.   The outside lawyers who work for contractor and subcontractor organizations are doing that work to be visible to industry members.   The surety and insurance company claim departments will know any number of contractor lawyers.   The legal reference text *Martindale Hubbell*, which can be found in law schools and other libraries, gives descriptions of law firm abilities. Major law firms almost always have partners who have a background in contractor disputes as well as men who are trained litigators or trial attorneys. Experience indicates that every lawyer fancies himself a good or great litigator but be warned that some are better litigators than others.   Fellow members of the bar are usually familiar with each other's practice. If counsel is not known as a good litigator or construction attorney, better to consider another or be prepared to reinforce the attorney with an expert or much of one's time.   Construction litigation is a specialty.   The importance of capable legal talent in a contract dispute cannot be overemphasized.   The range of ability and interest in your situation will vary widely from lawyer to lawyer. Counsel's skill at persuasion is extremely important. The best lawyer is simply the best lawyer you can find. One should first compile a list of likely firms or attorneys that might do the job. Once done, this list should be culled to perhaps four choices. Then the attorneys should be interviewed. The lawyer should be invited to the client's place of business for the interview. Questions of experience and performance should also be asked directly during the period in which counsel's employment is considered.

The potential client should not expect to pay for attorney's time spent in the initial interview. Just as a contractor, designer, or other is not reimbursed for any time spent in marketing efforts for a job, an attorney is not reimbursed for discussions which may lead to employment. The same process and questions should be applied to several firms. Find out who on the staff will really do your work. Meet him, or them, as well as the partner and see if you like him immediately. Can the lawyer communicate? Is he persuasive? Simple things like dress, language, personal habits are important.

The prospective client should never hesitate to discuss fees with the attorney. In fact, it is the inexperienced attorney who will not bring up the discussion of fees first. Experienced name partner level construction counsel on the East and West coasts may cost as much as $150 per hour. The fees for similarly experienced counsel in the Midwest and South are less, although the quality is not less. Policies on contingent fees vary from firm to firm. Contruction disputants are best served by attorneys on an hourly basis. The contingent method of retaining counsel can leave great discretion with counsel as to how it spends its time.

Payment of fees should be treated as the client treats all other obligations. The attorney will usually expect to be paid net 30 days. However, under certain circumstances, the attorney, like any other creditor, will understand postponement of payment. The client should not hesitate to call and discuss any problems with bills.

Attorneys can also be terminated; their performance should be monitored continuously just as any other service is measured. The client needs no reasons to change attorneys, but economy and efficiency will always be sacrificed. Start-up costs of a large law firm on major litigation can run $20,000. Changing attorneys is not inexpensive.

Certainly, if dissatisfied or unhappy, the client is well advised to discuss the matter with counsel. The attorney is well aware of his performance and may be able to divert greater resources or attention to the resolution of the client's problem and remove the causes of dissatisfaction if brought to his attention. If the reason for an apparent lack of performance is not satisfactorily explained, the client may always dismiss counsel. Remember, though, successful attorneys are always busy. Remember, too, the attorney did not create the dispute.

Performance in a law case can be compared to the preparation and completion of any construction project. An attorney needs as much time to review and familiarize himself with the facts as a contractor needs to make material takeoffs and analysis of performance specifications. Once the attorney is sufficiently familiar with the facts involved to proceed to the initial steps in the litigation process, he will usually file a complaint if the client is the complaining party or an answer and perhaps a counterclaim if client is the defendant named in the lawsuit. An attorney must not only consider the merits of the facts of a particular client's case but also must ensure compliance with the procedural requirements of the courts. Often legal procedure seems unnecessary to a client, and the attorney may well agree. However, the client must be aware that both it and the opposition have the same rights in due legal process and the weaker or defensive party may quite properly use every device possible to delay and possibly escape the responsibilities of payment. The client must be patient with the legal manipulation, motions, and answers to ensure conformance with the procedural rules.

An attorney should be on board as soon as possible in a dispute. The selection process should be done before problems occur and a single law firm does not have to get all the work from one client. If selected early, the time and money spent keeping the attorney familiar with the progress of the work and development of problems will be minimal and will enable the attorney to react with greater speed if and when a time for action comes. Legal advice is as much a management tool as is cost accounting. The skill of counsel cannot be overemphasized. Success in bringing disputes to a close has so often turned on the ability of counsel that the efforts in securing good counsel are perhaps the best efforts a party can make in a dispute process. Attorneys should look and act like lawyers; those who effect unusual clothing, facial hair, mannerisms, or other peculiarities often carry the hazard of poor reaction from the hearer of the facts. Remember, attorneys in themselves are very success-oriented; they feel exposed to measurement for they get a reputation by winning cases.

In construction litigation it is common to bring in counsel who specializes in construction from other locales. A case in California may well have lawyers from Atlanta, Georgia and Washington, D.C. for much of the process. This is not at all unusual and either party to a dispute should feel no restrictions or necessity on bringing in big name construction legal talent.

Costs

The direct costs of litigation are not easily determined since scope can vary over a wide range. Lawyers, paralegals, consultants, experts, and so on, all have hourly rates starting in the $20 an hour range and working up to over $100 per hour. The sum of these costs usually translates into some 15 percent of the money that crosses from party to party. Hard fought litigation is expensive to each side. The plaintiff generally spends a little more than the defendant but both are in for major expenses. Good legal talent is expensive. Its value cannot be overemphasized. Often counsel costs are misunderstood, thought too high, and resented. Lawyers, however, are concerned with the best solution to a dispute. Its creation and the overtones that went into the dispute are not the lawyer's concern. His concern is to see that the full advantage of due process accrues to the client and that resolution occurs at the point at which the economics make sense. The lawyer measures his value on a different scale. The client will learn to use lawyers best when it takes the time to find out how lawyers measure their value in their own minds. Be careful of lawyers who at times are worth $1000 for minutes, and at other times $10 for hours. A person who is worth $80 an hour is worth that amount on routine matters and on the most important matter. He is worth the same rate winning or losing, whether the stakes are $500, $500,000, or $5 million.

This sort of thinking in dealings with the lawyer will build loyalties and long-term relationships that will serve the client well for many situations. The client best served is the honest client. Legal fees will be minimized if clients are truthful with counsel early in the game. If a dispute or claim is phony, a client insisting it is real will not make it real. Sooner or later the truth will surface and evaluation will show that the dollars wasted were the client's indiscretion and insistence, not the lawyer's overcharge.

The claimant seeking to retain an attorney on contingency is often regarded as unable to pay for legal fees or as one who has little faith in its claim. Plaintiffs who have valid claims will do well to pursue them by financing the legal operation involved. Attorneys are like all other businessmen; they need inspection, direction, questioning, criticism, encouragement, and support. If a client will not attentively support the attorneys with needed files, answers, books, records, time, etc., the attorney will lose some degree of interest. If the client does not care, the attorney will not care.

THE PRACTICAL ROLE OF AN ATTORNEY IN A DISPUTE

A dispute begins when management of an owner or contractor first recognizes a deviation in anticipated contract performance. The usual dispute or claim takes form as a process of accumulation of costs and somewhere along the line a presentation of a request for more dollars to the contracting officer or design professional occurs. In the plaintiff situation, typically the attorney will not enter a construction dispute until the architect, engineer, or contracting officer has refused the contractor's request for an extra. Counsel's first step will be to review the facts which led to the deviation and to analyze the efforts which the contractor has expended in accumulating the cost and other backup information in support of the amount claimed. If the attorney finds the presentation inadequate, he may call in consultants for independent review of the merits of the claim from a technical, construction viewpoint. If the attorney finds the cost compilation inadequate, he may add to the consultant's assignments the computation of additional costs involved. The consultant or client may also assemble visual exhibits and reports for negotiation and ultimately court or panel display. During this period, the attorney will be responsible to ensure that all procedures in the contract for resolution of disputes are followed so that the disputes may be arbitrated or litigated without hindrance.

Once the review and/or computation of cost has been completed, the attorney may renew negotiations with the contracting officer. The advantage of spending additional time at the first level of owner involvement is the possibility of settlement without further investment in collection costs by the contractor.

If consultants are available participation in any negotiation with the owner is often valuable. A well-spoken persuasive consultant can make a substantial contribution to the dispute resolution by being able to review opposition arguments attempting to reduce or eliminate an adjustment as the adversary instead of the contractor. Attempts at negotiation are also advantageous in that the opponents' arguments will often be disclosed. This can enable the attorney and consultant to begin development of appropriate offenses.

If negotiations fail, the arbitration or litigation process begins. As when the claim was in the contractual disputes procedure, the attorney's responsibility again is to ensure that procedural rules of the forum are followed by both opposition and client. Failure to conform may artificially restrict recovery. Failure by opposition to conform may prejudice the proceeding. In addition, the attorney will ensure that his client takes advantage of all available rights. Assuming that all factual data in support of the client's position have been collated and analyzed, the attorney will then take primary responsibility for its presentation.

Arbitration and litigation are difficult and drawn-out procedures. The skill of counsel will invariably be one of the strongest points in the success of either side. Persuasive, energetic counsel is an essential key to success.

THE ROLE OF THE EXPERT

It is the rare construction dispute which does not involve at least one expert witness on each side. The purpose of an expert is to give opinionative testimony which will assist the trier of facts to understand the evidence or to determine a fact at issue.

Restated in somewhat looser terms, the purpose of an expert is to help state the facts, and the expert should be the best available person to do that. The location of good experts is somewhat similar to that for counsel. Experienced construction counsel will know a number of experts from previous cases. Chances are that counsel may already have several experts with whom he typically works and is extremely comfortable and satisfied. This does not mean that the client should not have an opportunity to interview and question the expert. The claimant is the ultimate judge of both which lawyer and which expert will best serve him.

Purpose

The expert in the construction litigation can be needed for a very narrow or a very broad role, depending on the particular problems involved. If the litigation involves structural cracking, the expert will need to analyze it, be able to explain its origin, and assign responsibility to the individual or firm whose lack of performance caused it. The expert's structural knowledge will have to withstand cross-examination, inspection, and in-depth probing by the other side. Therefore, keen talent and qualification is needed in the particular specialty on which the expert must speak. There are experts in all fields—roofing experts, heating and ventilating experts, structural experts, soil engineering experts, waterproofing experts. Virtually every specialty in construction can point to men across the country who have become recognized by their peers as more proficient in the nuances of that subject than the rest of the profession.

The scheduling expert is particularly important in construction litigation since time-oriented claims are the greatest in dollar damages. Therefore, both sides will need to engage someone who is able to speak to the construction process and equate the delays, interferences, and disruptions that occurred into a meaningful dialogue as to its effect on the contractor's economics, the end date, and all the myriad points of impact that have to be considered.

Qualifications

Some of the qualifications the expert requires are obvious, simple, everyday things. The expert must be persuasive. The expert must be judged on general physical characteristics. Is he physically well proportioned? Does he look nice? Is he warm in his personality? Does he act like an expert? Is his integrity and his humor proper and obvious? Is he confident or is he overbearing? Does he give you an uncomfortable devious feeling? Will he lie under oath? (If he says he will, one does not need that expert.) Experts come with a variety of personalities and backgrounds but they are all measured by one quality. Are they knowledgeable? Then, are they persuasive? Can they win an argument; would you like to have them on your side? Would you worry if they were on the other side? These questions are applied to virtually every person one meets but in the selection of an expert they are particularly important.

The expert and your attorney should coalesce easily. The expert should not upstage the attorney. Some experts become aware of the law and quote legal precedent, and cite law cases. This is not the expert's role, it is the attorney's role.

If the expert is being hired to write the claim or to put together a written opinion, a type of which is described in Chapter 8, obviously his ability to write is important. Ask for some previous reports written by the expert.

Look at basic things such as the physical presentation of the expert's work. Is it attractive? Are there typing errors? How quickly can he do your job? What type of word processing equipment does he have? Is he capable of writing a report? Can he produce under pressure? Can he reproduce it in-house? How does he do his artwork? Again, basic questions. The expert's financial ability is important. Experts, like attorneys, are hired for a long period of time. It is not uncommon for an expert to work on projects that are 15 years old. Experience indicates that there are generally one or two projects like that in an expert's shop. Thus the expert should have some longevity as a corporation or in practice in the field to get major consideration. If one hires an expert today, will he be there four years from now when needed in trial? Will he become discouraged with the job and go off to bigger and better things and no longer be available, so that all the work and effort has been for nought.

The expert must be a man of integrity. He will be paid anywhere from $7000 to over $100,000 for investigation and writing the report, and the verbal presentations that go with it. If he has a flaw in integrity and the flaw is brought out by the opposition at the right time, all that money and all that effort may suddenly turn against you. Experience indicates that some experts claim to be college graduates and are not. Some men pose as experts but turn out not to be. Experience shows that there have been experts in construction litigation in the past ten years who claimed to have Ph.D. degrees and did not. Some claimed to have 30 years experience in the construction industry and did not. Many who had a legitimate background had claimed to know what they were talking about but in the final aspect really did not. Experience has shown that the federal government, when duped by a man claiming to be an expert, will seek indictment and prosecution on a number of charges, but the typical plaintiff or private firm when duped quietly suffers the loss and embarrassment for its being hoodwinked.

Along with integrity, the expert should have some credentials. Experts are usually not whiz kids. Perhaps some young men are true experts in their field but most experts are men 35 years of age or older, who have gray hair or a look of maturity about them. Their appearance alone brings them respect. They have 15 to 40 years of participation in their field of endeavor, and thus speak from a position of experience. They have memberships in professional societies, and most have college degrees. Some are licensed as engineers or architects, but experience indicates that licensing in the technical professions is not a prerequisite to a man's being an excellent witness or expert. Men who go into the construction industry have become experts in many areas and yet do not pursue the engineer's professional license since it is not needed in most of the industry.

When Do You Hire the Expert?

The expert should be hired as soon as a question has arisen in a technical area. If a crack has appeared in the building or a construction process is not working properly, it is time to go to the expert. Good experts generally give good advice. However, many times it will be difficult for the expert's advice to be followed since in a disputed problem situation, the designer will be the one from whom advice on correction comes. The expert's advice, however, is invaluable to someone in a claim or dispute dilemma who is being given directions which may or may not help the situation. The expert's advice should be shared. If the claimant's expert says go left and the designer says go right, by all means share the knowledge of the expert with all so that if it turns out to be left, one can say "I told you so."

The scheduling expert is best hired before the dispute occurs. Contractors or owners are best served in delay situations when an outside scheduling consultant has been retained to schedule a job and help with its planning and scheduling before construction or even design starts. This way an independent expert is able to display and amalgamate the thinking of the participants into one document which shows the roles of all. The updating of that schedule can be a process by which problems come to the light, are understood, and dealt with as they crop up or later as the parties may choose. The scheduling expert's presence will give an independent view should a dispute occur.

In the construction process, experience indicates that the party who retains the scheduling consultant is best served by the consultant's efforts. If a contractor retains the scheduling consultant, the scheduling work speaks first to the contractor's problems and plans. The consultant's assessments and updates will intrinsically carry the contractor's point of view. If the owner hires the scheduling consultant, the scheduling consultant will see the project not so much from the contractor's problems as from the owner's performance requirements, and the viewpoint usually takes on a somewhat different tone. The sophisticated owner will hire the CPM consultant on each of its projects and will ascertain that the language in its contract makes the scheduling work part and parcel of the job. That owner will insist on an updating and review procedure on a monthly basis throughout the job. Experience indicates that owners who follow this policy are dollars ahead when it comes to the solution and evaluation of delay claims.

How Much Do Experts Cost?

Experts typically charge by the hour and their rates will vary widely. Fees of $30 to $40 an hour represent the low end of the scale and rates of $90 to $125 an hour represent the top end of the scale. It should be noted that an expert costs the same whether the problem is $5 million or $75,000. Qualified experts do not work on contingencies. As soon as they do, they lose their independent status, and are no longer able to make an independent evaluation of the case since their fee depends on winning the case. Experts get paid whether the case is won or lost or at least the obligation accrues.

It should be realized that in order to become an expert in the case at hand the person chosen must do the preparation work necessary to gain command of the project in order to persuade the hearer of the facts as to his knowledge of the case. Thus an expert report is almost always prepared by the man who delivers the testimony thereon. Occasionally, two men will work on an expert report, but it is not like a set of drawings where 10, 12, or 15 draftsmen and engineers might work together to produce a design. The expert must invest the time and the client the money to give the expert the knowledge needed to be persuasive.

Experience indicates that well over half the claims on which the author has worked were settled shortly after or during the expert's deposition or upon the release and review of the expert's opinion. Many times in construction litigation, the witnesses who can attest first-hand to facts have scattered to new and distant jobs and the expert becomes the key witness to tie the entire case together. His written opinion and his deposition thus become the key measurement device of opposition counsel to the viability of the plaintiff's case. A strong expert with obvious credentials who knows the subject and is convinced of his client's position will often mean a significant settlement in the construction dispute.

7

THE AVENUES AVAILABLE FOR DISPUTE RESOLUTION

The dispute has occurred and counsel and experts are on hand or identifiable when needed. What happens next? What does the claimant do, be it contractor or owner?

A number of activities are necessary. At a point in time the claim must be expressed in depth and in writing. This type of presentation is discussed in Chapter 8. After the claim is expressed, it must be reviewed by the other party who may counter with a claim of its own, refuse to recognize the claim at all, or give a written rejection which will require some sort of formal litigation process to be pursued. The litigative procedures are negotiation and settlement by counsel, arbitration, court and litigation, or government agency review board proceeding. This chapter deals with the process of perfecting the claim.

HOW LONG DOES A RESOLUTION TAKE?

Disputes are not instantly resolved. Just to start off takes a good length of time before the necessity of recognizing, let alone solving, the dispute becomes a hard fact in the minds of those involved. Once it becomes obvious that the other side means business there is always a chance, a hope in the minds of all, that something will happen to eliminate the need to have to go through the difficult and unpleasant process.

Usually, six months go by after the incident before it is decided that a dispute must or will be pursued. Even then, before the dispute can be pursued, it has to be presented, and the collection of facts necessary to do this is a time-consuming affair. Typically the dispute is first presented in very broad terms, two or three pages of standard buzz words, some random letters for backup, some Xeroxed pencil calculations on a columnar pad forwarded to the owner and designer, followed by several meetings at which the claimant says why it should be paid; and then the dispute sits, generally ignored, and nearly everybody involved hopes it will go away.

Before it is all done, the dispute has to be presented in a documented fashion according to certain rules, much backup is given to the designer for study, and a sufficient amount of paper has to go back and forth to enable those who make the decision to resist or accede to the demands of the other party feel that they have done their job. As silly as it is, one cannot have $1 million for a piece of paper, one needs a presentation for the million dollars.

This presentation takes four to five months to assemble and then it is formally presented. Its volume, size, and intimidating characteristics cause its recipient to need a similarly long period of study for reply. Two or three months of study go by, followed usually by complete denial in a short one-page letter, after which a period of absolutely no activity occurs.

Then talks begin to take place. Three or four months of talking and litigation or arbitration starts. A complaint broadly stating the issues is filed and 40 days later an answer of total general denial is given. The attorneys now take over the process totally and begin fact-finding, including discovery of the other party's files, depositions, and interrogatories. Preparation goes on slowly, and the case also moves slowly on the court lists. About a year after the complaint is filed, people begin to work in earnest on both sides. Enough discovery has been done to take important depositions. Experts surface their reports, which may be exchanged and studied, and are then deposed.

The case moves to trial, arbitration, or review board. The court case, if not before a jury, is never continuous, and takes months, even years to finish. The typical construction arbitration takes, if complicated, at least six months, plus another two or three months for the award to be made. The review board takes six months to hear the case and four more months for the decision.

The typical construction dispute is not easily or quickly solved. A seriously contested dispute takes three to four years and perhaps as long as five or six years. Sometimes, in a good situation, it is over in 18 months. The dispute resolved in less than a year has moved to a rapid conclusion.

HOW DOES THE GOVERNMENT VIEW A DISPUTE?

The view of the government is similar to that of the large corporate owner or the insurance company. Since government is involved in 80 percent of the construction disputes, its reaction is of note.

Its Reaction

Government's view of a dispute is best described as a "so what" attitude. It is common for the government to be confronted by contractors, suppliers, and designers making demands. People who work for governments are used to it, for disputes come along regularly in government.

The contractor bringing the dispute may find itself in this situation or in the big dollar claim for the first time. Its life may depend on success in the dispute, but with government this is not the case. In specific situations, certain governmental staff may have a great deal at stake as far as their personal careers in the specific government bureaucracy; but even this is unusual. So government's reaction to a dispute, with the exception of the initial emotion of those who might be fingered as a result of the claim of the contractor, is one of indifference and slow reaction.

The government's reaction no doubt will contain a certain amount of disbelief. Over the years all on the government staff have gotten used to contractors' overstatements of problems. Thus, every dispute that comes along is looked at as being overstated and less alarming. If government's reaction or lack of reaction causes contractor failure, government does not get upset. It has seen contractors come and go.

The claim is generally known, has been anticipated, and has been discussed before hand among the staff. Government has already formed a policy. The members of the staff who have to deal with the problem have already reviewed it and are ready to force the solution they believe is the right answer.

Generally, government has accepted total responsibility in disputes. It's only been in the last few years that government has thought to spread the blame for a disputed item. If a design error has been made, government almost never turns its own collection process on the designer or manager who might be at fault. Instead government takes the loss. This is very true in the federal cases. The disputes process does not allow for the inclusion of the designer as a defendant, so recovery from a designer becomes just another assignment for an already overworked government staff.

The same "go it alone" policy is found in most lower government agencies as well. Indeed at this level the designers are probably the only advocate available to prove the government's position. It's awfully hard to blame the only fellow that's defending you even if he's at fault.

Progressive government law departments recently though have been impleading or third-partying the architect, the engineer, the soil consultant, etc., to the claim action brought by the contractors. This gains government the position of being able to say to the hearer of facts: We hired the contractor to build, we hired the architect to design, and if anything went wrong, it is certainly not our fault.

Its Machinery and People Process

Government has a methodology in dealing with disputes. The larger the government body, generally the more sophisticated is the methodology.

Federal government has a claims procedure consisting of a number of contracting officer reviews, a review board, and a court of claims. It is a process that is in constant use. Day by day, contracting officers make decisions, decisions are protested and remanded to higher levels of authority, and finally the review boards hear the claim.

The state level of government can be the most difficult body to bring to account in a dispute situation. Many states have sovereign immunity statutes or common law holding that the state cannot be sued unless it consents, and it rarely does. Many states with sovereign immunity have set up quasi-judicial arbitration and review boards which will hear contractor claims against the state.

Many large cities have standard contract policies which prevent the city from being joined directly in court action until the administrative processes set forth in the contracts have been exhausted. Smaller communities who rely on architects and engineers who use the AIA, NSPE, or similar contract forms find themselves in arbitration proceedings.

State and large city governments generally try to resolve disputes with contractors by administrative process. In a typical claim, the Director of the Public Facilities Department, General State Authority, Office of Building and Planning, or whatever, will schedule meetings with the contractor as well as with the design professional, hear them out, and attempt to come up with a compromise which can be sold to his superiors and the contractor and thus satisfy the situation.

This administrative route requires the contractor to submit in advance of claim meetings a statement of facts, the backup, and such other information for those who were responsible for the project to study. Typically, the agency director will not review the claim until the meeting and then he will listen to verbal presentations and arguments during the meeting. Government employees always counter the contractor's arguments by disclosing all of the unrelated sins and failings of the contractor, pointing to how good the state was throughout the job and how bad the contractor was. With the air then poisoned, the director gets down to the specifics.

A great majority of claims are resolved in this fashion. The success of the contractor in this method of settlement depends on a number of things. Its performance record has some bearing. The number of previously successful appearances at the negotiating table also has some bearing, although some contractors can seemingly go back time and time again and always be successful. The historical review of the response of the contractor to particular state needs during the project has some bearing, as well as the legitimacy of the complaint and open asset to some responsibility for the dilemma by all those involved.

The presentation of the claim is relevant and documentation of the cost figures is important. The contractor is asking a public employee in a sensitive position to negotiate additional money. Thus, the more sketchy and less definitive the cost records, the more awkward and difficult a decision is forced on the administrative officer.

The people who present the claim at the meeting are important. In some preclaim meetings counsel is absolutely necessary, and in others counsel is the worst thing one can have. Sometimes, the owner of the firm should be there, sometimes not. Knowing the people involved is important.

The size of the claim and the financial condition of the particular project may be in the department head's mind. When money is available in the appropriation for a specific project, claims can be dealt with much more readily on an administrative level than if funds must be sought from the political review board that has jurisdiction over the particular department. Sometimes that political review board forces what could be a negotiated settlement into the legal route by pressures totally beyond the scope, influence, and interest of the particular disputants in the claim.

Failure to reach solution in this informal situation leads to the next step, perhaps a formal presentation to the Director of Public Works or some type of in-house review board who will somewhat more formally in a day or less hear both sides and then deliver an "impartial" verdict which, if unacceptable to the contractor, must be formally challenged. This type of administrative review process is lengthy. It takes time to get the meeting set up, and then it can take four or five months to get the reviewer to present written findings, which are almost always the same: decision against the contractor. Next, if there are no additional administrative requirements, the city law department takes over the case. The contractor files a complaint in court, the city answers it, and the claim moves through court procedures.

In the state case, the state review board, which is somewhat similar to an arbitration panel, must be formally petitioned by the claimant with a legal-style complaint which is then answered by the state law department. The case is listed on the review board's calendar for several years down the line. State review boards are generally politically appointed positions carrying annual salaries of a maximum of $15,000, which means that a week a month is all that is available for hearings. Moreover, since the members of the review board do not necessarily live in the state capital, scheduling is a difficult and slow process. The contractors' usual experience in front of review boards, however, is good. States never take a plaintiff role in a review board hearing by third-partying the design consultant or any codefendant.

Some state review boards have difficulties making awards of the scope that contractors need. For example, in one review board hearing, a contractor proved $700,000 damages and absolute entitlement. The review board found the entitlement but awarded only $200,000. It was learned afterwards that this was the largest single award ever made by the board and the members who heard the case felt they had been generous. In some states the review board award cannot be paid until the legislature votes the funds.

The typical state defense against a claim is of lower quality than the contractor's side of the case. Occasionally, some states will hire outside counsel for major claim proceedings. In such cases, outside counsel is generally extremely qualified though unfortunately not supported by the type of witness or expertise that is needed. The typical state is slow to hire an expert to help defend itself from a contractor's claim, although this is changing as claims become larger and the necessity to deal with these on a hard line basis becomes more of an economic must to the state.*

In federal government, the claim is formally presented to the contracting officer, who issues a final decision after a fact-finding period. If the decision is not acceptable, the contractor must appeal to a higher review within a specified time period. If the higher level review yields no solution, the dispute is presented to the designated government board. The boards are somewhat informal, with the atmosphere of the proceeding one of obvious desire by those hearing the case to learn the facts, understand the problems, and appreciate what's going on in addition to the letter of the law. The hearing day can require many months of waiting and the hearing is usually restricted to entitlement. Damages are not presented until the issue of liability is decided. Once hearings are concluded, a very technical opinion as to settlement is written. The informality of the hearing room is concluded, however, in a written opinion of unusual formality, with many references to previous cases. Then the contractor, if successful in proving entitlement, must negotiate the quantum position of its claim starting back with the contracting officer, and is faced with the whole process again.

Sometimes government is disposed to be genuine in its negotiation over quantum; sometimes it seems to be trying to get even for the loss before the board on entitlement. The audit problem comes up as well, for any claim or contract modification over $100,000 on a federal project must be audited. Sometimes a claim is audited during the initial contracting officer review before going before the board and quite often the claim is audited a second time after entitlement has been found by the board.

All or any of the findings of the review board can be appealed in the United States Court of Claims. This is a long drawn-out process, expensive, and highly technical, though in good cases usually well rewarding.

The Resources of Government

The resources of government are far beyond those of the contractor. Government usually has the building and several hundred thousand dollars of the contractor's. It also has all the time needed because it is not going anywhere and is not working on a profit or loss situation. Thus, government moves slowly, although in fairness, countless well-presented and fairly priced claims are settled daily by the government.

While sympathy in construction claims is generally expressed by the industry for the contractor's viewpoint, the problems experienced by government staff in having to force or beg contractors to do work on time, to produce acceptable quality, not to substitute inferior equipment for specified

*I now am retained by five state Attorney Generals.

products, and to bill equitably are so overwhelming that strong resistance to many good claims has to follow.

Contractors constantly try to pay for their mistakes with other people's money. They are often not well managed and many do not care about any part of the contract but the bottom line. The fact that somehow a building or bridge is produced in the process is often amazing. Contractors work too cheaply; their planning is too short run for their own good. Only a few do any research; development of new methods is almost nil.

Truth is much too scarce in the construction industry; it is often harder to find than profit. Many contractors do not police their suppliers or subcontractors, and little or no effort is made to motivate employees. The largest construction jobs in the country are typified by hordes of labor mechanics who are openly working against the contractor. The ultimate idea that government or insurer must pay a claim because a contractor loses money is absurd. Construction claims will thus be resisted until the industry solves its problems from within.

Settlement

Settlement is always to be sought in construction disputes. When the government is the defendant, a properly presented claim is understood to be subject to settlement by both parties. The government will almost always resist the claim until it believes it has brought the figure down to that which is reasonable. Many times, one needs to include outside analysis to assess the claim in order to justify government's decision to settle as a safeguard against whatever internal government review process or investigation that might come up in the future.

Settlement should be sought as soon as possible by the contractor, obviously before litigation starts, and then continually during the adjudication process. Settlement sometimes is not possible. Government may not have the funds to pay. Cities, typically, do not have an extra $2 million laying around to settle the lawsuit on City Hall. If that sort of settlement bargain comes along, city council has to meet, appropriate the money, and raise the tax rate. Governments sometimes have to sell bonds or borrow money to meet court settlements—they sometimes make settlements and do not deliver the cash. Awarding authorities sometimes cannot be forced to deliver cash. Experience has shown that contractors can win large judgments against sewer and water authorities which cannot be enforced until the bond holders are all satisfied. Most state and lower level governments will settle the case as soon as they believe it is to their financial advantage.

The contractor who puts a claim together properly has the better chance of settlement. Settlement will not come fast, for the government takes time to settle. Claims have to grow a certain amount of whiskers; they have to sit around, and be studied and analyzed. Everybody has to get used to the fact that they are there and will not go away.

Government generally reacts to the claim with about the same degree of interest and enthusiasm the contractor displays. If the contractor comes in and pushes the issue, government will do business with him. If the contractor is obviously overstating, fooling, and stalling, government fools and stalls. Government's reaction is largely determined by the pattern set by the contractor. Government is not generally disposed to taking advantage of a contractor on claims nor does it run giveaway programs.

ARENAS AVAILABLE FOR CLAIM PRESENTATION

Arbitration

Well over half of the current construction contracts call for arbitration of disputes and provide for disputes to be arbitrated under the rules of the American Arbitration Association. The inclusion of AAA rules is relatively new in construction. Up until the mid-1960s there was no tie between the AAA and the construction industry. The AAA dates back to the 1920s, but its primary purpose was not construction-related.

Until the mid-1960s, parties to contracts with arbitration provisions found themselves with procedures and processes that often did not work. The lack of rules and no supervision made arbitration requirements ineffective in many cases. Procedural problems of not knowing how to start an arbitration added to the problem. The confusion led to a joint study by AAA and industry representatives which resulted in the adoption of rules for construction arbitrations. At the present time, nine natiohal professional organizations refer to the AAA construction industry rules in their recommended contract documents. These include the American Institute of Architects, the American Consulting Engineers Council, the Associated General Contractors, the Mechanical Contractors Association, the National Society of Professional Engineers, the American Society of Civil Engineers, Construction Specifications Institute, and the American Subcontractors Association.

The typical construction contract or design contract will have a clause reading as follows:

> Any controversy or claim arising out or relating to this contract or the breach thereof shall be settled in accordance with the construction industry arbitration rules of the American Arbitration Association rules and judgment upon the award may be entered in any court having jurisdiction thereof.

The rules are short, simple, and give great discretion to the arbitrators. They are primarily simple rules of procedure without provision for penalty. The AAA has no authority. The latest rules are available from the AAA on request.

One of the advantages claimed for arbitration is that frequently one member of the panel is experienced in the issues being contested, which gives the disputants a forum helpfully knowledgeable in the technical areas of the dispute. Arbitration hearings are a great deal less formal than a court proceeding. They take place in a hotel room, someone's conference room, a public room available to the governmental agency involved and so on. In some cities the hearings take place at the offices of the AAA; its New York headquarters has a number of hearing rooms that are in constant use.

Arbitrators are not required to follow legal rules of evidence and great latitude is generally afforded both sides in the presentation of the case. The arbitrators are empowered to listen to all evidence they deem relevant and material.

The selection of the panel can be done in a number of ways. Each party to the arbitration can name its own arbitrator and have those two select an independent arbitrator from the AAA panel, or all three arbitrators can be selected from the AAA list of panelists. The AAA has endeavored since the mid-1960s to empanel members of the construction industry. Anyone qualified who would like to be a member of the construction industry panel can do so by writing a letter to the nearest local office of the AAA for an application form. The application form is simple and once executed with a letter of reference leads to an almost automatic inclusion on the panel.

Under the rules arbitration is opened when one of the parties writes a demand for arbitration to the nearest office of the AAA. Upon receipt the AAA automatically sends a copy of the demand to the defendant along with a list of arbitrators from which the parties are asked to select. The typical list may include builders, contractors, engineers, architects, businessmen familiar with real estate or construction, and attorneys who do construction litigation.

In small arbitration the AAA will often contact several members on the list and prequalify them as having no conflict and being available. Larger arbitrations generally use three arbitrators. Typically both parties have seven days to study the list of arbitrators, delete any objectionable names and number the remaining names in order of preference. Those appearing as a choice in both lists are designated as arbitrators. When there are no mutual choices, an additional list is submitted. If the parties do not come to any agreement at the end of the second attempt the AAA makes administrative appointments. The AAA publishes the names of the appointed men and both parties have the right of approval. The typical selection of arbitrators takes three to four months. It can take a little longer, and occasionally it is done faster.

On small arbitrations the arbitrator will serve a day or two without payment as a public service. On major arbitrations the arbitrators are paid. Where millions of dollars are involved the arbitrators are paid well. Arbitrators receive as much as $1000 a day, and on a major construction arbitration, an arbitrator would expect to receive at least $300–$600 a day plus traveling expenses. The AAA will sometimes suggest a fee to the arbitrators and will in certain cases make fee arrangements beforehand. Inexpensive arbitrators generally are inexperienced and are not good. Good arbitrators cost money.

There are a variety of ways in which to handle the time and location of the arbitration. Generally the AAA makes the arrangements, and correspondence between panel and parties is formalized by using the AAA as a clearinghouse. However, most arbitrations are held with only the initial arrangements made by the AAA and then all future scheduling is made at the hearing. If hotels are the space being used, the AAA will make the arrangements for the hotel room.

The arbitrators are not imposed upon to make arrangements for routine elements of the process and they are accorded the privileges of privacy during the hearings—good taste prevails; one simply uses courtesy and good sense in dealing with an arbitrator while he is sitting formally during the hearing and in the times between the hearings. Correspondence to the arbitrator is always sent to the AAA with copies forwarded meticulously to all attorneys and the others involved.

Some states have adopted the uniform arbitration code and some have additional law from decisions or cases involving previous arbitrations. In some states, discovery is part of arbitration, while in other cases discovery is up to the panel. Some states, e.g. Texas, have no arbitration statute at all and the duty to arbitrate or an arbitration award cannot be enforced.

The typical arbitration has the parties represented by counsel with the proceedings very much like that of the court, where the plaintiff will explain its case, bring in witnesses, documents, and so forth, and present them to the panel. The panel will then permit cross-examination by the defendant, and the panel itself will ask questions. A record may be kept, the cost of which is generally borne by the party requesting it. The court reporter will typically keep the exhibits or they are turned over to the chief arbitrator, who is appointed from the three arbitrators. The chief or chairman arbitrator speaks for the panel during the proceedings, confers with his fellow members, and then addresses the assembled group.

Some difficulties in arbitration are the scheduling of hearings and the length of the overall process. Arbitrators are always very careful to be courteous and nonprejudicial to all parties and arbitrations can go on and on. Some cases have lasted a year and even more. The difficulty with this is that those hearing the case are human and have memories which fail. The best way to arbitrate a problem is to arbitrate as continuously as possible.

The hearings are closed at the end of the verbal presentations, and AAA rules call for the award to be furnished within 30 days—which sometimes happens, but most times does not. It often does not happen when the stenographic record is not typed and distributed for six weeks. Occasionally the arbitrators can waive the record of the last proceeding and make judgment without it. Sometimes arbitrators ask for memorandums of law or closing written briefs summarizing positions. Occasionally arbitrators ask for written findings of fact from both sides. Quite often in major construction arbitration 30 days is inadequate to review the record, make calculations, have conferences between the arbitrators, and make an award. In any event, the award is generally made within 30–90 days after the hearings close. Once an award is made, it can be entered in the court as collected. It can be challenged by either party in the court, but few challenges are sustained.

The arbitration process is expensive. The payment of the arbitrators, court reporter, and expenses of the panel alone without counsel, witnesses, etc., can be $1200–$2000 a day. If 20 hearings are required, that is $40,000. The filing fee for the AAA can be considerable. Experience indicates that the fee can go as high as $20,000. Arbitration lacks the force of a court in bringing pressure to bear on counsel to make a settlement. With arbitration clauses part of so many professional society standard contract forms, arbitration is common and in small jobs is almost automatic. In many cases, it is an excellent way to resolve disputes without the length of time it takes in court.

A second method of arbitration is to go ahead and arbitrate without the aegis of the AAA. This saves the filing fees and the rules can be followed except for the one difficulty of naming the panelists. The panel is preferably composed of a named arbitrator from each party who then select, with or without the help of the AAA, an independent arbitrator. One member of the panel should be experienced as an arbitrator or an attorney. This gives the panel the ability to respond to the comments typically made by attorneys in presenting evidence and valid objections as to hearsay. The presence of that experienced arbitrator or lawyer gives the panel the balance that is needed.

Frequently parties will agree to have a single arbitrator hear a case. Certain arbitrations have the parties agreeing on a man of experience to hear the case. Regardless of tribunal panel policies, the parties can always agree to have one man who is a recognized expert in the field be the arbitrator. His fee could be substantial, but in return the hearing can be expedited, held on a continuous basis, be given publicity if needed, and serve the parties well. There are many ideas of procedure in arbitration which, if the parties involved take the time to work out, can be to the advantage of both.

Multiparty arbitrations must be mentioned as the conclusion to this topic. This is the subject of much maneuvering by those involved. In construction, the owner is best served by having all disputants to a problem in the same proceeding. The disputants, particularly the designer, are best served by insisting that each issue be separately arbitrated between separate parties.

Recent professional society updated contract formats specify that the designer cannot be a party to an arbitration between contractor and owner. This, if enforced, saves the designer great exposure and is to the owner's detriment. It most likely hurts the contractor as well.

The other method of avoiding multiparty arbitration is to go into court and argue against multiparty arbitration, although the consensus in developing common law is toward consolidation. A better device for the professional to protect itself is to exclude arbitration from its contract, and place arbitration in the contractor's contract. Then the owner has great difficulty compelling both parties to be in the same proceeding.

A little thought by the drafter of the contract can mean impressive problems or safeguards five years later.

The Court

If provision for arbitration or other dispute process is absent, the disputants will seek the courts for a forum to find relief. There is typically a courthouse in each county and a federal courthouse in various districts across the country. There are certain rules which govern whether the suit goes into state or federal court but usually, if diversity of domicile of the parties exists, federal court is the answer.

The contractor will generally seek to adjudicate the dispute where the construction site is located. Both the majority of the contractor's records and the witnesses necessary to state confirmation of the contractor's side of the dispute will initially be located at the field office. A view of the project by the court is also possible.

The extent of participation of the owner is also to be considered in the selection of the forum. If the dispute is against a political body which depends on taxation for revenue (and the owner can demand a jury trial), the contractor may want to consider alternate locations. This consideration is not present in an arbitration. In addition, newspaper and other media coverage of the claim may be to the contractor's disadvantage if the dispute involves a politically sensitive project.

Final determination of the choice for forum will be made after review of the legal issues involved by the attorney. If the law differs on a particular question, the contractor will want to consider the state in which the decisions have been most favorable. The federal court system will be considered also.

Experience shows that the court is the favored forum. The presentation and litigation of construction disputes always involve intricate and difficult factual questions. In some cases, the interrelationships of 10,000 construction activities may be an item of proof. The amount of paperwork, cost records, daily reports, and correspondence generated by the typical two-year construction project is overwhelming. Construction disputes lead to protracted discovery, extensive interrogatories, and numerous depositions. The formality of the court and its authority are much more beneficial to the factfinding process in a claim. However, there is a risk of an unsophisticated judge who is not interested or too busy, or a jury of people who can be simply overwhelmed with financial discussions in the millions.

The costs involved in resolution are also an important consideration. In a court of law, there are only attorney's fees and court costs. These costs are also present in arbitration hearings. However, AAA arbitration also requires the payment of a fee based on the amount in dispute. The arbitrators are the pleasure of the parties—the judge is the taxpayers'.

The typical case proceeds with the filing of a complaint by the complaining party. The complaint is answered by the defendant who has an opportunity to bring before the court failure to conform to contractual or legal procedural matters, denials, and/or counterclaims. The plaintiff then has an opportunity to answer the counterclaims in the form of a reply.

Once the initial pleadings have been filed, both parties begin discovery or specificity is sought by further pleadings. In construction matters, a discovery is extensive. The task is almost always to define the real issue. Typical interrogatories may reach 200 questions and require extensive work on both sides by attorney, client, and consultant. Sometimes interrogatories can be burdensome, for they take unbearable effort and months to answer. Some lawyers will attempt to drown the other side by forcing it to answer detailed questions with multiparts. Some prefer tactics of skimpy answers; others want to answer with such specificity and force that somehow the case will be won by the interrogatories alone. Depositions, which are probably the best area in which the other side's case begins to emerge, may be extensive owing to the number of individuals associated with any construction project. Once discovery is completed, a process of one to three years, trial begins.

For some construction cases, the judge may direct fact-finding by a Master and the trial becomes similar to a one-man arbitration hearing. The typical trial moves very slowly; all sorts of legal maneuvers occur to create a record, the purpose of which is beyond the understanding of the bystander and sometimes the lawyer. Direct and cross-examination move slowly but invariably the truth surfaces. Again the skill and preparation of counsel are paramount. There are varieties of skill in lawyers and good and bad days for everyone, including plaintiff and defendants.

The courts, however, do work. Those who do not use or understand the courts usually have a poor impression of the system. Experience shows that the system works and permits both parties equal fact-finding opportunities, the ability to hire talent, and a place to run or settle a problem. It is expensive at times and it is slow at times, and once in a while the result will be wrong.*

Contractors who have learned to use the courts are constantly in court—so are insurance companies. The courts are a business tool. Subcontractors who have learned to give notice under the Miller Act constantly do so. True, some use the courts unfairly, but sooner or later those who do run out of the money.

---

*Judges are experienced fact finders and are capable of exerting settlement pressure.

Settlement discussions may go on continuously from the initiation of discovery until the jury or court reaches its decision, depending on the parties involved. Insurance and surety companies typically participate in a complete discovery and initiate the trial before making any firm offers to settle the matter. A contractor may be willing to buy out early and for much less than the claim is worth for quick money since a contractor's business typically requires fluid capital. Since interest may not be assessed until the date of award, there is incentive for the owner to resist settlement as long as possible. Indeed the owner's usual object in any dispute proceeding will be to hold as many dollars and delay, confound, and obfuscate the entire process. Who knows? The claimant may go broke in the meantime and the day of reckoning never come. Defense and delay are synonymous. Architects or engineers are the most difficult opponents, for they are generally unwilling to participate in meaningful settlement discussions, and much more so than the others in the dispute, because of the principles involved, are willing to defend to the full extent of the law.

The Government Board of Review

There are several types of government boards; the text dealt with this area in some depth earlier. Some boards, such as the Armed Services Review Board and the Review Board for the General Services Administration, have members who function primarily in that role as full-time government employees whose job is to hear disputes and make the decisions that are required.

Other federal government agencies with less disputes or localized disputes will make up a part-time board from members of its staff to hear a dispute. They can be the same people used over and over again, but typically are members of the agency's law department or contracting department who are empaneled and have the authority to hear the argument "impartially." If either of the parties disagree with the review board finding, it has a right of appeal to the Court of Appeals.

Some claimants feel that the review boards have difficulty in maintaining the total impartiality needed since the board members sometimes work with the government advocate in the same case. The contractor, however, is aware of this when it bids the job. The risk of being in a dispute situation with a board which someone thinks is unfriendly is not thrust upon the contractor suddenly. Experience has shown that the board is generally more than fair to the claimant.

Government boards of review generally follow through on a case once presentation is started in good order. The evidence is presented and heard in a reasonable length of time. Sometimes, however, decisions take a long time to be written, and even worse review board members in charge of the case change quite frequently. Thus the decision can be written by a board member who did not even hear the case.

Finally, most boards have discovery and fact-finding provisions. Boards are generally busy, and the waiting list is long. Government counsel is in-house and generally far more overworked from a case load point of view than the claimant's lawyer. Therefore, it takes time to get before the board. Board work and appeals to the U.S. Court of Appeals are best handled by lawyers who are well versed in the procedures. Most of these attorneys are located in Washington, D.C. or are with the prominent construction law firms in the major cities.

8

THE CLAIM AND ITS CALCULATION

THE FORMAT OF THE CLAIM

The presentation of the claim has a number of extremely important aspects. The presentation must be persuasive at a glance. Obviously a request for several hundreds of thousands or millions of dollars in a dispute situation must ostensibly reflect the degree of interest and expectation its author or presenter holds.

Yet the typical construction claim is often so poorly or ingenuously presented that owners who want to make contract adjustments find themselves unable to respond to the situation because of inadequate or unrealistic presentations. Too often the contractor submits a one and one-half page letter with several Xerox copies of accounting tablet pages in which some meager approximation of costs that it thinks sound reasonable have been scratched out in pencil, along with copies of enough job correspondence to create a book deemed sufficiently thick and heavy by the contractor to prove its case.

The result is almost always the same: The owner cannot do anything but ignore the claim or ask for specificity and proof. It is hard for contractors to come to the realization that presentation and collection of a construction claim is as exacting as estimating, pricing, buying services, or selling a new customer. The perfection of a dispute from a claim to cash costs effort, money, and time.

No group politic is going to roll over and play dead when asked to give a contractor millions of dollars in today's sunshine era of government. The evidence for extra dollars must be set forth so well that settlement is the only way out for the defendant and so difficult for its representative to achieve that when done the media, politician, management board, or whatever will applaud.

Some of the elements of a claim presentation are so basic they really should not have to be stated, but their constant violation means that they cannot be overlooked. A well-done claim needs a good typewriter used by a person who understands margins, spacing, listing numbers in vertical rows, reasonable syntax, spelling, and punctuation. Good paper is necessary, extra copies reproduced with clarity and class which bespeak the sincerity and seriousness of the claimant. The contractor who has fought his way from laborer in the trench to president with the big Lincoln and the CB radio cannot rely totally on the verbal excellence of his persuasive personality that has engendered such success. When that personality and all its trappings have left the negotiating table, the claim presentation that remains has to be sufficient enough to persuade its reader that the jig is up, the day of reckoning is soon to be at hand.

When putting a claim together, consider its physical appearance. Chances are the document will be routed between many and reviewed by at least ten people. The cover must be strong and it must look expensive. Spend $25 on a leather jacket; remember how much money is being wrapped in the jacket.

Remember, too, that the solution or settlement to the claim will most likely require the assent of a high-positioned person who is or fancies himself too busy to study the claim in detail. He may only hold the presentation in his possession for 15 minutes. Will the claim stand the 15-minute riffle and spot perusal of the big executive asked to approve the $600,000 settlement, or will his reaction be negative? Less attractive young ladies properly presented have gone far in this world and so have claims given the same care.

The claim is almost never ruled upon by the people who are most knowledgeable of what went on. Chances are that those who make the decisions as to settlement or defense will not have the necessary knowledge of what the project looks like, let alone an understanding of what went wrong in the boiler room or what a pressure-reducing station does. It is the burden of the author of the claim statement to provide enough information in the presentation to teach the intermediate reactors and ultimate judges what they need to know to be able to work the situation out. The claim statement also becomes the basis for the trial manual and can give the same training to counsel. Too many claims founder because the client cannot tell its own lawyer the story in proper perspective. The best construction lawyer is really a lawyer with a construction vocabulary and some education and experience in the technical aspects, but the real meat of the dispute must be conveyed to him from the client or technical side.

The presentation should be in the degree of detail each reader needs so it can learn what its job level requires it to know. The following is a guide for the contents in a claim statement:

1. Title Page
2. Table of Contents

3. Preface
4. Summary
5. Introduction of the Participants
6. Description of the Project
7. Contract Terms
8. Intentions of Claimant
9. Actual Experiences and Why's
10. Conclusion
11. Damages

The preface is some two or three double-spaced pages which describe the purpose of the report and the way the report was prepared. It will establish the sources of fact consulted and describe the qualifications of the author and the claimant. After being read it should leave the impression that a highly qualified person of integrity has done all the study necessary to write what follows.

The summary follows, and it too is short in length, no more than five single-spaced pages. It will be read by everyone and must pack all the information into a concise wallop, which is strong, poignant, and to a sense piquant as well. The big boss may only read the summary and the writer must catch his eye with enough problems on his part to set permanent conclusion that much of the money demanded at the end of the summary is going to have to be paid sooner or later. Do not fill the summary with too many dollar figures, it will tend to confuse the real issue.

The body of the report follows double spaced and ideally is no more than 100 pages long. If more space is needed, give details of individual problems in appended minireports. A claim report that is too long will lose its persuasive ability. The details of who struck John fit better in an annex out of the main flow of the story. The body should start off with a description of who, when, where, and how all the players came to this project. Never deal in personalities or people, always refer to the firm; do not worry about the reader needing to know which individuals are to blame, the record will make that evident.

Once the main body of the report starts, the process of exhibiting each statement of fact begins. Any fact cited in the main body needs an exhibit which should be provided in a separate place in the presentation. A system of notation such as "(Exhibit 6)" will do the job and when one turns to Exhibit 6, there easily identified and ready to be totally read is the very document needed. The exhibits should be numbered consecutively.

After introducing the reader to the parties, describe the project, setting down enough so that a mental picture can be formed by the words written. Supplement (but do not substitute for) the written description with a few photographs. Get the reader comfortable with the physical aspects involved. Compare the building to a football field, for instance, for spatial relationships. State what the purpose of the project was supposed to be. Then describe the contract requirements, and cite and copy important contract clauses. Quote the relief clauses on which the decisions or abilities to pay the sums sought can be justified. Do not practice law and do not make legal conclusions; just focus on and make available the parts of the contract that must be reviewed by the person who has to make a recommendation and be able to answer questions from superiors.

Next, the report will set out the intention of the contractor before any or all of those horrible things happened. Maybe the original job schedule or critical path needs to be described, or the method of shoring or bracing, or the erection sequence which was intended. Then the actual experience is described along with the reasons the contractor was penalized. Each reason needs exhibits showing that all described is fact.

The report then reaches its conclusion. Some well-written 16,000 or fewer words designed to convey sufficient persuasion in layman's language have led up to one or two pages of conclusion so obvious to the reader that any argument over entitlement, if it ever existed, is no longer viable, and the claimant needs and deserves instant adjustment.

The skilled writer will not argue his case in each line nor will he gild the lily. Instead the simple progression of asserting fact upon fact upon fact will give the reader all the conclusions necessary to be ready to agree with the writer. Do not lie, always tell the truth; do not volunteer errors, they are the other side's duty to discover and prove; and do not demand or claim so much that the result is an insulted reader ready to battle all the way to the Supreme Court.

The final part of the claim presentation is the damages.

THE CALCULATION OF DAMAGE

A note of caution about damages is in order. Gossip states that 30 cents on the dollar is an average prelitigation settlement. This leads to everyone multiplying what they want by 5 to get a number high enough so that a third of it is more than needed, and thus will cover the expenses of making the claim. Experience indicates that absurd claims are resisted by absurd adamancy. Therefore, do not invent a Mickey Mouse number that no one believes; this will only make the defendant bristle and dig in to resist the claim. If one is patently absurd, the good parts of the claim will lose some luster, and perhaps the claimant will overshoot and lose everything. Instead, set forth the dollars lost and sought with a reasonable amount of over-statement. Thirty cents on the dollar may be the average settlement, but that means some settlements are higher. The claimant whose damages are undeniably real for the most part is a worthy opponent. No one wants to tangle with a worthy opponent for longer than necessary to establish that the defense has done its job and is not giving away money.

Damages must be calculated in special ways as well. The idea of being owed all one has spent plus its overhead and a reasonable profit as damages will be acceptable in a total breach entitlement, but in construction such is rare. This method of damage calculation is called total cost, a method generally not acceptable to most hearers of facts. The difference between selling price or estimated cost and actual cost, if a negative number or a positive one smaller than originally agreed upon, may well be the loss the contractor has felt and experienced. However, that negative number is not necessarily the result of the actions or lack of action by the owner or its agents.

Damages must be calculated and proven in other ways than total cost alone. Total cost will be the first check used by the defendant in measuring the claimant's credence once cost records are discovered. If damages exceed the total cost check, they will be suspect and characterized as phony or overstated by the defendant. Therefore, damages are best calculated by individual analysis and then discussed in the context of total cost. These analyses properly done will lend support to each approach.

Two sample damage calculations are discussed in the following sections. The first is for an electrical separate prime contractor on a sewage treatment plant in the upper Midwest which was not totally complete when the claim was filed, and the second for a transportation system concrete tunnel which was completed. The principles in the calculations will apply to almost all problems that arise.

Sample Damage Calculation Number 1

Labor Escalation

X estimated that the Big City Sewage Treatment Facility would require 12,580 man-hours to complete. These estimated man-hours were divided into two labor wage classifications: foreman and journeyman. Each labor classification's wage rate had escalated at fixed contract time periods. These wage increases for the original period contemplated were known by X and included in its estimate. X was able to determine its labor costs accurately by dividing the amount of work each labor classification was expected to perform among each period of wage increase.

Owing to the inability to maintain the expected job progress, labor costs greatly increased. Work was forced into higher wage periods, and even when work was performed within the originally planned wage period, it was done at drastically reduced rates.

The amount of wage escalation which X incurred is calculated by multiplying the difference between wage rates from one period to another (escalation) by the difference between the total hours that should have been worked and the total hours actually worked in each wage period:

FOREMAN

Pay Period	(Exhibit #22) Planned	(Exhibit #21) Actual	(Exhibit #23) Escalation
#1 6/1/72–6/1/73	1104 (–)	588 = 516 X 0.46 =	$237.36
#2 6/1/76–6/1/74	2208 (–)	1150 = 1058 X 0.32 =	$338.56
			$575.92

JOURNEYMAN

Pay Period	(Exhibit #22) Planned	(Exhibit #21) Actual	(Exhibit #23) Escalation
#1 6/1/72–6/1/73	7,210 (–)	465 = 6745 X 0.50 =	$3372.50
#2 6/1/73–6/1/74	10,372 (–)	2875 = 7497 X 0.37 =	$2773.89
#3 6/1/74–6/1/75	10,372 (–)	8548 = 1824 X 0.42 =	$766.08
			$6912.47

Foreman	$575.92
Journeyman	$6912.47
	$7488.39
24% Payroll Burden	$1797.21
	$9285.60

In addition, X is entitled to escalation on labor hours in excess of the estimate. This work would have been completed in the second pay period if the job had been performed within the original contract period. Also, if work is carried into any additional pay periods, X will be entitled to additional escalation.

To date, only foreman hours have exceeded its estimate. Hence, X is also due the following amount:

Pay Period	Hours	Escalation
#3 6/1/74–6/1/75	2107 X 0.32 =	$674.24
	24% Payroll Burden	$161.82
		$836.05

Additional escalation on hours in excess of the estimate can be projected by determining the percentage of job completion at a known date and extending the remaining hours into the next pay period. As of June 15, 1975 labor was 60 percent complete; 40 percent is the amount of labor to be completed:

Wage Category	Total Hours 100%	Hours to 6/15, 60%	Hours Remaining, 40%	Escalation
Foreman	7,853 –	4,712 =	3141 X 0.80 =	$2512.80
Journeyman	17,095 –	10,257 =	6838 X 0.69 =	$4718.22
				$7231.02
			24% Payroll Burden	$1735.44
			Total	$8966.46

Thus, the total amount due X for wage escalation is

$9,285.60
$836.05
$8,966.46
$19,088.11

Material Increase

Price escalation.   Because of inability to achieve timely performance, X was forced to purchase materials at increased costs. X is entitled to recover the additional sums paid for material purchased at increased prices. Through June 1975, X's material had escalated as follows:

Conduit (Exhibit #)*	$12,365.00
Equipment (Exhibit #)	$807.00
Fixtures (Exhibit #)	$8,952.00
Lamps (Exhibit #)	$1,267.00
Wire and Cable (Exhibit #)	$38,743.00
	$62,134.00
7% State Sales Tax	$ 4,349.38
	$66,483.38

*These exhibits typically would include invoices and backup documents.

It is estimated that X will incur an additional $24,672.00 in material escalation during the remainder of the job (Exhibit #). Thus,

$66,483.38
$24,672.00
$91,155.38 Total Material Escalation

Material storage.   X has further incurred additional storage charges. To date these charges are

$11,304.00   (Exhibit #)
$791.28   7% Sales Tax
$12,095.28

For material increase, therefore, X is due

$91,155.38   Price Escalation
$12,095.28   Storage
$103,250.66

Engineering and Supervision

Increased supervision.   The original work schedule called for 24 supervision months, 12 months in each pay period. The supervisory personnel to be utilized at the Big City Plant were to be working foremen. X estimated the supervision cost to be as follows:

Pay period #1, 6/1/72–6/1/73

12 foreman months $\times$ 92 hr = 1104 hr $\times$ 11.27 = $12,442.08

Pay Period #2, 6/1/73–6/1/74

12 foreman months $\times$ 92 hr = 1104 hr $\times$ 11.85 = $13,082.40
Total Estimated Supervision Costs $25,524.48

Because of its inability to obtain copies of all approved shop drawings from other contractors on the project and failure of the owner's representative to respond to legitimate questions of location, method, and sequence for installing electrical work, X's supervisory costs have greatly increased. Not only were supervisory personnel required for an extended period of time, but the problems at the Big City Plant were of such a magnitude that it was found necessary to employ a general foreman rather than just foremen to oversee job progress. In addition, it was necessary to use subforemen at various stages in the work schedule.

The total cost of supervision as of July 6, 1975 was $68,653.84. This includes the work of a general foreman, foremen, and subforemen. Thus the extended supervision costs are as follows:

Supervision Costs as of 7/6/75 $68,653.84
Estimated Supervision Costs $25,524.40
Extended Supervision Costs $43,129.36

X bid the job to require 2204 supervisory hours. The supervisory personnel to be used at the work site were to be working foremen who only devoted 36 percent of their time to nonproductive supervision activities.* The job actually took (as of 7/6/75) 4712 supervisory hours, and 2380 of these hours were worked by a general foreman who devotes 100 percent of his time to nonproductive supervision. X is entitled to recover 64 percent

---

*This is a standard percentage used in determining the nonproductive supervisory time of a working foreman. It is estimated that a working foreman will spend one hour in the morning, a half-hour before lunch, a half-hour after lunch, and one hour in the afternoon in nonproductive supervisory activities.

of the general foreman's hours representing lost production time:

$$
\begin{array}{rr}
\text{General Foreman Pay} & \$38,368.69 \\
\text{X} & 0.64 \\
\hline
& \$24,555.96
\end{array}
$$

Thus, the extended supervisory costs are as follows:

$$
\begin{array}{lr}
\text{Extended Supervision Costs} & \$43,129.36 \\
\text{Lost Production} & \$24,555.96 \\
\hline
& \$67,685.32 \\
24\% \text{ Payroll Burden} & \$16,244.47 \\
\hline
& \$83,929.79
\end{array}
$$

It is estimated that the job is only 60 percent completed. Additional extended supervision costs are projected to be

$$
\begin{array}{ll}
\$83,929.79 & 40\% \text{ Supervision Costs Remaining} \\
\$55,953.19 & \\
\hline
\$139,882.98 & \text{Total Extended Supervision Cost}
\end{array}
$$

Additional engineering.   In addition to causing increased supervision, the inability to obtain copies of approved shop drawings and answers to questions of electrical equipment location have caused additional engineering costs.

Prior to June 1974, X had spent $4746.00 on project engineering that should have been done by the design engineer or owner. X is entitled to recover these amounts:

$$
\begin{array}{ll}
\$4746.00 & \\
\$1139.04 & 24\% \text{ Payroll Burden} \\
\hline
\$5885.04 &
\end{array}
$$

Total

$$
\begin{array}{lr}
\text{Increased Supervision} & \$139,882.98 \\
\text{Additional Engineering} & \$5,885.04 \\
\hline
& \$145,768.02
\end{array}
$$

Interest Expense

An additional cost item necessitated by the extended construction period is interest on borrowings. Such interest is allowed as an equitable adjustment when a clear necessity for borrowings occasioned by change can be demonstrated. Such necessity can be demonstrated when actual prime costs exceed the estimated prime costs. Since the job will not be completed until February 1976, actual total prime costs are not now known.

It is possible, however, to project actual prime costs. As of 6/15/75, labor was 60 percent complete. The total projected prime cost at completion is equal to the total prime cost as of 6/15/75 plus 40 percent of the projected total:

Total Prime Costs as of 6/15/75	60%	$987,183.00
	40%	$658,122.00
	100%	$1,645,305.00

The projected total prime costs at completion will exceed the estimated prime costs by $350,519.00:

Projected Total Prime Cost	$1,645,305.00
Estimated Prime Cost	$1,294,786.00
	$350,519.00

The additional prime cost will deprive X of valuable working capital. X will have to borrow to replace this lost working capital, and is entitled to recover the interest necessitated by such borrowing.

During the period this money is lost, X's interest rate will average 10 percent. It is estimated that X will borrow $350,519.00 through its general line of credit for one year. Thus, X is due $35,052.00 in interest on this borrowing.

In addition, X estimated that it would be necessary to finance the retainage (5 percent of the value of the contract) at 6 percent for two years. Since the job has been extended, X is forced to finance this retainage for an additional two years at an escalated interest rate of 10 percent. The cost of this financing is as follows:

$1,401,000.00	Bid Contract Cost
5%	
$70,050.00	Retainage
10%	Interest
$7,005.00	
X2	Years
$14,010.00	

Thus, the total amount due X for interest is

$35,052.00
$14,010.00
$49,062.00

Lost Profits

As a result of the inability to achieve timely performance and the resulting increased costs, X has been denied its expected profit. X's bid summary sheet reveals that the Big City job was expected to generate $31,250.00.

This profit represents only 2.5 percent of the estimated net cost of construction, a reasonable and modest figure. X is entitled to recover the profit which it has been denied as a result of increased costs: $31,250.00.

Winter Inefficiency

The work schedule for the project at Big City has been significantly altered. Outside work originally scheduled for performance during the "construction season" has been forced into difficult winter conditions. X is entitled to recover the lost efficiency which it has experienced while performing outside work in winter conditions which should have been performed during warm weather. The value of the lost efficiency can be determined by multiplying an inefficiency factor by the labor value of activities performed during the period.

Although work was done during three winter periods only during the 1974–1975 winter period, there were a significant number of hours worked. The work done during this period is as follows:

Work Area	Job	Hours
Primary settling tank	Electrical hook-up	200
Outside lights and grounding	Site lights rough-in	34
Pipe gallery	Set motor control center	12
	Hook-up equipment	17
	Electrical fixtures	198
		461 hr

The labor value of the activities will be determined by applying the hourly wage rate of the journeyman to the actual hours worked for the activities within the time period. The total labor value of these activities becomes

$$461 \text{ hr} \times \$11.07 = \$5103.27$$

According to recognized and accepted bidding formulas, the labor correction factor for work performed in winter can be calculated at 30 percent. Accordingly, for work performed in adverse winter conditions, X is due

$$
\begin{array}{ll}
\$5103.27 & \\
\underline{30\%} & \text{Labor Efficiency} \\
\$1530.98 & \\
\underline{367.44} & 24\% \text{ Payroll Burden} \\
\$1898.42 &
\end{array}
$$

Extended Main Office Overhead

Because of the inability to achieve timely performance on the Big City Sewage Treatment Plant, the project has taken an additional year, and work is expected to continue until February 1976. During this extended period the project is supported by significant amounts of main office overhead, and X is entitled to recover such expenses.

The amount of extended main office overhead has been calculated by allocating the total recorded main office overhead to the Big City project on the ratio of contract billings to total billings for the extended period of performance:

$$\frac{\text{Extended Job Billings (6/1/74-6/30/75)}}{\text{Total X Billings (6/1/74-6/30/75)}} \times \frac{\text{Total X Overhead}}{(6/1/74-6/30/75)}$$

$$= \text{Extended Main Office Overhead Allocable to Big City Job}$$

As a result of the use of this formula, X has reported its extended main office overhead as $82,336.00. This $82,336.00 represents the cost of extended main office overhead for 13 months. There remains eight months until completion. Projecting the extended main office overhead until February 1976, $82,336 represents 13/21 of the extended overhead, or 62 percent. Therefore, the remaining 8/21 or 38 percent of all extended overhead equals $50,464.

Thus for extended main office 8/21 overhead X will be due

6/1/73–6/30/75	62%	$82,336.00
	38%	$50,464.00
	100%	$132,800.00

Extended Direct Job Expense

X's bid was based on contract completion within two years. As such, the bid included two years of direct job expense. Since the time of performance has been extended due to factors outside the control of X, it is entitled to recover amounts paid during this extended period for direct job expense.

X has reported the amount $17,384.00 for such expenses paid during the extended period of performance from June 1974 to June 1975. To project the direct job expense for the completed job, multiply the average monthly direct job expense by the eight remaining months until job completion:

$$\$1337.23 \times 8 \text{ months} = \$10,697.85$$

Thus, the total due X for extended direct job expense is

June 1974 to June 1975	$17,384.00
Projected to February 1976	$10,697.85
	$28,081.85

Out-of-Sequence Work

Changes, delays, and other problems so disrupted the work schedule at Big City that construction had to be done out of sequence. The out-of-sequence work caused X inefficiencies which substantially contributed to the increased cost of the project. The increased costs suffered by X due to out-of-sequence work are divided into (1) job access continuity, (2) inefficiency due to unavailability of work, and (3) congestion.

Job access continuity.   There are distinct and separate time-consuming operations required for all of the processes involved in the installation of electrical work. The labor estimate is composed of the man-hours required to perform each of the operations. In the preparation of the labor estimate, consideration is given to those operations which are nonrecurring and which must continue throughout the life of the project. Operations involved in moving in and moving out are nonrecurring operations. They will be reflected in the labor estimate only once.

At Big City, X has experienced delays which have resulted in the normally nonrepetitive activities of moving in and moving out to recur, causing substantial dollar penalties. X is allowed to recover the money spent for excessive moves.

The operation of moving out of an area when work is unavailable and back in when work becomes available is made up of the following items:

1. Studying and rechecking plans
2. Moving to point of usage
3. Tooling up
4. Measuring and layout
5. Cleanup

Recognized bidding formulas assign values between 4 and 30 percent for the sum of these items. For purposes of determining the losses incurred by X's excessive moves, 10 percent has been used.

X experienced the following additional moves (move in and move out) in the areas listed.

Raw Sewage Pumping Station	2 Extra Moves X 10% = 20%
Primary Settling Tanks	2 Extra Moves X 10% = 20%
Incinerator and Chemicals Storage Bldg.	3 Extra Moves X 10% = 30%
Sludge Thickener Bldg.	4 Extra Moves X 10% = 40%
Exisiting Administration Bldg.	2 Extra Moves X 10% = 20%
Blower Bldg.	4 Extra Moves X 10% = 40%
Pipe Gallery and Aeration Tank	1 Extra Move X 10% = 10%
Lime Sludge Pumping Station	2 Extra Moves X 10% = 20%

The cost of the excessive moves can be calculated by multiplying the additional percentages by the value of the labor in each year (determined by multiplying labor hours for each area by the estimated average wage rate of $10.45):

	Labor Hours	Estimated Average Wage ($)	Labor Value ($)
Raw Sewage Pumping Station	900 X	10.45	9,405.00
Primary Settling Tank	762 X	10.45	7,962.90
Incinerator and Chemicals Storage Bldg.	3487 X	10.45	36,439.15
Sludge Thickener Bldg.	502 X	10.45	5,245.90
Existing Administration Bldg.	376 X	10.45	3,929.20
Blower Bldg.	1991 X	10.45	20,805.95
Pipe Gallery and Aeration Tank	8267 X	10.45	86,390.15
Lime Sludge Pumping Station	1776 X	10.45	18,559.20

Raw Sewage Pumping Station	$9,405.00 X 20% =	$1,881.00
Primary Settling Tank	$7,962.90 X 20% =	$1,592.58
Incinerator and Chemicals Storage Bldg.	$36,439.15 X 30% =	$10,931.74
Sludge Thickner Bldg.	$5,245.90 X 40% =	$2,098.36
Existing Administration Bldg.	$3,929.20 X 20% =	$785.84
Blower Bldg.	$20,805.95 X 40% =	$8,322.38
Pipe Gallery and Aeration Tank	$86,390.15 X 10% =	$8,639.02
Lime Sludge Pumping Station	$18,559.20 X 20% =	$3,711.84
		$37,962.76

Thus for additional move in–move out expenses X is due

	$37,962.76
24% Payroll Burden	$9,111.06
	$47,073.82

Inefficiency due to unavailable work.   Since the beginning of the job X has been subjected to inefficiencies due to insufficient volumes of work in areas occupied by its crews. The nature of the work in those areas was such that the work product of several contractors was simultaneously incorporated into the finished result. Although X's presence was necessary, it was not supplied with sufficient quantities of work to fill the time of the crews. Labor durations increased while productivity fell. The areas that experienced additional moves (see the section on job access continuity) were also the areas in which X's crews were confronted with the lack of work due to the delays and slow progress of other contractors.

By comparing planned construction to actual construction in those work areas experiencing inefficiencies due to unavailable work the reader can see how failure to maintain planned volumes of work has caused construction to decelerate, thereby adversely affecting X's production.

The problem of unavailable work plagued X's crews until March 1975. Since that time the problem has not been unavailable work, but congestion (see the following section). Only the hours worked prior to March 1975 are subject to a labor correction factor.

	Hours Prior to 3/75	Estimated Average Wage ($)	Labor Value ($)
Raw Sewage Pumping Station	287 X	10.45	2,999.15
Primary Settling Tanks	645 X	10.45	6,740.25
Incinerator and Chemicals Storage Bldg.	1007 X	10.45	10,523.15
Sludge Thickener Building	432 X	10.45	4,514.40
Existing Administration Bldg.	387 X	10.45	4,044.15
Blower Bldg.	820 X	10.45	8,569.00
Pipe Gallery and Aeration	1083 X	10.45	11,317.35
Lime Sludge Pumping Station	1106 X	10.45	11,557.70
			60,265.15

The labor correction factor for inefficiency due to unavailable work has been observed by us to be between 40 and 110 percent. For purposes of determining losses incurred by X we use the minimal value:

$$\$60,265.15 \times 40\% = \quad \begin{array}{l} \$24,106.06 \\ \underline{\$5,785.45} \quad 24\% \text{ Payroll Burden} \\ \$29,891.51 \end{array}$$

Congestion. Rescheduling and interruptions in the individual elements of the construction schedule have transformed an orderly sequenced work plan into one in which many conflicting operations were performed concurrently. The workmen of several trades became stacked in a limited work area, creating a situation in which work could not be done efficiently. A labor correction factor of 40 to 110 percent is necessary in our experience for inefficiencies due to stacking of trades or concurrent operations.

X has experienced such inefficiency in the pipe gallery, raw sewage pumping station, and primary settling tanks since March 1975. Inefficiency costs in these areas can be determined by multiplying the number of hours worked therein since March 1975 by the estimated average wage rate to get the labor value. Then the labor value is multiplied by the inefficiency factor of 60 percent:

Pipe Gallery	2100 X 10.45 = $21,945.00
Raw Sewage Pumping Station	2407 X 10.45 = $25,153.15
Primary Settling Tanks	324 X 10.45 = $ 3,385.80
	$50,483.95

$50,483.95
<u>   60%</u> Inefficiency

$30,290.37
<u>   7,269.69</u> Payroll Burden
$37,560.05

Total due for cost as a result of out-of-sequence work:

1. Job Access Continuity	$47,073.82
2. Inefficiency Due to Unavailable Work	$29,891.51
3. Congestion	<u>$37,560.05</u>
	$114,525.38

Unapproved Change Orders

In addition to the damages calculated in the preceding sections, the design engineer has failed to recognize that the multiconductor twisted shielded cable noted on Drawing E-47 is low-energy wiring and has not issued appropriate direction.

By contract, low-energy wiring is to be installed by the general contractor. However, in some cases low-energy instrumentation wiring was designed into X's conduit schedule necessitating pulling both X and general contractor cable together. The installation of this low-energy wiring cost $21,764.

Summary of Damages

In summary, X has experienced the following increased costs in the performance of work at Big City Sewage Treatment Plant:

A. Labor Escalation	$19,088.11
B. Material Increase	$103,250.66
C. Engineering and Supervision	$145,768.02
D. Interest Expense	$49,062.00
E. Lost Profit	$31,250.00
F. Winter Inefficiency	$1,898.42
G. Extended Main Office Overhead	$132,800.00
H. Extended Direct Job Expense	$28,081.85
I. Out-of-Sequence Work	$114,525.38
J. Unapproved Change Orders	<u>$21,764.00</u>
	$647,488.44

Sample Damage Calculation Number 2

This section of the report quantifies the cost increases. The damage calculations are based on the books and records of Merit Construction Co., Inc.; all figures can be readily audited.

Total Costs

Direct field costs.   As of January 1977, the final contract amount to be paid to Merit exclusive of pending claims and modifications was $8,764,728.00. Merit maintained sophisticated systems of *cost control* by computer for all of its work. Our examination of these cost records shows that Merit spent a total of $10,648,724.00 in direct costs and field overhead.

Administrative overhead calculations.   Merit's administrative overhead is calculated from figures taken from its IRS Form 1120 tax return. Merit sales in those years are as follows:

Year	Total Sales	Sales This Contract ($)	Percent of Total Sales This Contract
1972	4,221,164.00	511,433.47	12.1
1973	4,762,443.11	3,117,744.43	65.5
1974	5,276,422.07	2,889,997.84	54.8
1975	1,776,441.71	1,471,289.36	82.8
1976[a]			45.0 (est)

Year	Administrative Overhead ($)	This Contract Administrative Overhead ($)
1972	176,428.47	21,272.44
1973	298,727.41	177,284.07
1974	342,176.22	182,330.17
1975	289,117.07	183,211.03
1976[a]	176,122.11	57,603.98
	1,282,571.20	621,701.69

[a]1976 figures are not yet final; however, the estimate is accurate for these purposes.

Profit calculation.   Merit had a history of profit. This is demonstrated by review of Merit's operations in the several years prior to its problems on this contract:

		PROFITS	
Year	Gross Receipts ($)	Dollars	Percent
1970	6,541,002.73	137,842.11	2.11
1971	1,774,889.77	161,476.03	9.10
1972	4,221,164.00	631,111.01	14.95
Total	12,537,056.00	930,429.15	7.42

$$100\% - (\$12,537 - 930)/12,537 = 7.42\%$$

Therefore, Merit's history indicates it would ordinarily realize a profit from this contract as follows:

$$\text{Cost of Sales} \qquad \text{X Historical \% Profit} = \quad \text{Profit}$$

$$(\$10,648,724.00 + \$621,701.69) \times \quad 7.42\% \quad = \$836,265.53$$

Thus, based on the costs incurred on the project, home office costs, plus historical profit Merit is due an adjustment as follows:

Direct Field Cost	$10,648,724.00
Administration Overhead	$621,701.69
Historical Profit	$836,265.53
Total Due	$12,106,691.22
Paid to Date	$8,578,942.00
Amount Now Due	$3,527,749.22

Individual Damage Calculations

In addition to the foregoing total cost calculation of damage, we have also examined the individual facets of damage which Merit suffered. These costs are in the following categories:

(1) stretch-out costs of field overhead,
(2) stretch-out costs of administration,
(3) escalation,
(4) backfill increase,
(5) excavation increase,
(6) inefficiency in concrete, and
(7) acceleration costs.

This section examines these costs.

Stretch-out costs of field overhead. Merit field activity extended from April 1972 through May 1976. The project was accepted as substantially complete in March 1976. Merit was required to maintain field or direct management costs for this project. The period of April 20, 1972 to May 25, 1976, or 1497 days, is used as job duration.

Merit recorded its field overhead costs as they related to labor, material, rentals, subcontractor, miscellaneous, and equipment in cost code category 003. The following figures are taken from Merit's records:

Through 1976 $1,274,649.10

Therefore, the cost per day of construction duration for general conditions or direct field overhead is

$$\frac{\$1,274,649.10}{1497 \text{ days}} = \$851.47 \text{ per day}$$

The job duration is 1497 days, and the length of delay is 1497 − 521 = 976 days. Earlier negotiations agreed 74 of these days to be noncompensable. Thus Merit is due $851.47 × 902 days = $768,025.94 for extended field overhead.

This cost of daily operation compares with Merit's estimated program for the project. Merit had estimated $412,000 for 521 days, or $790.78. Hence, this is one check that the cost is consistent and proper. The increase reflects degree of difficulty and escalation.

Stretch-out costs of administrative office work.   Under the calculations at the beginning of this section, home office administrative overhead charged to this project was $621,701.69. On a daily basis, this is $415.30 per day. Merit is due 902 days × $415.30 for extended field overhead, or $374,600.60.

Escalation

Labor escalation. The contract called for performance in 521 days. Further, the contract called for completion of a number of items within 376 days. Restraints held certain start times on the job to specified dates.

Review indicates that there were three distinct categories of labor usage.

Category A   April 20 to mid-July 1972 included primarily demolition and preparatory work.

Category B   Mid-July 1972 to May 1973 included the 376-day list— demolition, excavation, concrete for the tunnel, office building, and closure wall, underground utilities, etc.

Category C   May 1973 to job end in November 1973 included the balance.

Merit's contract was such that 12.75 percent of the labor would be spent during category A, 58.44 percent during category B, and 28.81 percent in category C. The labor rates which prevailed were as follows:

Trade	5/1/72–4/30/73	5/73–4/74	5/74–4/75	5/75–4/76
Carpenters	$8.38	$8.83	$ 9.31	$10.61
Rodmen	8.58	9.08	10.08	10.98
Concrete Finishers	8.19	8.66	9.91	10.64
Operating Engineers	9.09	9.53	10.87	11.62
Labor	6.65	7.65	8.61	9.16
Average Wage	$8.18	$8.75	$9.76	$10.60

1. Rodmen contract actually changes on July 1. An April date was used for this contract to make calculations.
2. The 10 days remaining on the labor contracts ending April 1972 were ignored.

Merit spent the following labor hours on the project exclusive of the field overhead previously noted:

To April 30, 1973	42,425 man hours
Year to April 30, 1974	177,242
Year to April 30, 1975	187,198
Year Plus to End of Job	101,263
	508,128 man-hours

By category, these hours should have been spent as follows:

Category A	12.75% × 508,128 = 64,786.32 man hours
Category B	58.44% × 508,128 = 296,950.00
Category C	28.81% × 508,128 = 146,391.67
	508,127.99 man hours

Unescalated wage costs are as follows:

Category A
64,786.32 man-hours × 8.18   = $529,952.09

Category B

$$296,950 \text{ man-hours} \times \frac{(9.5 \times 8.18 + 1 \times 8.75)}{10.5} = \$2,445,171.10$$

Category C
146,391.67 man-hours × 8.75   = $1,280,927.10

Total Unescalated Wages = $4,256,050.20

Escalation = $4,974,662.71   Actual Wage Costs
$4,256,050.20
$718,612.50

$$\text{Adjustment for Change Order Labor} = \frac{\$286,274}{\$4,974,662.71} \quad \frac{\text{Change Order Labor}}{\text{Total Labor}} = 5.75\%$$

Adjusted Total of Escalation without Change Orders:

$$718,612.50 \times 94.25\% = \$677,292.28$$

Wage and Fringe Escalation Cost = $677,292.28	
FICA, UC, & AL @ 10.13%	$68,609.71
	$745,901.99

74 days were noncompensable delay. Therefore, $902/976 \times 745,901.99 =$ $689,347.94.

Material escalation: Concrete.   Merit's concrete purchase order agreement typically spelled out an escalating program of concrete prices. The job schedule was such that concrete purchases, like labor, were also set into the same time categories.

Calculation of the breakdown is as follows:

4/20/72–07/15/72	Category A =	8.10% of Concrete
7/15/72–06/1/73	Category B = 71.60% of Concrete	
6/1/73–11/25/73	Category C = 20.30% of Concrete	

The calculation is based on gross yardage of concrete purchased. Calculate concrete escalation as follows:

A. $8.10 \times 24,677 = $ 1,998.80 cubic yards
B. $71.60 \times 24,677 = $ 17,668.70 cubic yards
C. $20.30 \times 24,677 = $ 5,009.40 cubic yards

UNESCALATED COST

Category A   Price 4/20/72–1/1/73–$19.87/cubic yard
1998.8 cubic yards @$19.87/cubic yard = $39,716.15

Category B   Price 4/20/72–1/1/73–$19.87/cubic yard
Span 7/15/72–1/1/73–4.5 months
Price 1/1/73–3/13/73–$20.99/cubic yard
Span 1/1/73–3/13/73–2.43 months
Price 3/14/73–7/14/73–$21.50/cubic yard
Span 3/13/73–6/1/73–2.57 months

Category B	Cu. Yd./Mo.		Months		Price/Cu. Yd.	
	1859.86	X	4.5	X	$19.87	= $166,299.38
	1859.86	X	2.43	X	$20.99	= $94,863.46
	1859.86	X	2.57	X	$21.50	= $102,766.56
					Category B Total	$363,929.40

Category C   Price 3/14/73–7/14/73–$21.50/cubic yard
Span 6/1/73–7/14/73–1.47 months
Price 7/14/73–1/1/74–$22.15/cubic yard
Span 7/14/73–11/25/73–3.36 months

Category C	Cu. Yd./Mo.		Months		Price/Cu. Yd.		
	1037.14	X	1.47	X	$21.50	=	$32,778.81
	1037.14	X	3.36	X	$22.15	=	$77,188.11

Category C Total $109,966.92

Total Cost Concrete (No Escalation) $513,612.47

ACTUAL COST

We have examined Merit's records and find a total of $764,255.70 was actually paid for concrete. Thus, Merit is due

Actual Cost	$764,355.70
Unescalated Cost	513,612.47
	$250,643.23

Based on the 74-day reduction, this is reduced by 902/976. Thus, Merit is due an increase of $231,644.47 for concrete escalation.

Backfill.   The bulk of backfill was to be placed in the line area between station 61 + 50 and 68 + 00. Other backfill was needed in different small quantities throughout the job. As the work progressed, the backfill was delayed and obstructed.

Lack of planned access forced Merit to haul this backfill on roads congested by all the other contractors. Further, restricted access and working area required some of the material to be dumped and bucketed into place. In addition to the production decrease, the cost of backfill increased in that a source of backfill material at $17.00 per truckload was lost as delays and lack of storage prevented use. Merit was forced to pay $37.00 per truckload for the major portion of its backfill.

Merit lost efficient use of equipment since placing material on an intermittent basis in a congested area is very difficult to schedule and control. As an example, a machine with a capacity to handle 500 cubic yards of material per day was necessary to place only a small portion of that amount on a given day, but due to the nature of the work, the machine had to be available to perform the work when needed. The cost of this machine was the same regardless of the quantity of material handled; thus the cost per cubic yard increased sharply. Merit's added material cost on backfill is $146,774.

During the period January 26, 1975 to February 23, 1975, Merit used 1704 man-hours to place 4953 cubic yards of bulk backfill, 0.344 man-hour per cubic yard. Contrast this to 6330 man-hours expended from April 14 to July 21, 1974 to place 420 cubic yards, or 15.07 man-hours per cubic yard.

After reviewing the file our opinion is that the restrictions and interferences on the backfill increased Merit's cost by $\frac{1}{2}$ man-hour per yard. This reduction in production relates to added manual handling, tight quarters requiring equipment guides and compaction details.

Added Labor Cost 23,808 cubic yards @$\frac{1}{2}$ man-hour per cubic yard
$$= 11,904 \text{ man-hours}$$

The unescalated labor rate fully burdened and adjusted for crew mix is $9.45 per hour. The added labor cost is $112,492.80.

Added equipment costs. The backfill extended over eight months. Our judgment is that Merit lost five days per month in bare equipment standby. Thus, 40 equipment days with a loader at $24 per hour is $7,680 equipment adjustment due. Based on a five-man crew, the extended equipment time to compensate for the added labor is 60 days at $192, or $11,520. Added tamper cost is estimated at $1500.

Thus, backfill adjustment is as follows:

Added Material	$146,774.00
Added Labor	112,492.80
Added Equipment	20,700.00
	$279,966.80

Excavation increase. Excavation operations were grossly changed as follows:

1. Access from the north to the site was delayed.
2. Demolition of the existing tunnel was delayed.

Merit excavated a total of 11,157 cubic yards.

Excavation was in part clammed though handled with a front end loader where possible. It was stop-and-go work due to the required demolition. Costs increased due to equipment time. Review indicates that the problems decreased the production normal to this work by 146 cubic yards per day. Therefore, machine time increased by 146/376 or 39 percent, or an additional 11.57 (12) days. Further, machine downtime occurred for 13 days due to demolition. Excavation cost increased by 25 equipment days @ $375 per day = $9375.00.

The added time of 25 days requires an adjustment in truck costs as well. An additional 130 truck days were required @ 255/day = $33,150.00. A foreman in the cut and a four-man crew for 12 days add $3917.04 ($326.42 per day X 12 days).

Merit damage on excavation is $46,442.04.

Inefficiencies in concrete. Demands and delays by the owner affected Merit's concrete work adversely. We have made a detailed study of concrete production which is summarized in Exhibit 109.

Review of the exhibit shows in detail the effects on Merit's production. The inefficiency is readily seen in the walls: 22.50 man-hours was required to perform that which normally requires 8 to 9 man-hours. The effects of the out-of-sequence work are evident in the concrete costs during that period. Production was far below the normal standards.

Our opinion is that Merit was penalized an average of 3.0 man-hours per cubic yard of concrete by the conditions under which it had to perform. Based on 21,706.4 cubic yards of concrete at an unescalated rate of $8.75 per man-hour, we find that Merit is due $21,706.4 × 3 × 8.75 = $569,793.00 for concrete inefficiency.

Acceleration. Merit spent specific sums of money in response to acceleration directives. Merit's books and records show that it is due the following for acceleration:

Overtime Premium	$364,187.91
Inefficiency	$213,771.01
Increased Form Materials	$22,434.11
Rent Storage Material	$7,770.32
Rehandling Rebar	$41,463.87
Rehandling Rebar	$3,707.84
Rehandling Track Materials	$1,000.15
Overtime to Subcontractor	$6,214.07
Total Acceleration Cost =	$660,549.28

Summary of individual damage. We find Merit damaged in the following categories:

1. Direct Field Overhead	$768,025.94
2. Home Office Overhead	$374,600.60
3. Escalation	
Labor	$689,347.94
Concrete	$250,643.23
4. Backfill Increase	$279,996.80
5. Excavation Increase	$46,442.04
6. Concrete Cost Increase	$569,793.00
7. Acceleration Increase	$660,549.28
	$3,639,398.83
Profit @ 7.42% Over Cost of Sales	$270,043.39
	$3,909,442.22

*   *   *   *   *   *   *   *

The damage calculations do not include speculations on dollars that might have been earned if not for loss of investment or operating capital or loss of bonding capacity. Experience indicates that such speculative calculations on what might have been are not recognized by the courts and are a waste of time. However, I have recently seen an argument in a state court by a contractor to recover projects which were diminished by lost bonding capacity and lower working capital. The merits were argued on an earlier case wherein the owner of a maltreated race horse was awarded winnings which the horse could have won were it not for the bad veterinary treatment. At this writing, the court has yet to rule on the contractors argument.

9

SOME SAMPLE DISPUTES AND CLOSING THOUGHTS

This chapter describes a number of claims which are either historical or fictional. The principles in each situation have been preserved for the benefit of the reader but the description of the project or dispute has been modified to prevent identification.

1. A general contractor brought a $350,000 delay claim against a federal agency on a $1.9 million job. The government had retained a consultant to analyze the delay and, based on the consultant's report, had denied the claim in total. The government had shared the consultant's report with the contractor during the claim process. Study showed that the consultant had made a number of basic errors in its analysis. The dispute went before the agency hearing board and the contractor began its case by severe attack on the errors in the consultant's report. The errors were so obvious at the end of the first day that the government's counsel was too embarrassed to go further and had lost all confidence in the government's position. Hearings were adjourned and the case settled a week later for $190,000. It took two and one-half years to get to the hearing board.

Experience indicates that federal agencies tend to buy expert witness services with restrictive policies on procurement which often make the level of experts it retains less than the best.

2. A contractor sued for $6 million in delay damages and $2.8 million for interference on a $25 million monumental building. A number of subcontractors sued the general contractor separately for the same reasons. The general contractor was at fault for much of the subcontractor's delay complaints but the plans were also poorly done by the architect, which hurt the entire job. The contractor was not local, had no other work in the area, and paid little attention to the suit for several years because of bigger problems

in other places. The result was that its local lawyer did little to push the case and only what was necessary to defend or stall against the suit action. All the subcontractors had the same lawyer, who was on a contingency and not pushy. After three years of relative languish, the contractor underwent an ownership change, key people left, and new management gave the lawsuit to another local law firm to clean up. Over the next year or so, a settlement was worked out wherein the government agency put up $100,000, the architect's insurance company $300,000, and the contractor $200,000, with the eight subcontractors working out the split of the $600,000 after their attorney took $180,000.

The contractor had made $216,000 clear on the job before counsel fees and the $200,000 contribution. Had the contractor management known how to assess or present its claims, retained knowledgeable counsel, or perhaps only policed the activities of its first counsel, it had every chance and right to recover several million dollars from the owner. Poor counsel, poor management, and a clever government in-house attorney turned the legitimate claim into a loss.

3. A city brought suit against a contractor for $2 million for late completion of a revenue-producing parking facility. The contractor just could not seem to do part of the work; although it finally did finish the job, it was only after three years of legal maneuvering. The cause was simple: The contractor was over its head and could not solve its problems. The city held $600,000 of the $6 million contract and instituted suit when it could get no further contractor action on punch lists. The Surety settled the suit by leaving the $600,000.

4. A Midwest general contractor brought a delay claim of $500,000 on a $16 million federal office building for late release of site. The contract award was made in midsummer but no work could start until November because the government did not own the site. Government disallowed almost all subcontractor claims since all subcontracts were dated after the late notice to proceed, and thus in the government's view were with knowledge and implied acceptance of the late start. The excavation translation to winter conditions, which obviously was an increased cost was also turned down by the federal auditor, who found that the subcontractor who did the work did not price the work until late October, and then offered a lower price for the work than used by the general contractor in bidding the job.

After 20 months of discussion, the government settled the claim for $125,000, purely by contracting officer negotiation with an audit by his staff as his basis. Contractor and government both had counsel involved to a minor extent.

5. A separate prime general contractor brought a $160,000 delay claim against a state building agency for costs that arose out of the failure and

bankruptcy of a separate plumbing and HVAC prime contractor. The state agency took over five months to replace the defaulted prime contractor on a $3 million college building. Poor contract language for recovery on delay damage coupled with the state holding $225,000 of the general contractor's $1.8 million contract forced the general contractor into taking a $25,000 settlement plus payment of its $225,000 in 30 days. Although the contractor was unquestionably due $40,000 more, it would have taken two to three years to get to the review board and cost $25,000 to pursue. The problem of establishing that the state failed to replace the defaulted subcontractor properly was considered.

6. A small general contractor failed to progress a $375,000 separate foundation contract on time and did not use good workmanship. The result was that the structural steel and completion contract were held up and much foundation rework was required. The government agency had paid all but $100,000 to the foundation firm and refused to make further progress payments until all was resolved. Shortly thereafter, the foundation contractor went broke. The government owner kept the $100,000, and although the trustee in bankruptcy made some noise initially, eventually it gave up and everyone forgot the problems. After two or three years, the government agency reallocated the funds.

7. A bonded local southern subcontractor was terminated for cause, and quite rightly, for although 80 percent of its work was complete, it was far behind and woefully over its head. Right after it was terminated it sued the general contractor for $1.5 million. The general contractor, who was perhaps the largest in that area, to its complete amazement was to spend ultimately $900,000 to complete the subcontractor's work and to redo much work which was found defective. The general contractor had $400,000 of contract balance when it terminated the subcontractor, and felt at first it would finish for less than the balance. When the job was finally completed, the subcontractor indicated it would settle for $500,000, which offended the general contractor. The subcontractor brought its case before the jury and told them it was broke. Its two owners, both of whom were of retirement age, had lost their houses and were having a difficult time. All reason was lost with the jury who simply saw the little guy versus the big guy. The subcontractor was awarded $1.5 million plus interest.

The subcontractor's lawyer had a contingent fee with certain costs borne by the subcontractor's surety. The case was appealed since the loss the subcontractor suffered was only $850,000. When all was said and done, both of the defunct firm's owners got $200,000 in cash and had their bank loan paid. The general contractor spent over $175,000 in legal and expert fees. This was a seemingly unfair fiasco to the general contractor, except that it knew the deal it signed with the subcontractor was $800,000 lower than any other solution available when the job started. Low prices, even if bonded, do not guarantee that windfall buys or unfair knowledge of economic duress in the other party will ultimately prevail.

8. A local general contractor who had a reputation for good work filed a claim against a state agency for $450,000 on a $2.6 million contract for state-caused interference on a new building. The state stopped progress on a separately contracted major underground steam utility line around the entire perimeter of the new building for a year which created a complete moat that the state refused to backfill because of cost. It was so insistent in its attitude that it refused to let the general contractor backfill the moat free, and finally the state refused an offer from the general contractor to backfill and then reexcavate key parts of the moat free of charge.

There was no question about the moat or the interference. When the claim got to someone in the legal department at the capitol, the reaction was one of total aghast. The state settled for $150,000 without any questions. Had the state bothered to inquire, it could have learned that the contractor, although annoyed by the moat, had subcontracted the entire job to the subcontractors with whom it had bid the work. It never had more than a superintendent and two or three laborers on the site and had not given the subcontractors any extra money because of the moat. The job was completed as bid and the $150,000 went right into the contractor's profit side of the ledger.

States almost never audit or do discovery on contractor's claims. Since most claims are tried before review boards, no discovery is done by the plaintiff and state legal staffs typically do not pursue the real accounting records. This claim under discussion was handled without a lawyer being involved by the contractor.

9. A general contractor filed a $900,000 delay claim against a state building agency on a $5 million contract. Separate prime contracts existed on HVAC, electrical and plumbing. A 100-day total suspension occurred in excavation when a tenant refused to vacate an existing building on the site. The suspension started in August and ended in December.

The excavation and foundations went through a difficult winter and a major strike the next spring held back the steel frame erection for three months. Had the delay not occurred at the job start, the strike would have occurred after the steel was erected. The 100 days turned into a nine-month delay. The state agency, as in the preceding case, settled by reacting to the contractor's claim statement and cut the money demanded to $300,000. The general had subcontracted the entire job, made $270,000 and had paid off and secured releases from the subcontractors who shared the delay claims before the claim was formally presented. The $300,000 settlement, after counsel and other costs, was added to the $270,000 fee already earned. Again, one sees that slow claim submittal and the state processing the claim by reaction instead of investigation led to an excellent settlement.

The question of undeserved dollars arises. Did this contractor pursue or get dollars to which it was not entitled? That is a question with more than one answer and up to the individuals involved. The basic approach to the free enterprise system is that a contractor can exercise its skills to earn as much as it can. There are skills in placing concrete and skills at the conference table. Only the federal government once in a great while exercises any authority to look at profits as excess. Good fishermen get more fish with the same equipment than bad fishermen. The same applies to contractors, evangelists, car salesmen, and ballroom dancers.

Claims are not an opportunity for a bonanza of undeserved recovery. While the skill of those presenting one side of the dispute can win more than someone else might feel is really due, the idea that some contractors are good at inventing claims and pursuing them is not really true.

Most contractors do not want to bother with claims. It diverts the attention and time of their people from the ordinary profitable lines of the business. A claim almost always moves too slowly to contractors to bank on. It cannot be spent or utilized in a surety or credit line, though experience is that at times many dollars are lent by banks on claims. Therefore, the pursuit of claims gets put on the back burner. The type of man needed to pursue the claim is of sufficient caliber to be needed more pressingly elsewhere, and since claims are slow and dull, not much reason is needed to direct the manager's attention to more dynamic situations.

Pursuit of undeserved dollars is unusual, and most will give up the dispute of such along the way. Most contractors do not win enough deserved dollars to chase those that are undeserved.

Do owners get undeserved dollars? Indeed they do, and probably more so than the contractors. Owners constantly get undeserved dollars, for work items and other things not necessarily due from contractors who simply do not have the time, cash, knowhow, or desire to fight. The process of dispute resolution is so difficult in the typical contract it is just not worth the effort. The designer almost always sides against the contractor in any dispute so the contractor must make its argument against more than one party.

10. A small general contractor brought claim for $170,000 on a $535,000 school job against a city. The city had unquestioned liability for $75,000 and a good argument existed on the whole amount. The city, on the advice of its private construction manager who had caused much of the loss, forced the contractor into a long series of administrative remedies which ultimately led to court. The contractor hired an indifferent lawyer who saw the process involved and the lack of contractor money available for its bills, and did little to pursue the issue. The contractor is back to curbs and sidewalks.

11. A miscellaneous iron subcontractor sued the owner through the general contractor for $60,000 for an installation error by the architect who corrected its mistake by marking up shop drawing submittals and forcing a

higher cost of performance method. The architect was the only witness the owner offered in court. The owner was ordered to pay the $60,000 extra— the architect went free. Indeed, no one in the owner camp really knew the architect was the responsible party.

12. A state agency sued a boring contractor for $1.4 million for errors it claimed were made in a $4000 boring contract the results of which were used to design the foundations for a $6 million hospital. The architect and its structural consultant were actually at fault for they had not checked the boring logs, locations, and elevations. The architect was not a party to the suit and could not be gotten into the lawsuit by the driller. The changed conditions which resulted from erroneous design, and not bad borings, led to the state settling with its building contractor for $600,000 more than the change was worth. The state simply was outplayed by the general contractor— that lack of discovery again. The driller could not open up this error either. The state's claim went before a jury, but the boring contractor's insurance company settled for $175,000 shortly after the judge charged the jury.

13. A large widely experienced general contractor sued a local govern- ment in county court for $2 million on an $8 million job, claiming a cardinal change had caused delay and costs. The local government hired a local law firm which filed an answer and impleaded a separately contracted out-of- state west coast equipment supplier and the local designer as third-party defendants.

Examination of records by the third-party equipment supplier defendant counsel revealed that the general contractor had suffered losses of $400,000 due to the default of several marginal subcontractors who could not perform, but no $2 million loss could be found. The outside government counsel made no investigation, did no discovery, slept during key depositions in Los Angeles and was totally unprepared for the trial.

On the third day of the trial, the government offered $350,000 to settle, the designer's insurance company put up $100,000, and the equipment sup- plier put up $50,000. Government counsel took credit for the $150,000 contribution from the third parties and acted like he had won the case.*

14. An experienced general contractor brought a claim against the govern- ment for $2 million for changed site conditions claiming they were the cause of a $1.3 million loss suffered on concrete work. The government acknow- ledged the loss by audit, but personalities involved resulted in total refusal to consider the contractor's claim. The contractor went to the board and won a significant partial award on entitlement but could not negotiate any relief. Finally after five years of struggle and pursuit government by chance changed contracting officers and the claim was settled for $650,000.

---

*The general contractor took the $500,000 after complaining bitterly.

15. A subcontractor from the United States entered into contract to do civil work on a refinery located in South America. The cost plus a fixed fee general contractor was from Europe and the contract called for payment to the U.S. subcontractor on a unit price and line item basis.

The American subcontractor found that the line item descriptions were ignored in its billing and the cheapest unit price for that category was applied. Some $640,000 was ultimately contested. The government owner sided with the general contractor's interpretation. The subcontractor demanded arbitration under the Paris rules of arbitration to recover its $640,000 plus costs and interest. Half-way through the arbitration, it took another similar subcontract from the same contractor for more plant expansion at the same site at reduced unit prices.

The arbitrator named spoke French. However, everyone else spoke English, Italian, or Spanish, so the arbitrator required an interpreter. It was immediately obvious that the arbitrator had a very high standard of ethics, enjoyed his role, and was a man of great skill. After three hearings, which were like attending an affair of state, the contractor received its $640,000, its costs, interest and an apology from the arbitrator.

16. A local government sued an architect who did its own structural design for $1.2 million to correct design defects and for delays on a recreation center which was in danger of collapse in the long-span area. The architect's insurance company settled for $400,000 on the third day of jury trial. The government finally spent $250,000 for repairs and had to pay a $300,000 negotiated settlement with the general contractor. The contractor had arbitration in its contract, the architect did not. The insurance company attorneys argued successfully against the government motion to have one proceeding to hear the facts.

17. A major private owner of a new $15 million 20-story downtown office building offered its construction manager, who obviously had encountered many problems, $100,000 over the balance due to close out the CM contract. The CM refused and sued the owner for $750,000 for delay damages. Four subcontractors to the CM on the project heard of the suit and each separately sued the CM. They all later advised that they would not have pursued this action but since the CM sued the owner they decided to get in.

The owner retained qualified experts and eminent counsel who reviewed the entire program. The result was a consolidated lawsuit with much legal expense and discovery. When the facts were set forth, the CM's deficiencies were obvious and a large settlement was made. The four subcontractors received a total of $650,000 from the CM, and the owner received $350,000 from the CM's insurer. The owner's fact-finding essentially won the money for the four subcontractors who did little themselves in the lawsuit. The CM who had turned down a gift of $100,000 is out of business today.

Very typically large corporate groups are not at all interested in asserting claims against poorly performing contractors. The preference is to work out the problems when it comes to award the next project. However, the large corporation when challenged becomes very difficult to defeat in a legal battle since its resources and motivation are different than those of the contracting world. Large corporations are usually fair and generous almost to a fault. When challenged they are formidable opponents who can and do spend large sums of money to preserve and establish images which are important.

Subcontractors are the poorest participants in construction disputes. Subcontractors almost never pursue delay claims or changed conditions claims. Most do not have the knowhow, money, or time to pursue the claims. Instead, subcontractors all learn to write the same kind of unacceptable letter describing losses which cannot be proved. General contractors will often bargain claims against owners down to palatable levels by giving away the subcontractors' claims.

Overton Currie, a noted contractor lawyer, often states that 39 percent of all contractors stay in business for only two and a half years, and 63 percent fail after five and a half years. Seven years is the average business life of a contractor. Due legal process takes too long for the typical contractor. Its life is too short for adjudication.

18. A general contractor who was far over its head and really fouled up on the project brought suit against a state agency for $885,000 for design error and delay on a $5 million hospital job. The building agency was offended and its attitude was so persuasive the state legal department concluded the contractor was not due an adjustment of substance. Limited funds meant the lawyers at hand could not hire any experts or outside help. The state did not do any discovery, and in the end relied on several low level state inspectors for its side of the case. The inspectors were not qualified to be witnesses and were afraid of the whole proceeding and intimidated by the contractor's lawyer.*

The state lost the case but appealed the decision through every available process. Finally, the state lost in the Supreme Court and paid $537,000 to the contractor. This is an instance of superior skills in the private sector prevailing.

19. A general contractor defaulted on a number of jobs it had in progress. Retrospective analysis showed it had in one month taken some $30 million in work on six condominium projects all of which was priced about $3 million too low. At the time of default, its surety retained a well-known construction lawyer who supervised completion of the work by continuing

*One inspector fell asleep and became enraged when his snooze was put in the record by the state lawyer.

the original contractor on some jobs and by bringing in another major contractor.

On one job in default the general contractor was working for a well-known developer, and since the work was nearly done and completion cost far less than the $600,000 contract balance, the surety attorney chose to finish the work with some minor surety financing and the existing staff. Then counsel somewhat haughtily filed and pursued a delay claim demanding arbitration to recover $2 million for alleged owner delay during construction. This demand preceded by a year or so a lawsuit by the condominium owners association and a lawsuit for breach by the developer. After much courtroom skirmishing the dispute between developer and contractor was arbitrated and the condominium owners association dispute was litigated with the contractor and developer as codefendants.

During the period it took for litigation to start, the developer voluntarily, but carefully, performed repairs and corrected all the owner association complaints except the parking lot; an essentially uncontested judgment in favor of the owners was entered for $65,000 against both codefendants for the parking lot. The developer paid the $65,000 in cash to the owners association. The developer counterclaimed some $1.5 million in delay costs and workmanship defects against the surety for the defaulted contractor. After eight days in front of the arbitration panel the case settled with the surety releasing all rights to the $600,000 balance on the contract and, further, it repaid the $65,000 for the parking lot repairs.

The developer had politely and skillfully documented its position throughout construction. The record showed it had time after time spent considerable sums of money to correct defects and satisfy the owners. It had advised and offered help during construction to the contractor on a weekly basis in a manner that could not be denied. The surety attorney, in spite of his opinion of himself, met much resistance in the local courts. The developer's local counsel could not be intimidated. The eight days in arbitration showed that the surety and its attorney were really not well prepared. It had relied on a lightweight expert as well. Sureties often rely heavily on counsel but do not police them to any degree. In this case the surety lawyer seemingly misled its client into pursuit of a delay claim that really had no foundation.

20. A major general contractor demanded arbitration of an $800,000 delay claim on a $6.5 million county hospital project on the West coast. It hired a national name construction lawyer. The hospital retained its usual local attorney who retained an expert consultant to analyze the claim. At completion of the expert analysis, the hospital offered $110,000 to settle the claim.

The general contractor's lawyer advised against settlement and 52 sessions before an arbitration panel followed over the next 16 months. The arbitration panel awarded the contractor $117,000. To some extent both parties lost: The hospital spent $85,000 in the process; the contractor spent $160,000.

Capable local counsel who is willing to listen and learn can frequently defeat the construction lawyers who have big names and reputations.

21. A sewer contractor demanded arbitration of a $2 million claim against a municipality for claims of changed conditions in that water-bearing soil was found as opposed to the findings of the borings, which indicated little or no water would be experienced.

The consulting engineer who had written the specification and contracted for the borings was the main witness for the municipality. During the hearing the engineer said it had no knowledge of a number of records in city hall describing wet underground conditions in other sewer projects in adjacent locations. The contractor produced a retired city inspector who said he had discussed and forwarded the documents to the engineer eight or nine years before. Within a month the municipality settled the claim for $700,000. The municipality then sued the engineer for the $700,000 but lost when the judge ruled that the municipality did not establish the engineer had been negligent.

Had the municipality been able to get the engineer into the arbitration as a third party it very possibly could have perfected its claim against the engineer.

22. A sewer contractor sued a design engineer for failure to perform the duties required by the engineer's contract with the county. The engineer had failed to answer correspondence, issue minutes of meetings, or process requisitions and change orders on a timely basis. The claim of $500,000 was settled for $75,000 by the designer's insurer. The engineer's project resident had been diverted in performing his duties and had fallen in love with a very attractive female who during discovery was found to be in the employ of the sewer contractor on a somewhat boldly stated unit price basis. There was much speculation about what her deposition might be like, but the case was settled rather abruptly when the situation was discovered.

These examples establish several points for all disputes: First, there is the probability of settlement after the fact-finding is done; few cases went all the way to an adjudicated decision, and those that did often produced results that were a surprise to both parties. The examples indicate that certain levels of government are not strong adversaries in a dispute. They show that the skill of counsel is extremely important and the time for solution is long and often exhausting.

Further, the examples show little pattern for settlement. Experience indicates that claims settled for 30 cents on the dollar are considered good settlements by government rule of thumb before litigation starts. Hard-fought litigation with clear entitlement developing during fact-finding leads to settlements of 60 cents on the dollar. Another seeming way of life in disputes is that insurers and sureties will use every legal defense to which they are entitled but invariably will work out a settlement if their assured is at fault.

Claim size has little to do with settlement. Some small claims go the full route, and some $5 million claims which should be resisted with all the fury man can muster get settled with little effort.

One seemingly sure point in construction disputes is that an argument, if presented honestly and properly, with enthusiasm and follow-through, will always get the attention of the other side.

Arguments not advanced cannot be won. Reasonable overstatement is expected in the dispute and can be to one's advantage. The ultimate result in a dispute will depend on the truth involved, on the desire to prevail in the parties in the argument, and on the persuasive ability of those who present the argument.

Experience indicates that systems available to resolve disputes usually work. Most of several hundreds of disputes on which the author has worked have been resolved in a manner which brought acceptable justice to those involved. The process is sometimes slow and expensive but it works if someone cares; besides, there is no other way to settle the dispute.

APPENDIX A

A.I.A. CONTRACT FORMS

THE AMERICAN INSTITUTE OF ARCHITECTS

AIA Document A101

Standard Form of Agreement Between Owner and Contractor

where the basis of payment is a

STIPULATED SUM

AGREEMENT

made this   day of   in the year of Nineteen
Hundred and

BETWEEN the Owner:

and the Contractor:

the Project:

the Architect:

The Owner and the Contractor agree as set forth below.

AIA DOCUMENT A101 • OWNER-CONTRACTOR AGREEMENT • JANUARY 1974 EDITION • AIA® • ©1974
THE AMERICAN INSTITUTE OF ARCHITECTS, 1735 NEW YORK AVE., N.W., WASHINGTON, D. C. 20006

1

ARTICLE 1

THE CONTRACT DOCUMENTS

The Contract Documents consist of this Agreement, Conditions of the Contract (General, Supplementary and other Conditions), Drawings, Specifications, all Addenda issued prior to execution of this Agreement and all Modifications issued subsequent thereto. These form the Contract, and all are as fully a part of the Contract as if attached to this Agreement or repeated herein. An enumeration of the Contract Documents appears in Article 7.

ARTICLE 2

THE WORK

The Contractor shall perform all the Work required by the Contract Documents for

(Here insert the caption descriptive of the Work as used on other Contract Documents.)

ARTICLE 3

TIME OF COMMENCEMENT AND COMPLETION

The Work to be performed under this Contract shall be commenced

and completed

(Here insert any special provisions for liquidated damages relating to failure to complete on time.)

ARTICLE 4

CONTRACT SUM

The Owner shall pay the Contractor for the performance of the Work, subject to additions and deductions by Change Order as provided in the Conditions of the Contract, in current funds, the Contract Sum of

(State here the lump sum amount, unit prices, or both, as desired.)

ARTICLE 5

PROGRESS PAYMENTS

Based upon Applications for Payment submitted to the Architect by the Contractor and Certificates for Payment issued by the Architect, the Owner shall make progress payments on account of the Contract Sum to the Contractor as provided in the Conditions of the Contract as follows:

On or about the   day of each month   per cent of the proportion of the Contract Sum properly allocable to labor, materials and equipment incorporated in the Work and   per cent of the portion of the Contract Sum properly allocable to materials and equipment suitably stored at the site or at some other location agreed upon in writing by the parties, up to   days prior to the date on which the Application for Payment is submitted, less the aggregate of previous payments in each case; and upon Substantial Completion of the entire Work, a sum sufficient to increase the total payments to   per cent of the Contract Sum, less such retainages as the Architect shall determine for all incomplete Work and unsettled claims.

(If not covered elsewhere in the Contract Documents, here insert any provision for limiting or reducing the amount retained after the Work reaches a certain stage of completion.)

Any moneys not paid when due to either party under this Contract shall bear interest at the legal rate in force at the place of the Project.

ARTICLE 6

FINAL PAYMENT

Final payment, constituting the entire unpaid balance of the Contract Sum, shall be paid by the Owner to the Contractor   days after Substantial Completion of the Work unless otherwise stipulated in the Certificate of Substantial Completion, provided the Work has then been completed, the Contract fully performed, and a final Certificate for Payment has been issued by the Architect.

ARTICLE 7

MISCELLANEOUS PROVISIONS

7.1   Terms used in this Agreement which are defined in the Conditions of the Contract shall have the meanings designated in those Conditions.

7.2   The Contract Documents, which constitute the entire agreement between the Owner and the Contractor, are listed in Article 1 and, except for Modifications issued after execution of this Agreement, are enumerated as follows:

(List below the Agreement, Conditions of the Contract (General, Supplementary, and other Conditions), Drawings, Specifications, Addenda and accepted Alternates, showing page or sheet numbers in all cases and dates where applicable.)

This Agreement executed the day and year first written above.

OWNER   CONTRACTOR

_____   _____

_____   _____

THE AMERICAN INSTITUTE OF ARCHITECTS

AIA DOCUMENT A101/CM

CONSTRUCTION MANAGEMENT EDITION

Standard Form of Agreement Between Owner and Contractor

where the basis of payment is a

STIPULATED SUM

*THIS DOCUMENT HAS IMPORTANT LEGAL CONSEQUENCES; CONSULTATION WITH
AN ATTORNEY IS ENCOURAGED WITH RESPECT TO ITS COMPLETION OR MODIFICATION*

Use only with the latest edition of AIA Document A201/CM, General Conditions of the Contract
for Construction, Construction Management Edition.

AGREEMENT

made this   day of   in the year of Nineteen
Hundred and

BETWEEN the Owner:

and the Contractor:

For the following Project:
(Include detailed description of Project location and scope)

the Construction Manager:

the Architect:

The Owner and the Contractor agree as set forth below.

AIA DOCUMENT A101/CM • OWNER-CONTRACTOR AGREEMENT • CONSTRUCTION MANAGEMENT EDITION • MAY 1975
EDITION • AIA® • ©1975 • THE AMERICAN INSTITUTE OF ARCHITECTS, 1735 NEW YORK AVE., N.W., WASHINGTON, D.C. 20006   **1**

ARTICLE 1

THE CONTRACT DOCUMENTS

The Contract Documents consist of this Agreement and any Exhibits attached hereto, Conditions of the Contract (General, Supplementary and other Conditions), Drawings, Specifications, all Addenda issued prior to execution of this Agreement and all Modifications issued subsequent thereto. These form the Contract, and all are as fully a part of the Contract as if attached to this Agreement or repeated herein. An enumeration of the Contract Documents appears in Article 8.

ARTICLE 2

THE WORK

The Contractor shall perform all the Work required by the Contract Documents for

(Here insert a precise description of the Work covered by this Contract and refer to numbers of Drawings and pages of Specifications including Addenda and accepted Alternates.)

ARTICLE 3

TIME OF COMMENCEMENT AND COMPLETION

(Here insert the specific provisions that are applicable to this Contract including any information pertaining to notice to proceed or other method of notification for commencement of Work starting and completion dates or duration and any provisions for liquidated damages relating to failure to complete on time.)

Time is of the essence of this Contract.

ARTICLE 4

CONTRACT SUM

The Owner shall pay the Contractor for the performance of the Work, subject to additions and deductions by Change Order as provided in the Conditions of the Contract, in current funds, the Contract Sum of

(State here the lump sum amount, unit prices, or both, as desired.)

ARTICLE 5

PROGRESS PAYMENTS

Based upon Applications for Payment submitted to the Construction Manager by the Contractor and Certificates for Payment issued by the Architect, the Owner shall make progress payments on account of the Contract Sum to the Contractor as provided in the Conditions of the Contract as follows:

On or about the   day of each month   per cent of the proportion of the Contract Sum properly allocable to labor, materials and equipment incorporated in the Work and   per cent of the portion of the Contract Sum properly allocable to materials and equipment suitably stored at the site or at some other location agreed upon in writing by the parties up to   days prior to the date on which the Application for Payment is submitted less the aggregate of previous payments in each case; and upon Substantial Completion of the entire Work, a sum sufficient to increase the total payments to   per cent of the Contract Sum, less such retainages as the Architect shall determine for all incomplete Work and unsettled claims.

(If not covered elsewhere in the Contract Documents, here insert any provision for limiting or reducing the amount retained after the Work reaches a certain stage of completion.)

Any moneys not paid when due to either party under this Contract shall bear interest at the legal rate in force at the place of the Project.

AIA DOCUMENT A101/CM • OWNER-CONTRACTOR AGREEMENT • CONSTRUCTION MANAGEMENT EDITION • MAY 1975 EDITION • AIA® • ©1975 • THE AMERICAN INSTITUTE OF ARCHITECTS, 1735 NEW YORK AVE., N.W., WASHINGTON, D.C. 20006   **3**

ARTICLE 6

FINAL PAYMENT

Final payment, constituting the entire unpaid balance of the Contract Sum, shall be paid by the Owner to the Contractor   days after Substantial Completion of the Work unless otherwise stipulated in the Certificate of Substantial Completion, provided the Work has then been completed, the Contract fully performed, and a final Certificate for Payment has been issued by the Architect.

ARTICLE 7

TEMPORARY FACILITIES AND SERVICES

(Here insert temporary facilities and services which are different from or in addition to those included elsewhere in the Contract Documents.)

ARTICLE 8

MISCELLANEOUS PROVISIONS

8.1 Terms used in this Agreement which are defined in the Conditions of the Contract shall have the meanings designated in those Conditions.

8.2 Working conditions:

(List below any special conditions affecting the Contract.)

AIA DOCUMENT A101/CM • OWNER-CONTRACTOR AGREEMENT • CONSTRUCTION MANAGEMENT EDITION • MAY 1975
EDITION • AIA® • ©1975 • THE AMERICAN INSTITUTE OF ARCHITECTS, 1735 NEW YORK AVE., N.W., WASHINGTON, D.C. 20006   **5**

8.3 The Contract Documents, which constitute the entire agreement between the Owner and the Contractor, are listed in Article 1 and, except for Modifications issued after execution of this Agreement, are enumerated as follows:

(List below the Agreement and any exhibits thereto, Conditions of the Contract (General, Supplementary, and other Conditions), Drawings, Specifications, Addenda and accepted Alternates, showing page or sheet numbers in all cases and dates where applicable.)

This Agreement executed the day and year first written above.

OWNER _____   CONTRACTOR _____

_____   _____

AIA DOCUMENT A101/CM • OWNER-CONTRACTOR AGREEMENT • CONSTRUCTION MANAGEMENT EDITION • MAY 1975
EDITION • AIA® • ©1975 • THE AMERICAN INSTITUTE OF ARCHITECTS, 1735 NEW YORK AVE., N.W., WASHINGTON, D.C. 20006   **6**

THE AMERICAN INSTITUTE OF ARCHITECTS

AIA Document A107

Standard Form of Agreement Between Owner and Contractor

Short Form Agreement for **Small Construction Contracts**

Where the Basis of Payment is a

STIPULATED SUM

THIS DOCUMENT HAS IMPORTANT LEGAL CONSEQUENCES; CONSULTATION WITH
AN ATTORNEY IS ENCOURAGED WITH RESPECT TO ITS COMPLETION OR MODIFICATION

For other contracts the AIA issues Standard Forms of Owner-Contractor Agreements and Standard General Conditions
of the Contract for Construction for use in connection therewith.

This document has been approved and endorsed by The Associated General Contractors of America.

AGREEMENT

made this   day of   in the year Nineteen
Hundred and

BETWEEN the Owner:

and the Contractor:

the Project:

the Architect:

The Owner and Contractor agree as set forth below.

AIA DOCUMENT A107 • SMALL CONSTRUCTION CONTRACT • JANUARY 1974 EDITION • AIA® • ©1974
THE AMERICAN INSTITUTE OF ARCHITECTS, 1735 NEW YORK AVE., N.W., WASHINGTON, D. C. 20006

1

ARTICLE 1
THE WORK

The Contractor shall perform all the Work required by the Contract Documents for
(Here insert the caption descriptive of the Work as used on other Contract Documents.)

ARTICLE 2
TIME OF COMMENCEMENT AND COMPLETION

The Work to be performed under this Contract shall be commenced

and completed

ARTICLE 3
CONTRACT SUM

The Owner shall pay the Contractor for the performance of the Work, subject to additions and deductions by Change Order as provided in the General Conditions, in current funds, the Contract Sum of
(State here the lump sum amount, unit prices, or both, as desired.)

ARTICLE 4
PROGRESS PAYMENTS

Based upon Applications for Payment submitted to the Architect by the Contractor and Certificates for Payment issued by the Architect, the Owner shall make progress payments on account of the Contract Sum to the Contractor as follows:

ARTICLE 5
FINAL PAYMENT

The Owner shall make final payment   days after completion of the Work, provided the Contract be then fully performed, subject to the provisions of Article 16 of the General Conditions.

ARTICLE 6
ENUMERATION OF CONTRACT DOCUMENTS

The Contract Documents are as noted in Paragraph 7.1 of the General Conditions and are enumerated as follows:
(List below the Agreement, Conditions of the Contract (General, Supplementary, and other Conditions), Drawings, Specifications, Addenda and accepted Alternates, showing page or sheet numbers in all cases and dates where applicable.)

ARTICLE 7
CONTRACT DOCUMENTS

7.1 The Contract Documents consist of this Agreement (which includes the General Conditions), Supplementary and other Conditions, the Drawings, the Specifications, all Addenda issued prior to the execution of this Agreement, all modifications, Change Orders, and written interpretations of the Contract Documents issued by the Architect. These form the Contract and what is required by any one shall be as binding as if required by all. The intention of the Contract Documents is to include all labor, materials, equipment and other items as provided in Paragraph 10.2 necessary for the proper execution and completion of the Work and the terms and conditions of payment therefor, and also to include all Work which may be reasonably inferable from the Contract Documents as being necessary to produce the intended results.

7.2 The Contract Documents shall be signed in not less than triplicate by the Owner and the Contractor. If either the Owner or the Contractor do not sign the Drawings, Specifications, or any of the other Contract Documents, the Architect shall identify them. By executing the Contract, the Contractor represents that he has visited the site and familiarized himself with the local conditions under which the Work is to be performed.

7.3 The term Work as used in the Contract Documents includes all labor necessary to produce the construction required by the Contract Documents, and all materials and equipment incorporated or to be incorporated in such construction.

ARTICLE 8
ARCHITECT

8.1 The Architect will provide general administration of the Contract and will be the Owner's representative during construction and until issuance of the final Certificate for Payment.

8.2 The Architect shall at all times have access to the Work wherever it is in preparation and progress.

8.3 The Architect will make periodic visits to the site to familiarize himself generally with the progress and quality of the Work and to determine in general if the Work is proceeding in accordance with the Contract Documents. On the basis of his on-site observations as an architect, he will keep the Owner informed of the progress of the Work, and will endeavor to guard the Owner against defects and deficiencies in the Work of the Contractor. The Architect will not be required to make exhaustive or continuous on-site inspections to check the quality or quantity of the Work. The Architect will not be responsible for construction means, methods, techniques, sequences or procedures, or for safety precautions and programs in connection with the Work, and he will not be responsible for the Contractor's

failure to carry out the Work in accordance with the Contract Documents.

8.4 Based on such observations and the Contractor's Applications for Payment, the Architect will determine the amounts owing to the Contractor and will issue Certificates for Payment in accordance with Article 16.

8.5 The Architect will be, in the first instance, the interpreter of the requirements of the Contract Documents. He will make decisions on all claims and disputes between the Owner and the Contractor. All his decisions are subject to arbitration.

8.6 The Architect will have authority to reject Work which does not conform to the Contract Documents.

ARTICLE 9
OWNER

9.1 The Owner shall furnish all surveys.

9.2 The Owner shall secure and pay for easements for permanent structures or permanent changes in existing facilities.

9.3 The Owner shall issue all instructions to the Contractor through the Architect.

ARTICLE 10
CONTRACTOR

10.1 The Contractor shall supervise and direct the work, using his best skill and attention. The Contractor shall be solely responsible for all construction means, methods, techniques, sequences and procedures and for coordinating all portions of the Work under the Contract.

10.2 Unless otherwise specifically noted, the Contractor shall provide and pay for all labor, materials, equipment, tools, construction equipment and machinery, water, heat, utilities, transportation, and other facilities and services necessary for the proper execution and completion of the Work.

10.3 The Contractor shall at all times enforce strict discipline and good order among his employees, and shall not employ on the Work any unfit person or anyone not skilled in the task assigned to him.

10.4 The Contractor warrants to the Owner and the Architect that all materials and equipment incorporated in the Work will be new unless otherwise specified, and that all Work will be of good quality, free from faults and defects and in conformance with the Contract Documents. All Work not so conforming to these standards may be considered defective.

10.5 The Contractor shall pay all sales, consumer, use and other similar taxes required by law and shall secure all permits, fees and licenses necessary for the execution of the Work.

10.6 The Contractor shall give all notices and comply with all laws, ordinances, rules, regulations, and orders of any public authority bearing on the performance of the Work, and shall notify the Architect if the Drawings and Specifications are at variance therewith.

10.7 The Contractor shall be responsible for the acts and omissions of all his employees and all Subcontractors, their agents and employees and all other persons performing any of the Work under a contract with the Contractor.

10.8 The Contractor shall review, stamp with his approval and submit all samples and shop drawings as directed for approval of the Architect for conformance with the design concept and with the information given in the Contract Documents. The Work shall be in accordance with approved samples and shop drawings.

10.9 The Contractor at all times shall keep the premises free from accumulation of waste materials or rubbish caused by his operations. At the completion of the Work he shall remove all his waste materials and rubbish from and about the Project as well as his tools, construction equipment, machinery and surplus materials, and shall clean all glass surfaces and shall leave the Work "broom clean" or its equivalent, except as otherwise specified.

10.10 The Contractor shall indemnify and hold harmless the Owner and the Architect and their agents and employees from and against all claims, damages, losses and expenses including attorneys' fees arising out of or resulting from the performance of the Work, provided that any such claim, damage, loss or expense (1) is attributable to bodily injury, sickness, disease or death, or to injury to or destruction of tangible property (other than the Work itself) including the loss of use resulting therefrom, and (2) is caused in whole or in part by any negligent act or omission of the Contractor, any Subcontractor, anyone directly or indirectly employed by any of them or anyone for whose acts any of them may be liable, regardless of whether or not it is caused in part by a party indemnified hereunder. In any and all claims against the Owner or the Architect or any of their agents or employees by any employee of the Contractor, any Subcontractor, anyone directly or indirectly employed by any of them or anyone for whose acts any of them may be liable, the indemnification obligation under this Paragraph 10.10 shall not be limited in any way by any limitation on the amount or type of damages, compensation or benefits payable by or for the Contractor or any Subcontractor under workmen's compensation acts, disability benefit acts or other employee benefit acts. The obligations of the Contractor under this Paragraph 10.10 shall not extend to the liability of the Architect, his agents or employees arising out of (1) the preparation or approval of maps, drawings, opinions, reports, surveys, Change Orders, designs or specifications, or (2) the giving of or the failure to give directions or instructions by the Architect, his agents or employees provided such giving or failure to give is the primary cause of the injury or damage.

ARTICLE 11
SUBCONTRACTS

11.1 A Subcontractor is a person who has a direct contract with the Contractor to perform any of the Work at the site.

11.2 Unless otherwise specified in the Contract Documents or in the Instructions to Bidders, the Contractor, as soon as practicable after the award of the Contract, shall furnish to the Architect in writing a list of the names of Subcontractors proposed for the principal portions of the Work. The Contractor shall not employ any Subcontractor to whom the Architect or the Owner may have a reasonable objection. The Contractor shall not be required to employ any Subcontractor to whom he has a reasonable objection. Contracts between the Contractor and the Subcontractor shall be in accordance with the terms of this Agreement and shall include the General Conditions of this Agreement insofar as applicable.

ARTICLE 12
SEPARATE CONTRACTS

12.1 The Owner reserves the right to award other contracts in connection with other portions of the Project or other work on the site under these or similar Conditions of the Contract.

12.2 The Contractor shall afford other contractors reasonable opportunity for the introduction and storage of their materials and equipment and the execution of their work, and shall properly connect and coordinate his Work with theirs.

12.3 Any costs caused by defective or ill-timed work shall be borne by the party responsible therefor.

ARTICLE 13
ROYALTIES AND PATENTS

The Contractor shall pay all royalties and license fees. The Contractor shall defend all suits or claims for infringement of any patent rights and shall save the Owner harmless from loss on account thereof.

ARTICLE 14
ARBITRATION

All claims or disputes arising out of this Contract or the breach thereof shall be decided by arbitration in accordance with the Construction Industry Arbitration Rules of the American Arbitration Association then obtaining unless the parties mutually agree otherwise. Notice of the demand for arbitration shall be filed in writing with the other party to the Contract and with the American Arbitration Association and shall be made within a reasonable time after the dispute has arisen.

ARTICLE 15
TIME

15.1 All time limits stated in the Contract Documents are of the essence of the Contract.

15.2 If the Contractor is delayed at any time in the progress of the Work by changes ordered in the Work, by labor disputes, fire, unusual delay in transportation, unavoidable casualties, causes beyond the Contractor's control, or by any cause which the Architect may determine justifies the delay, then the Contract Time shall be extended by Change Order for such reasonable time as the Architect may determine.

AIA DOCUMENT A107 • SMALL CONSTRUCTION CONTRACT • JANUARY 1974 EDITION • AIA® • ©1974
THE AMERICAN INSTITUTE OF ARCHITECTS, 1735 NEW YORK AVE., N.W., WASHINGTON, D. C. 20006

5

ARTICLE 16
PAYMENTS

16.1 Payments shall be made as provided in Article 4 of this Agreement.

16.2 Payments may be withheld on account of (1) defective Work not remedied, (2) claims filed, (3) failure of the Contractor to make payments properly to Subcontractors or for labor, materials, or equipment, (4) damage to another contractor, or (5) unsatisfactory prosecution of the Work by the Contractor.

16.3 Final payment shall not be due until the Contractor has delivered to the Owner a complete release of all liens arising out of this Contract or receipts in full covering all labor, materials and equipment for which a lien could be filed, or a bond satisfactory to the Owner indemnifying him against any lien.

16.4 The making of final payment shall constitute a waiver of all claims by the Owner except those arising from (1) unsettled liens, (2) faulty or defective Work appearing after Substantial Completion, (3) failure of the Work to comply with the requirements of the Contract Documents, or (4) terms of any special guarantees required by the Contract Documents. The acceptance of final payment shall constitute a waiver of all claims by the Contractor except those previously made in writing and still unsettled.

ARTICLE 17
PROTECTION OF PERSONS AND PROPERTY

The Contractor shall be responsible for initiating, maintaining, and supervising all safety precautions and programs in connection with the Work. He shall take all reasonable precautions for the safety of, and shall provide all reasonable protection to prevent damage, injury or loss to (1) all employees on the Work and other persons who may be affected thereby, (2) all the Work and all materials and equipment to be incorporated therein, and (3) other property at the site or adjacent thereto. He shall comply with all applicable laws, ordinances, rules, regulations and orders of any public authority having jurisdiction for the safety of persons or property or to protect them from damage, injury or loss. All damage or loss to any property caused in whole or in part by the Contractor, any Subcontractor, any Sub-subcontractor or anyone directly or indirectly employed by any of them, or by anyone for whose acts any of them may be liable, shall be remedied by the Contractor, except damage or loss attributable to faulty Drawings or Specifications or to the acts or omissions of the Owner or Architect or anyone employed by either of them or for whose acts either of them may be liable but which are not attributable to the fault or negligence of the Contractor.

ARTICLE 18
CONTRACTOR'S LIABILITY INSURANCE

The Contractor and each separate Contractor shall purchase and maintain such insurance as will protect him from claims under workmen's compensation acts and other employee benefit acts, from claims for damages because of bodily injury, including death, and from claims for damages to property which may arise out of or result from the Contractor's operations under this Contract, whether such operations be by himself or by any Subcontractor or anyone directly or indirectly employed by any of them. This insurance shall be written for not less than any limits of liability specified as part of this Contract, or required by law, whichever is the greater, and shall include contractual liability insurance as applicable to the Contractor's obligations under Paragraph 10.10. Certificates of such insurance shall be filed with the Owner and each separate Contractor.

ARTICLE 19
OWNER'S LIABILITY INSURANCE

The Owner shall be responsible for purchasing and maintaining his own liability insurance and, at his option, may maintain such insurance as will protect him against claims which may arise from operations under the Contract.

ARTICLE 20
PROPERTY INSURANCE

20.1 Unless otherwise provided, the Owner shall purchase and maintain property insurance upon the entire Work at the site to the full insurable value thereof. This insurance shall include the interests of the Owner, the Contractor, Subcontractors and Sub-subcontractors in the Work and shall insure against the perils of Fire, Extended Coverage, Vandalism and Malicious Mischief.

20.2 Any insured loss is to be adjusted with the Owner and made payable to the Owner as trustee for the insureds, as their interests may appear, subject to the requirements of any mortgagee clause.

20.3 The Owner shall file a copy of all policies with the Contractor prior to the commencement of the Work.

20.4 The Owner and Contractor waive all rights against each other for damages caused by fire or other perils to the extent covered by insurance provided under this paragraph. The Contractor shall require similar waivers by Subcontractors and Sub-subcontractors.

ARTICLE 21
CHANGES IN THE WORK

21.1 The Owner without invalidating the Contract may order Changes in the Work consisting of additions, deletions, or modifications, the Contract Sum and the Contract Time being adjusted accordingly. All such Changes in the Work shall be authorized by written Change Order signed by the Owner or the Architect as his duly authorized agent.

21.2 The Contract Sum and the Contract Time may be changed only by Change Order.

21.3 The cost or credit to the Owner from a Change in the Work shall be determined by mutual agreement.

ARTICLE 22
CORRECTION OF WORK

The Contractor shall correct any Work that fails to conform to the requirements of the Contract Documents where such failure to conform appears during the progress of the Work, and shall remedy any defects due to faulty materials, equipment or workmanship which appear within a period of one year from the Date of Substantial Completion of the Contract or within such longer period of time as may be prescribed by law or by the terms of any applicable special guarantee required by the Contract Documents. The provisions of this Article 22 apply to Work done by Subcontractors as well as to Work done by direct employees of the Contractor.

ARTICLE 23
TERMINATION BY THE CONTRACTOR

If the Architect fails to issue a Certificate of Payment for a period of thirty days through no fault of the Contractor, or if the Owner fails to make payment thereon for a period of thirty days, the Contractor may, upon seven days' written notice to the Owner and the Architect, terminate the Contract and recover from the Owner payment for all Work executed and for any proven loss sustained upon any materials, equipment, tools, and construction equipment and machinery, including reasonable profit and damages.

ARTICLE 24
TERMINATION BY THE OWNER

If the Contractor defaults or neglects to carry out the Work in accordance with the Contract Documents or fails to perform any provision of the Contract, the Owner may, after seven days' written notice to the Contractor and without prejudice to any other remedy he may have, make good such deficiencies and may deduct the cost thereof from the payment then or thereafter due the Contractor or, at his option, may terminate the Contract and take possession of the site and of all materials, equipment, tools, and construction equipment and machinery thereon owned by the Contractor and may finish the Work by whatever method he may deem expedient, and if the unpaid balance of the Contract Sum exceeds the expense of finishing the Work, such excess shall be paid to the Contractor, but if such expense exceeds such unpaid balance, the Contractor shall pay the difference to the Owner.

ARTICLE 25
MISCELLANEOUS PROVISIONS

AIA DOCUMENT A107 • SMALL CONSTRUCTION CONTRACT • JANUARY 1974 EDITION • AIA® • ©1974
THE AMERICAN INSTITUTE OF ARCHITECTS, 1735 NEW YORK AVE., N.W., WASHINGTON, D. C. 20006

This Agreement executed the day and year first written above.

OWNER _____   CONTRACTOR _____

_____   _____

THE AMERICAN INSTITUTE OF ARCHITECTS

AIA Document A111

Standard Form of Agreement Between Owner and Contractor

where the basis of payment is the

COST OF THE WORK PLUS A FEE

*THIS DOCUMENT HAS IMPORTANT LEGAL CONSEQUENCES; CONSULTATION WITH
AN ATTORNEY IS ENCOURAGED WITH RESPECT TO ITS COMPLETION OR MODIFICATION*

Use only with the latest Edition of AIA Document A201, General Conditions of the Contract for Construction.

This document has been approved and endorsed by The Associated General Contractors of America.

AGREEMENT

made this   day of   in the year of Nineteen
Hundred and

BETWEEN the Owner:

and the Contractor:

the Project:

the Architect:

The Owner and the Contractor agrees as set forth below.

ARTICLE 1

THE CONTRACT DOCUMENTS

The Contract Documents consist of this Agreement, Conditions of the Contract (General, Supplementary and other Conditions), Drawings, Specifications, all Addenda issued prior to execution of this Agreement and all Modifications issued subsequent thereto. These form the Contract, and all are as fully a part of the Contract as if attached to this Agreement or repeated herein. An enumeration of the Contract Documents appears in Article 16. If anything in the General Conditions is inconsistent with this Agreement, the Agreement shall govern.

ARTICLE 2

THE WORK

The Contractor shall perform all the Work required by the Contract Documents for

(Here insert the caption descriptive of the Work as used on other Contract Documents.)

ARTICLE 3

THE CONTRACTOR'S DUTIES AND STATUS

The Contractor accepts the relationship of trust and confidence established between him and the Owner by this Agreement. He covenants with the Owner to furnish his best skill and judgment and to cooperate with the Architect in furthering the interests of the Owner. He agrees to furnish efficient business administration and superintendence and to use his best efforts to furnish at all times an adequate supply of workmen and materials, and to perform the Work in the best way and in the most expeditious and economical manner consistent with the interests of the Owner.

ARTICLE 4

TIME OF COMMENCEMENT AND COMPLETION

The Work to be performed under this Contract shall be commenced

and completed

(Here insert any special provisions for liquidated damages relating to failure to complete on time.)

ARTICLE 5

COST OF THE WORK AND GUARANTEED MAXIMUM COST

5.1   The Owner agrees to reimburse the Contractor for the Cost of the Work as defined in Article 8. Such reimbursement shall be in addition to the Contractor's Fee stipulated in Article 6.

5.2   The maximum cost to the Owner, including the Cost of the Work and the Contractor's Fee, is guaranteed not to exceed the sum of
dollars ($); such Guaranteed Maximum Cost shall be increased or decreased for Changes in the Work as provided in Article 7.

(Here insert any provision for distribution of any savings. Delete Paragraph 5.2 if there is no Guaranteed Maximum Cost.)

ARTICLE 6

CONTRACTOR'S FEE

6.1 In consideration of the performance of the Contract, the Owner agrees to pay the Contractor in current funds as compensation for his services a Contractor's Fee as follows:

6.2 For Changes in the Work, the Contractor's Fee shall be adjusted as follows:

6.3 The Contractor shall be paid   per cent (%) of the proportionate amount of his Fee with each progress payment, and the balance of his Fee shall be paid at the time of final payment.

ARTICLE 7

CHANGES IN THE WORK

7.1 The Owner may make Changes in the Work in accordance with Article 12 of the General Conditions insofar as such Article is consistent with this Agreement. The Contractor shall be reimbursed for Changes in the Work on the basis of Cost of the Work as defined in Article 8.

7.2 The Contractor's Fee for Changes in the Work shall be as set forth in Paragraph 6.2, or in the absence of specific provisions therein, shall be adjusted by negotiation on the basis of the Fee established for the original Work.

ARTICLE 8

COSTS TO BE REIMBURSED

8.1 The term Cost of the Work shall mean costs necessarily incurred in the proper performance of the Work and paid by the Contractor. Such costs shall be at rates not higher than the standard paid in the locality of the Work except with prior consent of the Owner, and shall include the items set forth below in this Article 8.

8.1.1 Wages paid for labor in the direct employ of the Contractor in the performance of the Work under applicable collective bargaining agreements, or under a salary or wage schedule agreed upon by the Owner and Contractor, and including such welfare or other benefits, if any, as may be payable with respect thereto.

8.1.2 Salaries of Contractor's Personnel when stationed at the field office, in whatever capacity employed. Personnel engaged, at shops or on the road, in expediting the production or transportation of materials or equipment, shall be considered as stationed at the field office and their salaries paid for that portion of their time spent on this Work.

8.1.3 Cost of contributions, assessments or taxes for such items as unemployment compensation and social security, insofar as such cost is based on wages, salaries, or other remuneration paid to employees of the Contractor and included in the Cost of the Work under Subparagraphs 8.1.1 and 8.1.2.

8.1.4 The proportion of reasonable transportation, traveling and hotel expenses of the Contractor or of his officers or employees incurred in discharge of duties connected with the Work.

8.1.5 Cost of all materials, supplies and equipment incorporated in the Work, including costs of transportation thereof.

8.1.6 Payments made by the Contractor to Subcontractors for Work performed pursuant to subcontracts under this Agreement.

8.1.7 Cost, including transportation and maintenance, of all materials, supplies, equipment, temporary facilities and hand tools not owned by the workmen, which are consumed in the performance of the Work, and cost less salvage value on such items used but not consumed which remain the property of the Contractor.

8.1.8 Rental charges of all necessary machinery and equipment, exclusive of hand tools, used at the site of the Work, whether rented from the Contractor or others, including installation, minor repairs and replacements, dismantling, removal, transportation and delivery costs thereof, at rental charges consistent with those prevailing in the area.

8.1.9 Cost of premiums for all bonds and insurance which the Contractor is required by the Contract Documents to purchase and maintain.

8.1.10 Sales, use or similar taxes related to the Work and for which the Contractor is liable imposed by any governmental authority.

8.1.11 Permit fees, royalties, damages for infringement of patents and costs of defending suits therefor, and deposits lost for causes other than the Contractor's negligence.

8.1.12 Losses and expenses, not compensated by insurance or otherwise, sustained by the Contractor in connection with the Work, provided they have resulted from causes other than the fault or neglect of the Contractor. Such losses shall include settlements made with the written consent and approval of the Owner. No such losses and expenses shall be included in the Cost of the Work for the purpose of determining the Contractor's Fee. If, however, such loss requires reconstruction and the Contractor is placed in charge thereof, he shall be paid for his services a Fee proportionate to that stated in Paragraph 6.1.

8.1.13 Minor expenses such as telegrams, long distance telephone calls, telephone service at the site, expressage, and similar petty cash items in connection with the Work.

8.1.14 Cost of removal of all debris.

8.1.15 Costs incurred due to an emergency affecting the safety of persons and property.

8.1.16 Other costs incurred in the performance of the Work if and to the extent approved in advance in writing by the Owner.

ARTICLE 9

COSTS NOT TO BE REIMBURSED

9.1   The term Cost of the Work shall not include any of the items set forth below in this Article 9.

9.1.1   Salaries or other compensation of the Contractor's Personnel at the Contractor's principal office and branch offices.

9.1.2   Expenses of the Contractor's Principal and Branch Offices other than the Field Office.

9.1.3   Any part of the Contractor's capital expenses, including interest on the Contractor's capital employed for the Work.

9.1.4   Overhead or general expenses of any kind, except as may be expressly included in Article 8.

9.1.5   Costs due to the negligence of the Contractor, any Subcontractor, anyone directly or indirectly employed by any of them, or for whose acts any of them may be liable, including but not limited to the correction of defective or nonconforming Work, disposal of materials and equipment wrongly supplied, or making good any damage to property.

9.1.6   The cost of any item not specifically and expressly included in the items described in Article 8.

9.1.7   Costs in excess of the Guaranteed Maximum Cost, if any, as set forth in Article 5 and adjusted pursuant to Article 7.

ARTICLE 10

DISCOUNTS, REBATES AND REFUNDS

All cash discounts shall accrue to the Contractor unless the Owner deposits funds with the Contractor with which to make payments, in which case the cash discounts shall accrue to the Owner. All trade discounts, rebates and refunds, and all returns from sale of surplus materials and equipment shall accrue to the Owner, and the Contractor shall make provisions so that they can be secured.

(Here insert any provisions relating to deposits by the Owner to permit the Contractor to obtain cash discounts.)

ARTICLE 11

SUBCONTRACTS

11.1   All portions of the Work that the Contractor's organization has not been accustomed to perform shall be performed under subcontracts. The Contractor shall request bids from subcontractors and shall deliver such bids to the Architect. The Architect will then determine, with the advice of the Contractor and subject to the approval of the Owner, which bids will be accepted.

11.2   All Subcontracts shall conform to the requirements of Paragraph 5.3 of the General Conditions. Subcontracts awarded on the basis of the cost of such work plus a fee shall also be subject to the provisions of this Agreement insofar as applicable.

ARTICLE 12

ACCOUNTING RECORDS

The Contractor shall check all materials, equipment and labor entering into the Work and shall keep such full and detailed accounts as may be necessary for proper financial management under this Agreement, and the system shall be satisfactory to the Owner. The Owner shall be afforded access to all the Contractor's records, books, correspondence, instructions, drawings, receipts, vouchers, memoranda and similar data relating to this Contract, and the Contractor shall preserve all such records for a period of three years after the final payment.

ARTICLE 13

APPLICATIONS FOR PAYMENT

The Contractor shall, at least ten days before each progress payment falls due, deliver to the Architect a statement, sworn to if required, showing in complete detail all moneys paid out or costs incurred by him on account of the Cost of the Work during the previous month for which he is to be reimbursed under Article 5 and the amount of the Contractor's Fee due as provided in Article 6, together with payrolls for all labor and all receipted bills for which payment has been received.

ARTICLE 14

PAYMENTS TO THE CONTRACTOR

14.1 The Architect will review the Contractor's statement of moneys due as provided in Article 13 and will promptly issue a Certificate for Payment to the Owner for such amount as he approves, which Certificate shall be payable on or about the   day of the month.

14.2 Final payment, constituting the unpaid balance of the Cost of the Work and of the Contractor's Fee, shall be paid by the Owner to the Contractor when the Work has been completed, the Contract fully performed and a final Certificate for Payment has been issued by the Architect. Final payment shall be due
days after the date of issuance of the final Certificate for Payment.

ARTICLE 15

TERMINATION OF CONTRACT

15.1 The Contract may be terminated by the Contractor as provided in Article 14 of the General Conditions.

15.2 If the Owner terminates the Contract as provided in Article 14 of the General Conditions, he shall reimburse the Contractor for any unpaid Cost of the Work due him under Article 5, plus (1) the unpaid balance of the Fee computed upon the Cost of the Work to the date of termination at the rate of the percentage named in Article 6, or (2) if the Contractor's Fee be stated as a fixed sum, such an amount as will increase the payments on account of his Fee to a sum which bears the same ratio to the said fixed sum as the Cost of the Work at the time of termination bears to the adjusted Guaranteed Maximum Cost, if any, otherwise to a reasonable estimated Cost of the Work when completed. The Owner shall also pay to the Contractor fair compensation, either by purchase or rental at the election of the Owner, for any equipment retained. In case of such termination of the Contract the Owner shall further assume and become liable for obligations, commitments and unsettled claims that the Contractor has previously undertaken or incurred in good faith in connection with said Work. The Contractor shall, as a condition of receiving the payments referred to in this Article 15, execute and deliver all such papers and take all such steps, including the legal assignment of his contractual rights, as the Owner may require for the purpose of fully vesting in him the rights and benefits of the Contractor under such obligations or commitments.

ARTICLE 16

MISCELLANEOUS PROVISIONS

16.1 Terms used in this Agreement which are defined in the Conditions of the Contract shall have the meanings designated in those Conditions.

16.2 The Contract Documents, which constitute the entire agreement between the Owner and the Contractor, are listed in Article 1 and, except for Modifications issued after execution of this Agreement, are enumerated as follows:

(List below the Agreement, Conditions of the Contract, (General, Supplementary, other Conditions), Drawings, Specifications, Addenda and accepted Alternates, showing page or sheet numbers in all cases and dates where applicable.)

This Agreement executed the day and year first written above.

OWNER

_____

_____

CONTRACTOR

_____

_____

THE AMERICAN INSTITUTE OF ARCHITECTS

AIA Document A201

General Conditions of the Contract for Construction

THIS DOCUMENT HAS IMPORTANT LEGAL CONSEQUENCES; CONSULTATION WITH AN ATTORNEY IS ENCOURAGED WITH RESPECT TO ITS MODIFICATION

1976 EDITION
TABLE OF ARTICLES

This document has been approved and endorsed by The Associated General Contractors of America.

INDEX

AIA DOCUMENT A201 • GENERAL CONDITIONS OF THE CONTRACT FOR CONSTRUCTION • THIRTEENTH EDITION • AUGUST 1976
AIA® • © 1976 • THE AMERICAN INSTITUTE OF ARCHITECTS, 1735 NEW YORK AVENUE, N.W., WASHINGTON, D.C. 20006

GENERAL CONDITIONS OF THE CONTRACT FOR CONSTRUCTION

ARTICLE 1

CONTRACT DOCUMENTS

1.1 DEFINITIONS

1.1.1 THE CONTRACT DOCUMENTS

The Contract Documents consist of the Owner-Contractor Agreement, the Conditions of the Contract (General, Supplementary and other Conditions), the Drawings, the Specifications, and all Addenda issued prior to and all Modifications issued after execution of the Contract. A Modification is (1) a written amendment to the Contract signed by both parties, (2) a Change Order, (3) a written interpretation issued by the Architect pursuant to Subparagraph 2.2.8, or (4) a written order for a minor change in the Work issued by the Architect pursuant to Paragraph 12.3. The Contract Documents do not include Bidding Documents such as the Advertisement or Invitation to Bid, the Instructions to Bidders, sample forms, the Contractor's Bid or portions of Addenda relating to any of these, or any other documents, unless specifically enumerated in the Owner-Contractor Agreement.

1.1.2 THE CONTRACT

The Contract Documents form the Contract for Construction. This Contract represents the entire and integrated agreement between the parties hereto and supersedes all prior negotiations, representations, or agreements, either written or oral. The Contract may be amended or modified only by a Modification as defined in Subparagraph 1.1.1. The Contract Documents shall not be construed to create any contractual relationship of any kind between the Architect and the Contractor, but the Architect shall be entitled to performance of obligations intended for his benefit, and to enforcement thereof. Nothing contained in the Contract Documents shall create any contractual relationship between the Owner or the Architect and any Subcontractor or Sub-subcontractor.

1.1.3 THE WORK

The Work comprises the completed construction required by the Contract Documents and includes all labor necessary to produce such construction, and all materials and equipment incorporated or to be incorporated in such construction.

1.1.4 THE PROJECT

The Project is the total construction of which the Work performed under the Contract Documents may be the whole or a part.

1.2 EXECUTION, CORRELATION AND INTENT

1.2.1 The Contract Documents shall be signed in not less than triplicate by the Owner and Contractor. If either the Owner or the Contractor or both do not sign the Conditions of the Contract, Drawings, Specifications, or any of the other Contract Documents, the Architect shall identify such Documents.

1.2.2 By executing the Contract, the Contractor represents that he has visited the site, familiarized himself with the local conditions under which the Work is to be performed, and correlated his observations with the requirements of the Contract Documents.

1.2.3 The intent of the Contract Documents is to include all items necessary for the proper execution and completion of the Work. The Contract Documents are complementary, and what is required by any one shall be as binding as if required by all. Work not covered in the Contract Documents will not be required unless it is consistent therewith and is reasonably inferable therefrom as being necessary to produce the intended results. Words and abbreviations which have well-known technical or trade meanings are used in the Contract Documents in accordance with such recognized meanings.

1.2.4 The organization of the Specifications into divisions, sections and articles, and the arrangement of Drawings shall not control the Contractor in dividing the Work among Subcontractors or in establishing the extent of Work to be performed by any trade.

1.3 OWNERSHIP AND USE OF DOCUMENTS

1.3.1 All Drawings, Specifications and copies thereof furnished by the Architect are and shall remain his property. They are to be used only with respect to this Project and are not to be used on any other project. With the exception of one contract set for each party to the Contract, such documents are to be returned or suitably accounted for to the Architect on request at the completion of the Work. Submission or distribution to meet official regulatory requirements or for other purposes in connection with the Project is not to be construed as publication in derogation of the Architect's common law copyright or other reserved rights.

ARTICLE 2

ARCHITECT

2.1 DEFINITION

2.1.1 The Architect is the person lawfully licensed to practice architecture, or an entity lawfully practicing architecture identified as such in the Owner-Contractor Agreement, and is referred to throughout the Contract Documents as if singular in number and masculine in gender. The term Architect means the Architect or his authorized representative.

2.2 ADMINISTRATION OF THE CONTRACT

2.2.1 The Architect will provide administration of the Contract as hereinafter described.

2.2.2 The Architect will be the Owner's representative during construction and until final payment is due. The Architect will advise and consult with the Owner. The Owner's instructions to the Contractor shall be forwarded through the Architect. The Architect will have authority to act on behalf of the Owner only to the extent provided in the Contract Documents, unless otherwise modified by written instrument in accordance with Subparagraph 2.2.18.

2.2.3 The Architect will visit the site at intervals appropriate to the stage of construction to familiarize himself generally with the progress and quality of the Work and to determine in general if the Work is proceeding in accordance with the Contract Documents. However, the Architect will not be required to make exhaustive or continuous on-site inspections to check the quality or quantity of the Work. On the basis of his on-site observations as an architect, he will keep the Owner informed of the progress of the Work, and will endeavor to guard the Owner against defects and deficiencies in the Work of the Contractor.

2.2.4 The Architect will not be responsible for and will not have control or charge of construction means, methods, techniques, sequences or procedures, or for safety precautions and programs in connection with the Work, and he will not be responsible for the Contractor's failure to carry out the Work in accordance with the Contract Documents. The Architect will not be responsible for or have control or charge over the acts or omissions of the Contractor, Subcontractors, or any of their agents or employees, or any other persons performing any of the Work.

2.2.5 The Architect shall at all times have access to the Work wherever it is in preparation and progress. The Contractor shall provide facilities for such access so the Architect may perform his functions under the Contract Documents.

2.2.6 Based on the Architect's observations and an evaluation of the Contractor's Applications for Payment, the Architect will determine the amounts owing to the Contractor and will issue Certificates for Payment in such amounts, as provided in Paragraph 9.4.

2.2.7 The Architect will be the interpreter of the requirements of the Contract Documents and the judge of the performance thereunder by both the Owner and Contractor.

2.2.8 The Architect will render interpretations necessary for the proper execution or progress of the Work, with reasonable promptness and in accordance with any time limit agreed upon. Either party to the Contract may make written request to the Architect for such interpretations.

2.2.9 Claims, disputes and other matters in question between the Contractor and the Owner relating to the execution or progress of the Work or the interpretation of the Contract Documents shall be referred initially to the Architect for decision which he will render in writing within a reasonable time.

2.2.10 All interpretations and decisions of the Architect shall be consistent with the intent of and reasonably inferable from the Contract Documents and will be in writing or in the form of drawings. In his capacity as interpreter and judge, he will endeavor to secure faithful performance by both the Owner and the Contractor, will not

show partiality to either, and will not be liable for the result of any interpretation or decision rendered in good faith in such capacity.

2.2.11 The Architect's decisions in matters relating to artistic effect will be final if consistent with the intent of the Contract Documents.

2.2.12 Any claim, dispute or other matter in question between the Contractor and the Owner referred to the Architect, except those relating to artistic effect as provided in Subparagraph 2.2.11 and except those which have been waived by the making or acceptance of final payment as provided in Subparagraphs 9.9.4 and 9.9.5, shall be subject to arbitration upon the written demand of either party. However, no demand for arbitration of any such claim, dispute or other matter may be made until the earlier of (1) the date on which the Architect has rendered a written decision, or (2) the tenth day after the parties have presented their evidence to the Architect or have been given a reasonable opportunity to do so, if the Architect has not rendered his written decision by that date. When such a written decision of the Architect states (1) that the decision is final but subject to appeal, and (2) that any demand for arbitration of a claim, dispute or other matter covered by such decision must be made within thirty days after the date on which the party making the demand receives the written decision, failure to demand arbitration within said thirty days' period will result in the Architect's decision becoming final and binding upon the Owner and the Contractor. If the Architect renders a decision after arbitration proceedings have been initiated, such decision may be entered as evidence but will not supersede any arbitration proceedings unless the decision is acceptable to all parties concerned.

2.2.13 The Architect will have authority to reject Work which does not conform to the Contract Documents. Whenever, in his opinion, he considers it necessary or advisable for the implementation of the intent of the Contract Documents, he will have authority to require special inspection or testing of the Work in accordance with Subparagraph 7.7.2 whether or not such Work be then fabricated, installed or completed. However, neither the Architect's authority to act under this Subparagraph 2.2.13, nor any decision made by him in good faith either to exercise or not to exercise such authority, shall give rise to any duty or responsibility of the Architect to the Contractor, any Subcontractor, any of their agents or employees, or any other person performing any of the Work.

2.2.14 The Architect will review and approve or take other appropriate action upon Contractor's submittals such as Shop Drawings, Product Data and Samples, but only for conformance with the design concept of the Work and with the information given in the Contract Documents. Such action shall be taken with reasonable promptness so as to cause no delay. The Architect's approval of a specific item shall not indicate approval of an assembly of which the item is a component.

2.2.15 The Architect will prepare Change Orders in accordance with Article 12, and will have authority to order minor changes in the Work as provided in Subparagraph 12.4.1.

AIA DOCUMENT A201 • GENERAL CONDITIONS OF THE CONTRACT FOR CONSTRUCTION • THIRTEENTH EDITION • AUGUST 1976
AIA® • © 1976 • THE AMERICAN INSTITUTE OF ARCHITECTS, 1735 NEW YORK AVENUE, N.W., WASHINGTON, D.C. 20006

2.2.16 The Architect will conduct inspections to determine the dates of Substantial Completion and final completion, will receive and forward to the Owner for the Owner's review written warranties and related documents required by the Contract and assembled by the Contractor, and will issue a final Certificate for Payment upon compliance with the requirements of Paragraph 9.9.

2.2.17 If the Owner and Architect agree, the Architect will provide one or more Project Representatives to assist the Architect in carrying out his responsibilities at the site. The duties, responsibilities and limitations of authority of any such Project Representative shall be as set forth in an exhibit to be incorporated in the Contract Documents.

2.2.18 The duties, responsibilities and limitations of authority of the Architect as the Owner's representative during construction as set forth in the Contract Documents will not be modified or extended without written consent of the Owner, the Contractor and the Architect.

2.2.19 In case of the termination of the employment of the Architect, the Owner shall appoint an architect against whom the Contractor makes no reasonable objection whose status under the Contract Documents shall be that of the former architect. Any dispute in connection with such appointment shall be subject to arbitration.

ARTICLE 3

OWNER

3.1 DEFINITION

3.1.1 The Owner is the person or entity identified as such in the Owner-Contractor Agreement and is referred to throughout the Contract Documents as if singular in number and masculine in gender. The term Owner means the Owner or his authorized representative.

3.2 INFORMATION AND SERVICES REQUIRED OF THE OWNER

3.2.1 The Owner shall, at the request of the Contractor, at the time of execution of the Owner-Contractor Agreement, furnish to the Contractor reasonable evidence that he has made financial arrangements to fulfill his obligations under the Contract. Unless such reasonable evidence is furnished, the Contractor is not required to execute the Owner-Contractor Agreement or to commence the Work.

3.2.2 The Owner shall furnish all surveys describing the physical characteristics, legal limitations and utility locations for the site of the Project, and a legal description of the site.

3.2.3 Except as provided in Subparagraph 4.7.1, the Owner shall secure and pay for necessary approvals, easements, assessments and charges required for the construction, use or occupancy of permanent structures or for permanent changes in existing facilities.

3.2.4 Information or services under the Owner's control shall be furnished by the Owner with reasonable promptness to avoid delay in the orderly progress of the Work.

3.2.5 Unless otherwise provided in the Contract Documents, the Contractor will be furnished, free of charge, all copies of Drawings and Specifications reasonably necessary for the execution of the Work.

3.2.6 The Owner shall forward all instructions to the Contractor through the Architect.

3.2.7 The foregoing are in addition to other duties and responsibilities of the Owner enumerated herein and especially those in respect to Work by Owner or by Separate Contractors, Payments and Completion, and Insurance in Articles 6, 9 and 11 respectively.

3.3 OWNER'S RIGHT TO STOP THE WORK

3.3.1 If the Contractor fails to correct defective Work as required by Paragraph 13.2 or persistently fails to carry out the Work in accordance with the Contract Documents, the Owner, by a written order signed personally or by an agent specifically so empowered by the Owner in writing, may order the Contractor to stop the Work, or any portion thereof, until the cause for such order has been eliminated; however, this right of the Owner to stop the Work shall not give rise to any duty on the part of the Owner to exercise this right for the benefit of the Contractor or any other person or entity, except to the extent required by Subparagraph 6.1.3.

3.4 OWNER'S RIGHT TO CARRY OUT THE WORK

3.4.1 If the Contractor defaults or neglects to carry out the Work in accordance with the Contract Documents and fails within seven days after receipt of written notice from the Owner to commence and continue correction of such default or neglect with diligence and promptness, the Owner may, after seven days following receipt by the Contractor of an additional written notice and without prejudice to any other remedy he may have, make good such deficiencies. In such case an appropriate Change Order shall be issued deducting from the payments then or thereafter due the Contractor the cost of correcting such deficiencies, including compensation for the Architect's additional services made necessary by such default, neglect or failure. Such action by the Owner and the amount charged to the Contractor are both subject to the prior approval of the Architect. If the payments then or thereafter due the Contractor are not sufficient to cover such amount, the Contractor shall pay the difference to the Owner.

ARTICLE 4

CONTRACTOR

4.1 DEFINITION

4.1.1 The Contractor is the person or entity identified as such in the Owner-Contractor Agreement and is referred to throughout the Contract Documents as if singular in number and masculine in gender. The term Contractor means the Contractor or his authorized representative.

4.2 REVIEW OF CONTRACT DOCUMENTS

4.2.1 The Contractor shall carefully study and compare the Contract Documents and shall at once report to the Architect any error, inconsistency or omission he may discover. The Contractor shall not be liable to the Owner or the Architect for any damage resulting from any such errors, inconsistencies or omissions in the Contract Documents. The Contractor shall perform no portion of the Work at any time without Contract Documents or, where required, approved Shop Drawings, Product Data or Samples for such portion of the Work.

4.3 SUPERVISION AND CONSTRUCTION PROCEDURES

4.3.1 The Contractor shall supervise and direct the Work, using his best skill and attention. He shall be solely responsible for all construction means, methods, techniques, sequences and procedures and for coordinating all portions of the Work under the Contract.

4.3.2 The Contractor shall be responsible to the Owner for the acts and omissions of his employees, Subcontractors and their agents and employees, and other persons performing any of the Work under a contract with the Contractor.

4.3.3 The Contractor shall not be relieved from his obligations to perform the Work in accordance with the Contract Documents either by the activities or duties of the Architect in his administration of the Contract, or by inspections, tests or approvals required or performed under Paragraph 7.7 by persons other than the Contractor.

4.4 LABOR AND MATERIALS

4.4.1 Unless otherwise provided in the Contract Documents, the Contractor shall provide and pay for all labor, materials, equipment, tools, construction equipment and machinery, water, heat, utilities, transportation, and other facilities and services necessary for the proper execution and completion of the Work, whether temporary or permanent and whether or not incorporated or to be incorporated in the Work.

4.4.2 The Contractor shall at all times enforce strict discipline and good order among his employees and shall not employ on the Work any unfit person or anyone not skilled in the task assigned to him.

4.5 WARRANTY

4.5.1 The Contractor warrants to the Owner and the Architect that all materials and equipment furnished under this Contract will be new unless otherwise specified, and that all Work will be of good quality, free from faults and defects and in conformance with the Contract Documents. All Work not conforming to these requirements, including substitutions not properly approved and authorized, may be considered defective. If required by the Architect, the Contractor shall furnish satisfactory evidence as to the kind and quality of materials and equipment. This warranty is not limited by the provisions of Paragraph 13.2.

4.6 TAXES

4.6.1 The Contractor shall pay all sales, consumer, use and other similar taxes for the Work or portions thereof provided by the Contractor which are legally enacted at the time bids are received, whether or not yet effective.

4.7 PERMITS, FEES AND NOTICES

4.7.1 Unless otherwise provided in the Contract Documents, the Contractor shall secure and pay for the building permit and for all other permits and governmental fees, licenses and inspections necessary for the proper execution and completion of the Work which are customarily secured after execution of the Contract and which are legally required at the time the bids are received.

4.7.2 The Contractor shall give all notices and comply with all laws, ordinances, rules, regulations and lawful orders of any public authority bearing on the performance of the Work.

4.7.3 It is not the responsibility of the Contractor to make certain that the Contract Documents are in accordance with applicable laws, statutes, building codes and regulations. If the Contractor observes that any of the Contract Documents are at variance therewith in any respect, he shall promptly notify the Architect in writing, and any necessary changes shall be accomplished by appropriate Modification.

4.7.4 If the Contractor performs any Work knowing it to be contrary to such laws, ordinances, rules and regulations, and without such notice to the Architect, he shall assume full responsibility therefor and shall bear all costs attributable thereto.

4.8 ALLOWANCES

4.8.1 The Contractor shall include in the Contract Sum all allowances stated in the Contract Documents. Items covered by these allowances shall be supplied for such amounts and by such persons as the Owner may direct, but the Contractor will not be required to employ persons against whom he makes a reasonable objection.

4.8.2 Unless otherwise provided in the Contract Documents:

  .1 these allowances shall cover the cost to the Contractor, less any applicable trade discount, of the materials and equipment required by the allowance delivered at the site, and all applicable taxes;

  .2 the Contractor's costs for unloading and handling on the site, labor, installation costs, overhead, profit and other expenses contemplated for the original allowance shall be included in the Contract Sum and not in the allowance;

  .3 whenever the cost is more than or less than the allowance, the Contract Sum shall be adjusted accordingly by Change Order, the amount of which will recognize changes, if any, in handling costs on the site, labor, installation costs, overhead, profit and other expenses.

4.9 SUPERINTENDENT

4.9.1 The Contractor shall employ a competent superintendent and necessary assistants who shall be in attendance at the Project site during the progress of the Work. The superintendent shall represent the Contractor and all communications given to the superintendent shall be as binding as if given to the Contractor. Important communications shall be confirmed in writing. Other communications shall be so confirmed on written request in each case.

4.10 PROGRESS SCHEDULE

4.10.1 The Contractor, immediately after being awarded the Contract, shall prepare and submit for the Owner's and Architect's information an estimated progress sched-

AIA DOCUMENT A201 • GENERAL CONDITIONS OF THE CONTRACT FOR CONSTRUCTION • THIRTEENTH EDITION • AUGUST 1976
AIA® • © 1976 • THE AMERICAN INSTITUTE OF ARCHITECTS, 1735 NEW YORK AVENUE, N.W., WASHINGTON, D.C. 20006

ule for the Work. The progress schedule shall be related to the entire Project to the extent required by the Contract Documents, and shall provide for expeditious and practicable execution of the Work.

4.11 DOCUMENTS AND SAMPLES AT THE SITE

4.11.1 The Contractor shall maintain at the site for the Owner one record copy of all Drawings, Specifications, Addenda, Change Orders and other Modifications, in good order and marked currently to record all changes made during construction, and approved Shop Drawings, Product Data and Samples. These shall be available to the Architect and shall be delivered to him for the Owner upon completion of the Work.

4.12 SHOP DRAWINGS, PRODUCT DATA AND SAMPLES

4.12.1 Shop Drawings are drawings, diagrams, schedules and other data specially prepared for the Work by the Contractor or any Subcontractor, manufacturer, supplier or distributor to illustrate some portion of the Work.

4.12.2 Product Data are illustrations, standard schedules, performance charts, instructions, brochures, diagrams and other information furnished by the Contractor to illustrate a material, product or system for some portion of the Work.

4.12.3 Samples are physical examples which illustrate materials, equipment or workmanship and establish standards by which the Work will be judged.

4.12.4 The Contractor shall review, approve and submit, with reasonable promptness and in such sequence as to cause no delay in the Work or in the work of the Owner or any separate contractor, all Shop Drawings, Product Data and Samples required by the Contract Documents.

4.12.5 By approving and submitting Shop Drawings, Product Data and Samples, the Contractor represents that he has determined and verified all materials, field measurements, and field construction criteria related thereto, or will do so, and that he has checked and coordinated the information contained within such submittals with the requirements of the Work and of the Contract Documents.

4.12.6 The Contractor shall not be relieved of responsibility for any deviation from the requirements of the Contract Documents by the Architect's approval of Shop Drawings, Product Data or Samples under Subparagraph 2.2.14 unless the Contractor has specifically informed the Architect in writing of such deviation at the time of submission and the Architect has given written approval to the specific deviation. The Contractor shall not be relieved from responsibility for errors or omissions in the Shop Drawings, Product Data or Samples by the Architect's approval thereof.

4.12.7 The Contractor shall direct specific attention, in writing or on resubmitted Shop Drawings, Product Data or Samples, to revisions other than those requested by the Architect on previous submittals.

4.12.8 No portion of the Work requiring submission of a Shop Drawing, Product Data or Sample shall be commenced until the submittal has been approved by the Architect as provided in Subparagraph 2.2.14. All such

portions of the Work shall be in accordance with approved submittals.

4.13 USE OF SITE

4.13.1 The Contractor shall confine operations at the site to areas permitted by law, ordinances, permits and the Contract Documents and shall not unreasonably encumber the site with any materials or equipment.

4.14 CUTTING AND PATCHING OF WORK

4.14.1 The Contractor shall be responsible for all cutting, fitting or patching that may be required to complete the Work or to make its several parts fit together properly.

4.14.2 The Contractor shall not damage or endanger any portion of the Work or the work of the Owner or any separate contractors by cutting, patching or otherwise altering any work, or by excavation. The Contractor shall not cut or otherwise alter the work of the Owner or any separate contractor except with the written consent of the Owner and of such separate contractor. The Contractor shall not unreasonably withhold from the Owner or any separate contractor his consent to cutting or otherwise altering the Work.

4.15 CLEANING UP

4.15.1 The Contractor at all times shall keep the premises free from accumulation of waste materials or rubbish caused by his operations. At the completion of the Work he shall remove all his waste materials and rubbish from and about the Project as well as all his tools, construction equipment, machinery and surplus materials.

4.15.2 If the Contractor fails to clean up at the completion of the Work, the Owner may do so as provided in Paragraph 3.4 and the cost thereof shall be charged to the Contractor.

4.16 COMMUNICATIONS

4.16.1 The Contractor shall forward all communications to the Owner through the Architect.

4.17 ROYALTIES AND PATENTS

4.17.1 The Contractor shall pay all royalties and license fees. He shall defend all suits or claims for infringement of any patent rights and shall save the Owner harmless from loss on account thereof, except that the Owner shall be responsible for all such loss when a particular design, process or the product of a particular manufacturer or manufacturers is specified, but if the Contractor has reason to believe that the design, process or product specified is an infringement of a patent, he shall be responsible for such loss unless he promptly gives such information to the Architect.

4.18 INDEMNIFICATION

4.18.1 To the fullest extent permitted by law, the Contractor shall indemnify and hold harmless the Owner and the Architect and their agents and employees from and against all claims, damages, losses and expenses, including but not limited to attorneys' fees, arising out of or resulting from the performance of the Work, provided that any such claim, damage, loss or expense (1) is attributable to bodily injury, sickness, disease or death, or to injury to or destruction of tangible property (other than the Work itself) including the loss of use resulting therefrom, and (2) is caused in whole or in part by any negligent act or omission of the Contractor, any Subcontractor, anyone directly or indirectly employed by any of them or anyone for whose acts any of them may be liable, regardless of whether or not it is caused in part by a party indemnified hereunder. Such obligation shall not be construed to negate, abridge, or otherwise reduce any other right or obligation of indemnity which would otherwise exist as to any party or person described in this Paragraph 4.18.

4.18.2 In any and all claims against the Owner or the Architect or any of their agents or employees by any employee of the Contractor, any Subcontractor, anyone directly or indirectly employed by any of them or anyone for whose acts any of them may be liable, the indemnification obligation under this Paragraph 4.18 shall not be limited in any way by any limitation on the amount or type of damages, compensation or benefits payable by or for the Contractor or any Subcontractor under workers' or workmen's compensation acts, disability benefit acts or other employee benefit acts.

4.18.3 The obligations of the Contractor under this Paragraph 4.18 shall not extend to the liability of the Architect, his agents or employees, arising out of (1) the preparation or approval of maps, drawings, opinions, reports, surveys, change orders, designs or specifications, or (2) the giving of or the failure to give directions or instructions by the Architect, his agents or employees providing such giving or failure to give is the primary cause of the injury or damage.

ARTICLE 5

SUBCONTRACTORS

5.1 DEFINITION

5.1.1 A Subcontractor is a person or entity who has a direct contract with the Contractor to perform any of the Work at the site. The term Subcontractor is referred to throughout the Contract Documents as if singular in number and masculine in gender and means a Subcontractor or his authorized representative. The term Subcontractor does not include any separate contractor or his subcontractors.

5.1.2 A Sub-subcontractor is a person or entity who has a direct or indirect contract with a Subcontractor to perform any of the Work at the site. The term Sub-subcontractor is referred to throughout the Contract Documents as if singular in number and masculine in gender and means a Sub-subcontractor or an authorized representative thereof.

5.2 AWARD OF SUBCONTRACTS AND OTHER CONTRACTS FOR PORTIONS OF THE WORK

5.2.1 Unless otherwise required by the Contract Documents or the Bidding Documents, the Contractor, as soon as practicable after the award of the Contract, shall furnish to the Owner and the Architect in writing the names of the persons or entities (including those who are to furnish materials or equipment fabricated to a special design) proposed for each of the principal portions of the Work. The Architect will promptly reply to the Contractor in writing stating whether or not the Owner or the Architect, after due investigation, has reasonable objection to any

such proposed person or entity. Failure of the Owner or Architect to reply promptly shall constitute notice of no reasonable objection.

5.2.2 The Contractor shall not contract with any such proposed person or entity to whom the Owner or the Architect has made reasonable objection under the provisions of Subparagraph 5.2.1. The Contractor shall not be required to contract with anyone to whom he has a reasonable objection.

5.2.3 If the Owner or the Architect has reasonable objection to any such proposed person or entity, the Contractor shall submit a substitute to whom the Owner or the Architect has no reasonable objection, and the Contract Sum shall be increased or decreased by the difference in cost occasioned by such substitution and an appropriate Change Order shall be issued; however, no increase in the Contract Sum shall be allowed for any such substitution unless the Contractor has acted promptly and responsively in submitting names as required by Subparagraph 5.2.1.

5.2.4 The Contractor shall make no substitution for any Subcontractor, person or entity previously selected if the Owner or Architect makes reasonable objection to such substitution.

5.3 SUBCONTRACTUAL RELATIONS

5.3.1 By an appropriate agreement, written where legally required for validity, the Contractor shall require each Subcontractor, to the extent of the Work to be performed by the Subcontractor, to be bound to the Contractor by the terms of the Contract Documents, and to assume toward the Contractor all the obligations and responsibilities which the Contractor, by these Documents, assumes toward the Owner and the Architect. Said agreement shall preserve and protect the rights of the Owner and the Architect under the Contract Documents with respect to the Work to be performed by the Subcontractor so that the subcontracting thereof will not prejudice such rights, and shall allow to the Subcontractor, unless specifically provided otherwise in the Contractor-Subcontractor agreement, the benefit of all rights, remedies and redress against the Contractor that the Contractor, by these Documents, has against the Owner. Where appropriate, the Contractor shall require each Subcontractor to enter into similar agreements with his Sub-subcontractors. The Contractor shall make available to each proposed Subcontractor, prior to the execution of the Subcontract, copies of the Contract Documents to which the Subcontractor will be bound by this Paragraph 5.3, and identify to the Subcontractor any terms and conditions of the proposed Subcontract which may be at variance with the Contract Documents. Each Subcontractor shall similarly make copies of such Documents available to his Sub-subcontractors.

ARTICLE 6

WORK BY OWNER OR BY SEPARATE CONTRACTORS

6.1 OWNER'S RIGHT TO PERFORM WORK AND TO AWARD SEPARATE CONTRACTS

6.1.1 The Owner reserves the right to perform work related to the Project with his own forces, and to award separate contracts in connection with other portions of the Project or other work on the site under these or similar Conditions of the Contract. If the Contractor claims that delay or additional cost is involved because of such action by the Owner, he shall make such claim as provided elsewhere in the Contract Documents.

6.1.2 When separate contracts are awarded for different portions of the Project or other work on the site, the term Contractor in the Contract Documents in each case shall mean the Contractor who executes each separate Owner-Contractor Agreement.

6.1.3 The Owner will provide for the coordination of the work of his own forces and of each separate contractor with the Work of the Contractor, who shall cooperate therewith as provided in Paragraph 6.2.

6.2 MUTUAL RESPONSIBILITY

6.2.1 The Contractor shall afford the Owner and separate contractors reasonable opportunity for the introduction and storage of their materials and equipment and the execution of their work, and shall connect and coordinate his Work with theirs as required by the Contract Documents.

6.2.2 If any part of the Contractor's Work depends for proper execution or results upon the work of the Owner or any separate contractor, the Contractor shall, prior to proceeding with the Work, promptly report to the Architect any apparent discrepancies or defects in such other work that render it unsuitable for such proper execution and results. Failure of the Contractor so to report shall constitute an acceptance of the Owner's or separate contractors' work as fit and proper to receive his Work, except as to defects which may subsequently become apparent in such work by others.

6.2.3 Any costs caused by defective or ill-timed work shall be borne by the party responsible therefor.

6.2.4 Should the Contractor wrongfully cause damage to the work or property of the Owner, or to other work on the site, the Contractor shall promptly remedy such damage as provided in Subparagraph 10.2.5.

6.2.5 Should the Contractor wrongfully cause damage to the work or property of any separate contractor, the Contractor shall upon due notice promptly attempt to settle with such other contractor by agreement, or otherwise to resolve the dispute. If such separate contractor sues or initiates an arbitration proceeding against the Owner on account of any damage alleged to have been caused by the Contractor, the Owner shall notify the Contractor who shall defend such proceedings at the Owner's expense, and if any judgment or award against the Owner arises therefrom the Contractor shall pay or satisfy it and shall reimburse the Owner for all attorneys' fees and court or arbitration costs which the Owner has incurred.

6.3 OWNER'S RIGHT TO CLEAN UP

6.3.1 If a dispute arises between the Contractor and separate contractors as to their responsibility for cleaning up as required by Paragraph 4.15, the Owner may clean up and charge the cost thereof to the contractors responsible therefor as the Architect shall determine to be just.

ARTICLE 7

MISCELLANEOUS PROVISIONS

7.1 GOVERNING LAW

7.1.1 The Contract shall be governed by the law of the place where the Project is located.

7.2 SUCCESSORS AND ASSIGNS

7.2.1 The Owner and the Contractor each binds himself, his partners, successors, assigns and legal representatives to the other party hereto and to the partners, successors, assigns and legal representatives of such other party in respect to all covenants, agreements and obligations contained in the Contract Documents. Neither party to the Contract shall assign the Contract or sublet it as a whole without the written consent of the other, nor shall the Contractor assign any moneys due or to become due to him hereunder, without the previous written consent of the Owner.

7.3 WRITTEN NOTICE

7.3.1 Written notice shall be deemed to have been duly served if delivered in person to the individual or member of the firm or entity or to an officer of the corporation for whom it was intended, or if delivered at or sent by registered or certified mail to the last business address known to him who gives the notice.

7.4 CLAIMS FOR DAMAGES

7.4.1 Should either party to the Contract suffer injury or damage to person or property because of any act or omission of the other party or of any of his employees, agents or others for whose acts he is legally liable, claim shall be made in writing to such other party within a reasonable time after the first observance of such injury or damage.

7.5 PERFORMANCE BOND AND LABOR AND MATERIAL PAYMENT BOND

7.5.1 The Owner shall have the right to require the Contractor to furnish bonds covering the faithful performance of the Contract and the payment of all obligations arising thereunder if and as required in the Bidding Documents or in the Contract Documents.

7.6 RIGHTS AND REMEDIES

7.6.1 The duties and obligations imposed by the Contract Documents and the rights and remedies available thereunder shall be in addition to and not a limitation of any duties, obligations, rights and remedies otherwise imposed or available by law.

7.6.2 No action or failure to act by the Owner, Architect or Contractor shall constitute a waiver of any right or duty afforded any of them under the Contract, nor shall any such action or failure to act constitute an approval of or acquiescence in any breach thereunder, except as may be specifically agreed in writing.

7.7 TESTS

7.7.1 If the Contract Documents, laws, ordinances, rules, regulations or orders of any public authority having jurisdiction require any portion of the Work to be inspected, tested or approved, the Contractor shall give the Architect timely notice of its readiness so the Architect may observe such inspection, testing or approval. The Contractor shall bear all costs of such inspections, tests or approvals conducted by public authorities. Unless otherwise provided, the Owner shall bear all costs of other inspections, tests or approvals.

7.7.2 If the Architect determines that any Work requires special inspection, testing, or approval which Subparagraph 7.7.1 does not include, he will, upon written authorization from the Owner, instruct the Contractor to order such special inspection, testing or approval, and the Contractor shall give notice as provided in Subparagraph 7.7.1. If such special inspection or testing reveals a failure of the Work to comply with the requirements of the Contract Documents, the Contractor shall bear all costs thereof, including compensation for the Architect's additional services made necessary by such failure; otherwise the Owner shall bear such costs, and an appropriate Change Order shall be issued.

7.7.3 Required certificates of inspection, testing or approval shall be secured by the Contractor and promptly delivered by him to the Architect.

7.7.4 If the Architect is to observe the inspections, tests or approvals required by the Contract Documents, he will do so promptly and, where practicable, at the source of supply.

7.8 INTEREST

7.8.1 Payments due and unpaid under the Contract Documents shall bear interest from the date payment is due at such rate as the parties may agree upon in writing or, in the absence thereof, at the legal rate prevailing at the place of the Project.

7.9 ARBITRATION

7.9.1 All claims, disputes and other matters in question between the Contractor and the Owner arising out of, or relating to, the Contract Documents or the breach thereof, except as provided in Subparagraph 2.2.11 with respect to the Architect's decisions on matters relating to artistic effect, and except for claims which have been waived by the making or acceptance of final payment as provided by Subparagraphs 9.9.4 and 9.9.5, shall be decided by arbitration in accordance with the Construction Industry Arbitration Rules of the American Arbitration Association then obtaining unless the parties mutually agree otherwise. No arbitration arising out of or relating to the Contract Documents shall include, by consolidation, joinder or in any other manner, the Architect, his employees or consultants except by written consent containing a specific reference to the Owner-Contractor Agreement and signed by the Architect, the Owner, the Contractor and any other person sought to be joined. No arbitration shall include by consolidation, joinder or in any other manner, parties other than the Owner, the Contractor and any other persons substantially involved in a common question of fact or law, whose presence is required if complete relief is to be accorded in the arbitration. No person other than the Owner or Contractor shall be included as an original third party or additional third party to an arbitration whose interest or responsibility is insubstantial. Any consent to arbitration involving an additional person or persons shall not constitute consent to arbitration of any dispute not described therein or with any person not named or described therein. The foregoing agreement to arbitrate and any other agreement to arbitrate with an additional person or persons duly consented to by the parties to the Owner-Contractor Agreement shall be specifically enforceable under the prevailing arbitration law. The award rendered by the arbitrators shall be final, and judgment may be entered upon it in accordance with applicable law in any court having jurisdiction thereof.

7.9.2 Notice of the demand for arbitration shall be filed in writing with the other party to the Owner-Contractor Agreement and with the American Arbitration Association, and a copy shall be filed with the Architect. The demand for arbitration shall be made within the time limits specified in Subparagraph 2.2.12 where applicable, and in all other cases within a reasonable time after the claim, dispute or other matter in question has arisen, and in no event shall it be made after the date when institution of legal or equitable proceedings based on such claim, dispute or other matter in question would be barred by the applicable statute of limitations.

7.9.3 Unless otherwise agreed in writing, the Contractor shall carry on the Work and maintain its progress during any arbitration proceedings, and the Owner shall continue to make payments to the Contractor in accordance with the Contract Documents.

ARTICLE 8

TIME

8.1 DEFINITIONS

8.1.1 Unless otherwise provided, the Contract Time is the period of time allotted in the Contract Documents for Substantial Completion of the Work as defined in Subparagraph 8.1.3, including authorized adjustments thereto.

8.1.2 The date of commencement of the Work is the date established in a notice to proceed. If there is no notice to proceed, it shall be the date of the Owner-Contractor Agreement or such other date as may be established therein.

8.1.3 The Date of Substantial Completion of the Work or designated portion thereof is the Date certified by the Architect when construction is sufficiently complete, in accordance with the Contract Documents, so the Owner can occupy or utilize the Work or designated portion thereof for the use for which it is intended.

8.1.4 The term day as used in the Contract Documents shall mean calendar day unless otherwise specifically designated.

8.2 PROGRESS AND COMPLETION

8.2.1 All time limits stated in the Contract Documents are of the essence of the Contract.

8.2.2 The Contractor shall begin the Work on the date of commencement as defined in Subparagraph 8.1.2. He shall carry the Work forward expeditiously with adequate forces and shall achieve Substantial Completion within the Contract Time.

8.3 DELAYS AND EXTENSIONS OF TIME

8.3.1 If the Contractor is delayed at any time in the progress of the Work by any act or neglect of the Owner or the Architect, or by any employee of either, or by any separate contractor employed by the Owner, or by changes ordered in the Work, or by labor disputes, fire, unusual delay in transportation, adverse weather conditions not reasonably anticipatable, unavoidable casualties, or any causes beyond the Contractor's control, or by delay authorized by the Owner pending arbitration, or by any other cause which the Architect determines may justify the delay, then the Contract Time shall be extended by Change Order for such reasonable time as the Architect may determine.

8.3.2 Any claim for extension of time shall be made in writing to the Architect not more than twenty days after the commencement of the delay; otherwise it shall be waived. In the case of a continuing delay only one claim is necessary. The Contractor shall provide an estimate of the probable effect of such delay on the progress of the Work.

8.3.3 If no agreement is made stating the dates upon which interpretations as provided in Subparagraph 2.2.8 shall be furnished, then no claim for delay shall be allowed on account of failure to furnish such interpretations until fifteen days after written request is made for them, and not then unless such claim is reasonable.

8.3.4 This Paragraph 8.3 does not exclude the recovery of damages for delay by either party under other provisions of the Contract Documents.

ARTICLE 9

PAYMENTS AND COMPLETION

9.1 CONTRACT SUM

9.1.1 The Contract Sum is stated in the Owner-Contractor Agreement and, including authorized adjustments thereto, is the total amount payable by the Owner to the Contractor for the performance of the Work under the Contract Documents.

9.2 SCHEDULE OF VALUES

9.2.1 Before the first Application for Payment, the Contractor shall submit to the Architect a schedule of values allocated to the various portions of the Work, prepared in such form and supported by such data to substantiate its accuracy as the Architect may require. This schedule, unless objected to by the Architect, shall be used only as a basis for the Contractor's Applications for Payment.

9.3 APPLICATIONS FOR PAYMENT

9.3.1 At least ten days before the date for each progress payment established in the Owner-Contractor Agreement, the Contractor shall submit to the Architect an itemized Application for Payment, notarized if required, supported by such data substantiating the Contractor's right to payment as the Owner or the Architect may require, and reflecting retainage, if any, as provided elsewhere in the Contract Documents.

9.3.2 Unless otherwise provided in the Contract Documents, payments will be made on account of materials or equipment not incorporated in the Work but delivered and suitably stored at the site and, if approved in advance by the Owner, payments may similarly be made for materials or equipment suitably stored at some other location agreed upon in writing. Payments for materials or equipment stored on or off the site shall be conditioned upon submission by the Contractor of bills of sale or such other procedures satisfactory to the Owner to establish the Owner's title to such materials or equipment or otherwise protect the Owner's interest, including applicable insurance and transportation to the site for those materials and equipment stored off the site.

9.3.3 The Contractor warrants that title to all Work, materials and equipment covered by an Application for Payment will pass to the Owner either by incorporation in the construction or upon the receipt of payment by the Contractor, whichever occurs first, free and clear of all liens, claims, security interests or encumbrances, hereinafter referred to in this Article 9 as "liens"; and that no Work, materials or equipment covered by an Application for Payment will have been acquired by the Contractor, or by any other person performing Work at the site or furnishing materials and equipment for the Project, subject to an agreement under which an interest therein or an encumbrance thereon is retained by the seller or otherwise imposed by the Contractor or such other person.

9.4 CERTIFICATES FOR PAYMENT

9.4.1 The Architect will, within seven days after the receipt of the Contractor's Application for Payment, either issue a Certificate for Payment to the Owner, with a copy to the Contractor, for such amount as the Architect determines is properly due, or notify the Contractor in writing his reasons for withholding a Certificate as provided in Subparagraph 9.6.1.

9.4.2 The issuance of a Certificate for Payment will constitute a representation by the Architect to the Owner, based on his observations at the site as provided in Subparagraph 2.2.3 and the data comprising the Application for Payment, that the Work has progressed to the point indicated; that, to the best of his knowledge, information and belief, the quality of the Work is in accordance with the Contract Documents (subject to an evaluation of the Work for conformance with the Contract Documents upon Substantial Completion, to the results of any subsequent tests required by or performed under the Contract Documents, to minor deviations from the Contract Documents correctable prior to completion, and to any specific qualifications stated in his Certificate); and that the Contractor is entitled to payment in the amount certified. However, by issuing a Certificate for Payment, the Architect shall not thereby be deemed to represent that he has made exhaustive or continuous on-site inspections to check the quality or quantity of the Work or that he has reviewed the construction means, methods, techniques, sequences or procedures, or that he has made any examination to ascertain how or for what purpose the Contractor has used the moneys previously paid on account of the Contract Sum.

9.5 PROGRESS PAYMENTS

9.5.1 After the Architect has issued a Certificate for Payment, the Owner shall make payment in the manner and within the time provided in the Contract Documents.

9.5.2 The Contractor shall promptly pay each Subcontractor, upon receipt of payment from the Owner, out of the amount paid to the Contractor on account of such Subcontractor's Work, the amount to which said Subcontractor is entitled, reflecting the percentage actually retained, if any, from payments to the Contractor on account of such Subcontractor's Work. The Contractor shall, by an appropriate agreement with each Subcontractor, require each Subcontractor to make payments to his Subsubcontractors in similar manner.

9.5.3 The Architect may, on request and at his discretion, furnish to any Subcontractor, if practicable, information regarding the percentages of completion or the amounts applied for by the Contractor and the action taken thereon by the Architect on account of Work done by such Subcontractor.

9.5.4 Neither the Owner nor the Architect shall have any obligation to pay or to see to the payment of any moneys to any Subcontractor except as may otherwise be required by law.

9.5.5 No Certificate for a progress payment, nor any progress payment, nor any partial or entire use or occupancy of the Project by the Owner, shall constitute an acceptance of any Work not in accordance with the Contract Documents.

9.6 PAYMENTS WITHHELD

9.6.1 The Architect may decline to certify payment and may withhold his Certificate in whole or in part, to the extent necessary reasonably to protect the Owner, if in his opinion he is unable to make representations to the Owner as provided in Subparagraph 9.4.2. If the Architect is unable to make representations to the Owner as provided in Subparagraph 9.4.2 and to certify payment in the amount of the Application, he will notify the Contractor as provided in Subparagraph 9.4.1. If the Contractor and the Architect cannot agree on a revised amount, the Architect will promptly issue a Certificate for Payment for the amount for which he is able to make such representations to the Owner. The Architect may also decline to certify payment or, because of subsequently discovered evidence or subsequent observations, he may nullify the whole or any part of any Certificate for Payment previously issued, to such extent as may be necessary in his opinion to protect the Owner from loss because of:

  .1 defective work not remedied,

  .2 third party claims filed or reasonable evidence indicating probable filing of such claims,

  .3 failure of the Contractor to make payments properly to Subcontractors or for labor, materials or equipment,

  .4 reasonable evidence that the Work cannot be completed for the unpaid balance of the Contract Sum,

  .5 damage to the Owner or another contractor,

  .6 reasonable evidence that the Work will not be completed within the Contract Time, or

  .7 persistent failure to carry out the Work in accordance with the Contract Documents.

9.6.2 When the above grounds in Subparagraph 9.6.1 are removed, payment shall be made for amounts withheld because of them.

9.7 FAILURE OF PAYMENT

9.7.1 If the Architect does not issue a Certificate for Payment, through no fault of the Contractor, within seven days after receipt of the Contractor's Application for Payment, or if the Owner does not pay the Contractor within seven days after the date established in the Contract Documents any amount certified by the Architect or awarded by arbitration, then the Contractor may, upon seven additional days' written notice to the Owner and the Architect, stop the Work until payment of the amount owing has been received. The Contract Sum shall be increased by the amount of the Contractor's reasonable costs of shut-down, delay and start-up, which shall be effected by appropriate Change Order in accordance with Paragraph 12.3.

9.8 SUBSTANTIAL COMPLETION

9.8.1 When the Contractor considers that the Work, or a designated portion thereof which is acceptable to the Owner, is substantially complete as defined in Subparagraph 8.1.3, the Contractor shall prepare for submission to the Architect a list of items to be completed or corrected. The failure to include any items on such list does not alter the responsibility of the Contractor to complete all Work in accordance with the Contract Documents. When the Architect on the basis of an inspection determines that the Work or designated portion thereof is substantially complete, he will then prepare a Certificate of Substantial Completion which shall establish the Date of Substantial Completion, shall state the responsibilities of the Owner and the Contractor for security, maintenance, heat, utilities, damage to the Work, and insurance, and shall fix the time within which the Contractor shall complete the items listed therein. Warranties required by the Contract Documents shall commence on the Date of Substantial Completion of the Work or designated portion thereof unless otherwise provided in the Certificate of Substantial Completion. The Certificate of Substantial Completion shall be submitted to the Owner and the Contractor for their written acceptance of the responsibilities assigned to them in such Certificate.

9.8.2 Upon Substantial Completion of the Work or designated portion thereof and upon application by the Contractor and certification by the Architect, the Owner shall make payment, reflecting adjustment in retainage, if any, for such Work or portion thereof, as provided in the Contract Documents.

9.9 FINAL COMPLETION AND FINAL PAYMENT

9.9.1 Upon receipt of written notice that the Work is ready for final inspection and acceptance and upon receipt of a final Application for Payment, the Architect will promptly make such inspection and, when he finds the Work acceptable under the Contract Documents and the Contract fully performed, he will promptly issue a final Certificate for Payment stating that to the best of his knowledge, information and belief, and on the basis of his observations and inspections, the Work has been completed in accordance with the terms and conditions of the Contract Documents and that the entire balance found to be due the Contractor, and noted in said final Certificate, is due and payable. The Architect's final Certificate for Payment will constitute a further representation that the conditions precedent to the Contractor's being entitled to final payment as set forth in Subparagraph 9.9.2 have been fulfilled.

9.9.2 Neither the final payment nor the remaining retained percentage shall become due until the Contractor submits to the Architect (1) an affidavit that all payrolls, bills for materials and equipment, and other indebtedness connected with the Work for which the Owner or his property might in any way be responsible, have been paid or otherwise satisfied, (2) consent of surety, if any, to final payment and (3), if required by the Owner, other data establishing payment or satisfaction of all such obligations, such as receipts, releases and waivers of liens arising out of the Contract, to the extent and in such form as may be designated by the Owner. If any Subcontractor refuses to furnish a release or waiver required by the Owner, the Contractor may furnish a bond satisfactory to the Owner to indemnify him against any such lien. If any such lien remains unsatisfied after all payments are made, the Contractor shall refund to the Owner all moneys that the latter may be compelled to pay in discharging such lien, including all costs and reasonable attorneys' fees.

9.9.3 If, after Substantial Completion of the Work, final completion thereof is materially delayed through no fault of the Contractor or by the issuance of Change Orders affecting final completion, and the Architect so confirms, the Owner shall, upon application by the Contractor and certification by the Architect, and without terminating the Contract, make payment of the balance due for that portion of the Work fully completed and accepted. If the remaining balance for Work not fully completed or corrected is less than the retainage stipulated in the Contract Documents, and if bonds have been furnished as provided in Paragraph 7.5, the written consent of the surety to the payment of the balance due for that portion of the Work fully completed and accepted shall be submitted by the Contractor to the Architect prior to certification of such payment. Such payment shall be made under the terms and conditions governing final payment, except that it shall not constitute a waiver of claims.

9.9.4 The making of final payment shall constitute a waiver of all claims by the Owner except those arising from:
.1 unsettled liens,
.2 faulty or defective Work appearing after Substantial Completion,
.3 failure of the Work to comply with the requirements of the Contract Documents, or
.4 terms of any special warranties required by the Contract Documents.

9.9.5 The acceptance of final payment shall constitute a waiver of all claims by the Contractor except those previously made in writing and identified by the Contractor as unsettled at the time of the final Application for Payment.

ARTICLE 10
PROTECTION OF PERSONS AND PROPERTY

10.1 SAFETY PRECAUTIONS AND PROGRAMS

10.1.1 The Contractor shall be responsible for initiating, maintaining and supervising all safety precautions and programs in connection with the Work.

10.2 SAFETY OF PERSONS AND PROPERTY

10.2.1 The Contractor shall take all reasonable precautions for the safety of, and shall provide all reasonable protection to prevent damage, injury or loss to:
.1 all employees on the Work and all other persons who may be affected thereby;
.2 all the Work and all materials and equipment to be incorporated therein, whether in storage on or off the site, under the care, custody or control of the Contractor or any of his Subcontractors or Sub-subcontractors; and
.3 other property at the site or adjacent thereto, including trees, shrubs, lawns, walks, pavements, roadways, structures and utilities not designated for removal, relocation or replacement in the course of construction.

10.2.2 The Contractor shall give all notices and comply with all applicable laws, ordinances, rules, regulations and lawful orders of any public authority bearing on the safety of persons or property or their protection from damage, injury or loss.

10.2.3 The Contractor shall erect and maintain, as required by existing conditions and progress of the Work, all reasonable safeguards for safety and protection, including posting danger signs and other warnings against hazards, promulgating safety regulations and notifying owners and users of adjacent utilities.

10.2.4 When the use or storage of explosives or other hazardous materials or equipment is necessary for the execution of the Work, the Contractor shall exercise the utmost care and shall carry on such activities under the supervision of properly qualified personnel.

10.2.5 The Contractor shall promptly remedy all damage or loss (other than damage or loss insured under Paragraph 11.3) to any property referred to in Clauses 10.2.1.2 and 10.2.1.3 caused in whole or in part by the Contractor, any Subcontractor, any Sub-subcontractor, or anyone directly or indirectly employed by any of them, or by anyone for whose acts any of them may be liable and for which the Contractor is responsible under Clauses 10.2.1.2 and 10.2.1.3, except damage or loss attributable to the acts or omissions of the Owner or Architect or anyone directly or indirectly employed by either of them, or by anyone for whose acts either of them may be liable, and not attributable to the fault or negligence of the Contractor. The foregoing obligations of the Contractor are in addition to his obligations under Paragraph 4.18.

10.2.6 The Contractor shall designate a responsible member of his organization at the site whose duty shall be the prevention of accidents. This person shall be the Contractor's superintendent unless otherwise designated by the Contractor in writing to the Owner and the Architect.

10.2.7 The Contractor shall not load or permit any part of the Work to be loaded so as to endanger its safety.

10.3 EMERGENCIES

10.3.1 In any emergency affecting the safety of persons or property, the Contractor shall act, at his discretion, to prevent threatened damage, injury or loss. Any additional compensation or extension of time claimed by the Contractor on account of emergency work shall be determined as provided in Article 12 for Changes in the Work.

ARTICLE 11

INSURANCE

11.1 CONTRACTOR'S LIABILITY INSURANCE

11.1.1 The Contractor shall purchase and maintain such insurance as will protect him from claims set forth below which may arise out of or result from the Contractor's operations under the Contract, whether such operations be by himself or by any Subcontractor or by anyone directly or indirectly employed by any of them, or by anyone for whose acts any of them may be liable:

.1 claims under workers' or workmen's compensation, disability benefit and other similar employee benefit acts;

.2 claims for damages because of bodily injury, occupational sickness or disease, or death of his employees;

.3 claims for damages because of bodily injury, sickness or disease, or death of any person other than his employees;

.4 claims for damages insured by usual personal injury liability coverage which are sustained (1) by any person as a result of an offense directly or indirectly related to the employment of such person by the Contractor, or (2) by any other person;

.5 claims for damages, other than to the Work itself, because of injury to or destruction of tangible property, including loss of use resulting therefrom; and

.6 claims for damages because of bodily injury or death of any person or property damage arising out of the ownership, maintenance or use of any motor vehicle.

11.1.2 The issuance required by Subparagraph 11.1.1 be written for not less than any limits of liability specified in the Contract Documents, or required by law, whichever is greater.

11.1.3 The insurance required by Subparagraph 11.1.1 shall include contractual liability insurance applicable to the Contractor's obligations under Paragraph 4.18.

11.1.4 Certificates of Insurance acceptable to the Owner shall be filed with the Owner prior to commencement of the Work. These Certificates shall contain a provision that coverages afforded under the policies will not be cancelled until at least thirty days' prior written notice has been given to the Owner.

11.2 OWNER'S LIABILITY INSURANCE

11.2.1 The Owner shall be responsible for purchasing and maintaining his own liability insurance and, at his option, may purchase and maintain such insurance as will protect him against claims which may arise from operations under the Contract.

11.3 PROPERTY INSURANCE

11.3.1 Unless otherwise provided, the Owner shall purchase and maintain property insurance upon the entire Work at the site to the full insurable value thereof. This insurance shall include the interests of the Owner, the Contractor, Subcontractors and Sub-subcontractors in the Work and shall insure against the perils of fire and extended coverage and shall include "all risk" insurance for physical loss or damage including, without duplication of coverage, theft, vandalism and malicious mischief. If the Owner does not intend to purchase such insurance for the full insurable value of the entire Work, he shall inform the Contractor in writing prior to commencement of the Work. The Contractor may then effect insurance which will protect the interests of himself, his Subcontractors and the Sub-subcontractors in the Work, and by appropriate Change Order the cost thereof shall be charged to the Owner. If the Contractor is damaged by failure of the Owner to purchase or maintain such insurance and to so notify the Contractor, then the Owner shall bear all reasonable costs properly attributable thereto. If not covered under the all risk insurance or otherwise provided in the Contract Documents, the Contractor shall effect and maintain similar property insurance on portions of the Work stored off the site or in transit when such portions of the Work are to be included in an Application for Payment under Subparagraph 9.3.2.

11.3.2 The Owner shall purchase and maintain such boiler and machinery insurance as may be required by the Contract Documents or by law. This insurance shall include the interests of the Owner, the Contractor, Subcontractors and Sub-subcontractors in the Work.

11.3.3 Any loss insured under Subparagraph 11.3.1 is to be adjusted with the Owner and made payable to the Owner as trustee for the insureds, as their interests may appear, subject to the requirements of any applicable mortgagee clause and of Subparagraph 11.3.8. The Contractor shall pay each Subcontractor a just share of any insurance moneys received by the Contractor, and by appropriate agreement, written where legally required for validity, shall require each Subcontractor to make payments to his Sub-subcontractors in similar manner.

11.3.4 The Owner shall file a copy of all policies with the Contractor before an exposure to loss may occur.

11.3.5 If the Contractor requests in writing that insurance for risks other than those described in Subparagraphs 11.3.1 and 11.3.2 or other special hazards be included in the property insurance policy, the Owner shall, if possible, include such insurance, and the cost thereof shall be charged to the Contractor by appropriate Change Order.

AIA DOCUMENT A201 • GENERAL CONDITIONS OF THE CONTRACT FOR CONSTRUCTION • THIRTEENTH EDITION • AUGUST 1976
AIA® • © 1976 • THE AMERICAN INSTITUTE OF ARCHITECTS, 1735 NEW YORK AVENUE, N.W., WASHINGTON, D.C. 20006

11.3.6 The Owner and Contractor waive all rights against (1) each other and the Subcontractors, Sub-subcontractors, agents and employees each of the other, and (2) the Architect and separate contractors, if any, and their sub-contractors, sub-subcontractors, agents and employees, for damages caused by fire or other perils to the extent covered by insurance obtained pursuant to this Paragraph 11.3 or any other property insurance applicable to the Work, except such rights as they may have to the pro-ceeds of such insurance held by the Owner as trustee. The foregoing waiver afforded the Architect, his agents and employees shall not extend to the liability imposed by Subparagraph 4.18.3. The Owner or the Contractor, as appropriate, shall require of the Architect, separate con-tractors, Subcontractors and Sub-subcontractors by ap-propriate agreements, written where legally required for validity, similar waivers each in favor of all other parties enumerated in this Subparagraph 11.3.6.

11.3.7 If required in writing by any party in interest, the Owner as trustee shall, upon the occurrence of an insured loss, give bond for the proper performance of his duties. He shall deposit in a separate account any money so re-ceived, and he shall distribute it in accordance with such agreement as the parties in interest may reach, or in ac-cordance with an award by arbitration in which case the procedure shall be as provided in Paragraph 7.9. If after such loss no other special agreement is made, replace-ment of damaged work shall be covered by an appropri-ate Change Order.

11.3.8 The Owner as trustee shall have power to adjust and settle any loss with the insurers unless one of the parties in interest shall object in writing within five days after the occurrence of loss to the Owner's exercise of this power, and if such objection be made, arbitrators shall be chosen as provided in Paragraph 7.9. The Owner as trustee shall, in that case, make settlement with the insurers in accordance with the directions of such arbitrators. If dis-tribution of the insurance proceeds by arbitration is re-quired, the arbitrators will direct such distribution.

11.3.9 If the Owner finds it necessary to occupy or use a portion or portions of the Work prior to Substantial Com-pletion thereof, such occupancy shall not commence prior to a time mutually agreed to by the Owner and Contrac-tor and to which the insurance company or companies providing the property insurance have consented by en-dorsement to the policy or policies. This insurance shall not be cancelled or lapsed on account of such partial occupancy. Consent of the Contractor and of the insur-ance company or companies to such occupancy or use shall not be unreasonably withheld.

11.4 LOSS OF USE INSURANCE

11.4.1 The Owner, at his option, may purchase and main-tain such insurance as will insure him against loss of use of his property due to fire or other hazards, however caused. The Owner waives all rights of action against the Contractor for loss of use of his property, including con-sequential losses due to fire or other hazards however caused, to the extent covered by insurance under this Paragraph 11.4.

ARTICLE 12

CHANGES IN THE WORK

12.1 CHANGE ORDERS

12.1.1 A Change Order is a written order to the Contrac-tor signed by the Owner and the Architect, issued after execution of the Contract, authorizing a change in the Work or an adjustment in the Contract Sum or the Con-tract Time. The Contract Sum and the Contract Time may be changed only by Change Order. A Change Order signed by the Contractor indicates his agreement there-with, including the adjustment in the Contract Sum or the Contract Time.

12.1.2 The Owner, without invalidating the Contract, may order changes in the Work within the general scope of the Contract consisting of additions, deletions or other revisions, the Contract Sum and the Contract Time being adjusted accordingly. All such changes in the Work shall be authorized by Change Order, and shall be performed under the applicable conditions of the Contract Docu-ments.

12.1.3 The cost or credit to the Owner resulting from a change in the Work shall be determined in one or more of the following ways:

   .1 by mutual acceptance of a lump sum properly itemized and supported by sufficient substantiating data to permit evaluation;

   .2 by unit prices stated in the Contract Documents or subsequently agreed upon;

   .3 by cost to be determined in a manner agreed upon by the parties and a mutually acceptable fixed or percentage fee; or

   .4 by the method provided in Subparagraph 12.1.4.

12.1.4 If none of the methods set forth in Clauses 12.1.3.1, 12.1.3.2 or 12.1.3.3 is agreed upon, the Contrac-tor, provided he receives a written order signed by the Owner, shall promptly proceed with the Work involved. The cost of such Work shall then be determined by the Architect on the basis of the reasonable expenditures and savings of those performing the Work attributable to the change, including, in the case of an increase in the Con-tract Sum, a reasonable allowance for overhead and profit. In such case, and also under Clauses 12.1.3.3 and 12.1.3.4 above, the Contractor shall keep and present, in such form as the Architect may prescribe, an itemized account-ing together with appropriate supporting data for inclu-sion in a Change Order. Unless otherwise provided in the Contract Documents, cost shall be limited to the fol-lowing: cost of materials, including sales tax and cost of delivery; cost of labor, including social security, old age and unemployment insurance, and fringe benefits re-quired by agreement or custom; workers' or workmen's compensation insurance; bond premiums; rental value of equipment and machinery; and the additional costs of supervision and field office personnel directly attributable to the change. Pending final determination of cost to the Owner, payments on account shall be made on the Archi-tect's Certificate for Payment. The amount of credit to be allowed by the Contractor to the Owner for any deletion or change which results in a net decrease in the Contract Sum will be the amount of the actual net cost as confirmed by the Architect. When both additions and credits covering related Work or substitutions are involved in any one change, the allowance for overhead and profit shall be figured on the basis of the net increase, if any, with respect to that change.

12.1.5 If unit prices are stated in the Contract Documents or subsequently agreed upon, and if the quantities originally contemplated are so changed in a proposed Change Order that application of the agreed unit prices to the quantities of Work proposed will cause substantial inequity to the Owner or the Contractor, the applicable unit prices shall be equitably adjusted.

12.2 CONCEALED CONDITIONS

12.2.1 Should concealed conditions encountered in the performance of the Work below the surface of the ground or should concealed or unknown conditions in an existing structure be at variance with the conditions indicated by the Contract Documents, or should unknown physical conditions below the surface of the ground or should concealed or unknown conditions in an existing structure of an unusual nature, differing materially from those ordinarily encountered and generally recognized as inherent in work of the character provided for in this Contract, be encountered, the Contract Sum shall be equitably adjusted by Change Order upon claim by either party made within twenty days after the first observance of the conditions.

12.3 CLAIMS FOR ADDITIONAL COST

12.3.1 If the Contractor wishes to make a claim for an increase in the Contract Sum, he shall give the Architect written notice thereof within twenty days after the occurrence of the event giving rise to such claim. This notice shall be given by the Contractor before proceeding to execute the Work, except in an emergency endangering life or property in which case the Contractor shall proceed in accordance with Paragraph 10.3. No such claim shall be valid unless so made. If the Owner and the Contractor cannot agree on the amount of the adjustment in the Contract Sum, it shall be determined by the Architect. Any change in the Contract Sum resulting from such claim shall be authorized by Change Order.

12.3.2 If the Contractor claims that additional cost is involved because of, but not limited to, (1) any written interpretation pursuant to Subparagraph 2.2.8, (2) any order by the Owner to stop the Work pursuant to Paragraph 3.3 where the Contractor was not at fault, (3) any written order for a minor change in the Work issued pursuant to Paragraph 12.4, or (4) failure of payment by the Owner pursuant to Paragraph 9.7, the Contractor shall make such claim as provided in Subparagraph 12.3.1.

12.4 MINOR CHANGES IN THE WORK

12.4.1 The Architect will have authority to order minor changes in the Work not involving an adjustment in the Contract Sum or an extension of the Contract Time and not inconsistent with the intent of the Contract Documents. Such changes shall be effected by written order, and shall be binding on the Owner and the Contractor.

The Contractor shall carry out such written orders promptly.

ARTICLE 13

UNCOVERING AND CORRECTION OF WORK

13.1 UNCOVERING OF WORK

13.1.1 If any portion of the Work should be covered contrary to the request of the Architect or to requirements specifically expressed in the Contract Documents, it must, if required in writing by the Architect, be uncovered for his observation and shall be replaced at the Contractor's expense.

13.1.2 If any other portion of the Work has been covered which the Architect has not specifically requested to observe prior to being covered, the Architect may request to see such Work and it shall be uncovered by the Contractor. If such Work be found in accordance with the Contract Documents, the cost of uncovering and replacement shall, by appropriate Change Order, be charged to the Owner. If such Work be found not in accordance with the Contract Documents, the Contractor shall pay such costs unless it be found that this condition was caused by the Owner or a separate contractor as provided in Article 6, in which event the Owner shall be responsible for the payment of such costs.

13.2 CORRECTION OF WORK

13.2.1 The Contractor shall promptly correct all Work rejected by the Architect as defective or failing to conform to the Contract Documents whether observed before or after Substantial Completion and whether or not fabricated, installed or completed. The Contractor shall bear all costs of correcting such rejected Work, including compensation for the Architect's additional services made necessary thereby.

13.2.2 If, within one year after the Date of Substantial Completion of the Work or designated portion thereof or within one year after acceptance by the Owner of designated equipment or within such longer period of time as may be prescribed by law or by the terms of any applicable special warranty required by the Contract Documents, any of the Work is found to be defective or not in accordance with the Contract Documents, the Contractor shall correct it promptly after receipt of a written notice from the Owner to do so unless the Owner has previously given the Contractor a written acceptance of such condition. This obligation shall survive termination of the Contract. The Owner shall give such notice promptly after discovery of the condition.

13.2.3 The Contractor shall remove from the site all portions of the Work which are defective or nonconforming and which have not been corrected under Subparagraphs 4.5.1, 13.2.1 and 13.2.2, unless removal is waived by the Owner.

13.2.4 If the Contractor fails to correct defective or nonconforming Work as provided in Subparagraphs 4.5.1, 13.2.1 and 13.2.2, the Owner may correct it in accordance with Paragraph 3.4.

13.2.5 If the Contractor does not proceed with the correction of such defective or non-conforming Work within a reasonable time fixed by written notice from the Architect, the Owner may remove it and may store the materials or equipment at the expense of the Contractor. If the Contractor does not pay the cost of such removal and storage within ten days thereafter, the Owner may upon ten additional days' written notice sell such Work at auction or at private sale and shall account for the net proceeds thereof, after deducting all the costs that should have been borne by the Contractor, including compensation for the Architect's additional services made necessary thereby. If such proceeds of sale do not cover all costs which the Contractor should have borne, the difference shall be charged to the Contractor and an appropriate Change Order shall be issued. If the payments then or thereafter due the Contractor are not sufficient to cover such amount, the Contractor shall pay the difference to the Owner.

13.2.6 The Contractor shall bear the cost of making good all work of the Owner or separate contractors destroyed or damaged by such correction or removal.

13.2.7 Nothing contained in this Paragraph 13.2 shall be construed to establish a period of limitation with respect to any other obligation which the Contractor might have under the Contract Documents, including Paragraph 4.5 hereof. The establishment of the time period of one year after the Date of Substantial Completion or such longer period of time as may be prescribed by law or by the terms of any warranty required by the Contract Documents relates only to the specific obligation of the Contractor to correct the Work, and has no relationship to the time within which his obligation to comply with the Contract Documents may be sought to be enforced, nor to the time within which proceedings may be commenced to establish the Contractor's liability with respect to his obligations other than specifically to correct the Work.

13.3 ACCEPTANCE OF DEFECTIVE OR NON-CONFORMING WORK

13.3.1 If the Owner prefers to accept defective or non-conforming Work, he may do so instead of requiring its removal and correction, in which case a Change Order will be issued to reflect a reduction in the Contract Sum where appropriate and equitable. Such adjustment shall be effected whether or not final payment has been made.

ARTICLE 14

TERMINATION OF THE CONTRACT

14.1 TERMINATION BY THE CONTRACTOR

14.1.1 If the Work is stopped for a period of thirty days under an order of any court or other public authority having jurisdiction, or as a result of an act of government, such as a declaration of a national emergency making materials unavailable, through no act or fault of the Contractor or a Subcontractor or their agents or employees or any other persons performing any of the Work under a contract with the Contractor, or if the Work should be stopped for a period of thirty days by the Contractor because the Architect has not issued a Certificate for Payment as provided in Paragraph 9.7 or because the Owner has not made payment thereon as provided in Paragraph 9.7, then the Contractor may, upon seven additional days' written notice to the Owner and the Architect, terminate the Contract and recover from the Owner payment for all Work executed and for any proven loss sustained upon any materials, equipment, tools, construction equipment and machinery, including reasonable profit and damages.

14.2 TERMINATION BY THE OWNER

14.2.1 If the Contractor is adjudged a bankrupt, or if he makes a general assignment for the benefit of his creditors, or if a receiver is appointed on account of his insolvency, or if he persistently or repeatedly refuses or fails, except in cases for which extension of time is provided, to supply enough properly skilled workmen or proper materials, or if he fails to make prompt payment to Subcontractors or for materials or labor, or persistently disregards laws, ordinances, rules, regulations or orders of any public authority having jurisdiction, or otherwise is guilty of a substantial violation of a provision of the Contract Documents, then the Owner, upon certification by the Architect that sufficient cause exists to justify such action, may, without prejudice to any right or remedy and after giving the Contractor and his surety, if any, seven days' written notice, terminate the employment of the Contractor and take possession of the site and of all materials, equipment, tools, construction equipment and machinery thereon owned by the Contractor and may finish the Work by whatever method he may deem expedient. In such case the Contractor shall not be entitled to receive any further payment until the Work is finished.

14.2.2 If the unpaid balance of the Contract Sum exceeds the costs of finishing the Work, including compensation for the Architect's additional services made necessary thereby, such excess shall be paid to the Contractor. If such costs exceed the unpaid balance, the Contractor shall pay the difference to the Owner. The amount to be paid to the Contractor or to the Owner, as the case may be, shall be certified by the Architect, upon application, in the manner provided in Paragraph 9.4, and this obligation for payment shall survive the termination of the Contract.

THE AMERICAN INSTITUTE OF ARCHITECTS

AIA Document A201/SC

Federal Edition

Supplementary Conditions of the Contract for Construction

THIS DOCUMENT HAS IMPORTANT LEGAL CONSEQUENCES; CONSULTATION WITH
AN ATTORNEY IS ENCOURAGED WITH RESPECT TO ITS COMPLETION OR MODIFICATION

TABLE OF ARTICLES

This document may be used for U.S. Department of Health, Education, and Welfare
Federally Assisted Construction Projects.

AIA DOCUMENTS A201/SC • SUPPLEMENTARY CONDITIONS OF THE CONTRACT FOR CONSTRUCTION • FEDERAL EDITION •
AIA® • AUGUST 1972 EDITION • THE AMERICAN INSTITUTE OF ARCHITECTS, 1735 NEW YORK AVE., N.W., WASHINGTON, D. C. 20006

1

SUPPLEMENTARY CONDITIONS OF THE CONTRACT FOR CONSTRUCTION

ARTICLE 15

MODIFICATIONS OF THE GENERAL CONDITIONS

15.1 MODIFICATION OF PARAGRAPH 1.1, DEFINITIONS

15.1.1 Revise the first sentence of Subparagraph 1.1.1 as set forth below:

The Contract Documents consist of the Agreement, the Conditions of the Contract (General, Supplementary and other Conditions), Performance Bond, Labor and Material Payment Bond, the Drawings, the Specifications, all Addenda issued prior to execution of the Contract, and all Modifications thereto.

15.2 MODIFICATION OF PARAGRAPH 4.8, CASH ALLOWANCES

15.2.1 Substitute the following for Subparagraph 4.8.1 as set forth below:

4.8.1 The Contractor shall include in his proposal the cash allowances stated in the Specifications. These stated allowances represent the net cost estimate of the materials and equipment delivered and unloaded at the site, and all applicable taxes. The Contractor's handling costs on the site, labor, installation costs, overhead, profit and other expenses contemplated for the cash allowance material and equipment shall be included in the Contract Sum since they are not included in the cash allowance estimates. The Contractor shall purchase the cash allowance materials and equipment as directed by the Architect on the basis of the lowest responsive bid of at least three competitive bids. If the actual cost of the materials and equipment delivered and unloaded at the site and all applicable taxes is more or less than the cash allowance estimates, the Contract Sum will be adjusted accordingly by Change Order.

15.3 MODIFICATION OF PARAGRAPH 5.1, DEFINITIONS

15.3.1 Add Subparagraph 5.1.4 as set forth below:

5.1.4 The Contractor may utilize the services of only those Subcontractors who have not been disqualified under existing Federal laws and regulations from participating in Federally assisted construction projects.

15.4 MODIFICATION OF PARAGRAPH 5.2, AWARD OF SUBCONTRACTS AND OTHER CONTRACTS FOR PORTIONS OF THE WORK

15.4.1 Substitute the following for Subparagraph 5.2.1 as set forth below:

5.2.1 Unless otherwise specified in the Contract Documents or in the Instructions to Bidders, the Contractor, as soon as practicable after the award

of the Contract, shall furnish to the Architect in writing for acceptance by the Owner and the Architect a list of the names of the Subcontractors proposed for all portions of the Work. The Architect shall promptly notify the Contractor in writing if either the Owner or the Architect, after due investigation, has reasonable objection to any Subcontractor on such list and does not accept him. Failure of the Owner or Architect to make objection promptly to any Subcontractor on the list shall constitute acceptance of such Subcontractor.

15.5 MODIFICATION OF PARAGRAPH 7.5, PERFORMANCE BOND AND LABOR AND MATERIAL PAYMENT BOND

15.5.1 Substitute the following for Subparagraph 7.5.1 as set forth below:

7.5.1 The Contractor shall furnish a Performance Bond in an amount equal to one hundred percent (100%) of the Contract Sum as security for the faithful performance of this Contract and also a Labor and Material Payment Bond in an amount not less than one hundred percent (100%) of the Contract Sum or in a penal sum not less than that prescribed by State, Territorial or local law, as security for the payment of all persons performing labor on the Project under this Contract and furnishing materials in connection with this Contract. The Performance Bond and the Labor and Material Payment Bond may be in one or in separate instruments in accordance with local law and shall be delivered to the Owner not later than the date of execution of the Contract.

15.6 MODIFICATION OF PARAGRAPH 11.1, CONTRACTOR'S LIABILITY INSURANCE

15.6.1 Add the following Subparagraph 11.1.4 at the end of Paragraph 11.1, Contractor's Liability Insurance:

11.1.4 The Contractor's Comprehensive General Liability Insurance and Automobile Liability Insurance required by Subparagraph 11.1.1 shall be in an amount not less than _____

_____ dollars ($) for injuries, including accidental death, to any one person and subject to the same limit for each person, and in an amount not less than _____

_____ dollars ($) on account of one occurrence. The Contractor's Property Damage Liability Insurance shall be in an amount not less than _____

_____ dollars ($). The Contractor shall either (1) require each of his Subcontractors to procure and to maintain during the life of his Sub-

AIA DOCUMENTS A201/SC • SUPPLEMENTARY CONDITIONS OF THE CONTRACT FOR CONSTRUCTION • FEDERAL EDITION •
AIA® • AUGUST 1972 EDITION • THE AMERICAN INSTITUTE OF ARCHITECTS, 1735 NEW YORK AVE., N.W., WASHINGTON, D. C. 20006

2

214

contract, Subcontractors' Comprehensive General Liability, Automobile Liability, and Property Damage Liability Insurance of the type and in the same amounts as specified in this Subparagraph, or (2) insure the activity of his Subcontractors in his own policy. The Contractor's and his Subcontractors' liability insurance shall include adequate protection against the following special hazards:

(List above special hazards, if any.)

15.7 MODIFICATION OF PARAGRAPH 11.3, PROPERTY INSURANCE

15.7.1 Revise the first sentence of Subparagraph 11.3.1 as set forth below:

11.3.1 Until the Work is completed and accepted by the Owner, the _____* shall purchase and maintain property insurance upon the entire Work at the site to the full insurable value thereof.

(The Owner, at his option, must insert either 'Owner' or 'Contractor').

ARTICLE 16

ADDITIONAL CONDITIONS

16.1 SUBSTITUTION OF MATERIALS AND EQUIPMENT

16.1.1 Whenever a material, article or piece of equipment is identified on the Drawings or in the Specifications by reference to manufacturers' or vendors' names, trade names, catalog numbers, or the like, it is so identified for the purpose of establishing a standard, and any material, article, or piece of equipment of other manufacturers or vendors which will perform adequately the duties imposed by the general design will be considered equally acceptable provided the material, article, or piece of equipment so proposed is, in the opinion of the Architect, of equal substance, appearance and function. It shall not be purchased or installed by the Contractor without the Architect's written approval.

16.2 FEDERAL INSPECTION

16.2.1 The authorized representatives and agents of the Federal Government shall be permitted to inspect all Work, materials, payrolls, records of personnel, invoices of materials, and other relevant data and records.

16.3 LANDS AND RIGHTS-OF-WAY

16.3.1 Prior to the start of construction, the Owner shall obtain all lands and rights-of-way necessary for the execution and completion of Work to be performed under this Contract.

16.4 EQUAL OPPORTUNITY

16.4.1 During the performance of this Contract the Contractor agrees as follows:

.1 The Contractor will not discriminate against any employee or applicant for employment because of race, religion, color, sex or national origin. The Contractor will take affirmative action to insure that applicants are employed and that employees are treated during employment without regard to their race, religion, color, sex or national origin. Such action shall include, but not be limited to, the following: employment, upgrading, demotion or transfer; recruitment or recruitment advertising; layoff or termination; rates of pay or other forms of compensation; and selection for training, including apprenticeship. The Contractor agrees to post in conspicuous places, available to employees and applicants for employment, notices to be provided by an appropriate agency of the Federal Government setting forth the requirements of this Equal Opportunity clause.

.2 The Contractor will, in all solicitations or advertisements for employees placed by or on behalf of the Contractor, state that all qualified applicants will receive consideration for employment without regard to race, religion, color, sex or national origin.

.3 The Contractor will send to each labor union or representative of workers with which he has a collective bargaining agreement or other contract or understanding a notice to be provided by the Owner, advising the labor union or workers' representative of the Contractor's commitments under Section 202 of Executive Order No. 11246 of September 24, 1965, and shall post copies of the notice in conspicuous places available to employees and applicants for employment.

.4 The Contractor will comply with all provisions of Executive Order No. 11246 of September 24, 1965, and of the rules, regulations and relevant orders of the Secretary of Labor.

.5 The Contractor will furnish all information and reports required by Executive Order No. 11246 of September 24, 1965, and by the rules, regulations and orders of the Secretary of Labor, or pursuant thereto, and will permit access to his books, records, and accounts by an appropriate agency of the Federal Government and by the Secretary of Labor for purposes of investigation to ascertain compliance with such rules, regulations and orders.

AIA DOCUMENTS A201/SC • SUPPLEMENTARY CONDITIONS OF THE CONTRACT FOR CONSTRUCTION • FEDERAL EDITION • AIA® • AUGUST 1972 EDITION • THE AMERICAN INSTITUTE OF ARCHITECTS, 1735 NEW YORK AVE., N.W., WASHINGTON, D. C. 20006   **3**

215

.6 In the event of the Contractor's noncompliance with the Equal Opportunity conditions of this Contract or with any of such rules, regulations or orders, this Contract may be cancelled, terminated or suspended in whole or in part, and the Contractor may be declared ineligible for further Government contracts or Federally assisted construction contracts, in accordance with procedures authorized in Executive Order No. 11246 of September 24, 1965, and such other sanctions may be imposed and remedies invoked as provided in said Executive Order, or by rule, regulation or order of the Secretary of Labor, or as provided by law.

.7 The Contractor will include all of Clauses 16.4.1.1 through 16.4.1.7 inclusive in every Subcontract or purchase order unless exempted by rules, regulations or orders of the Secretary of Labor issued pursuant to Section 204 of Executive Order No. 11246 of September 24, 1965, so that such provisions will be binding upon each Subcontractor or vendor. The Contractor will take such action with respect to any Subcontract or purchase order as the appropriate agency of the Federal Government may direct as a means of enforcing such provisions, including sanctions for non-compliance: provided, however, that in the event the Contractor becomes involved in, or is threatened with, litigation with a Subcontractor or vendor as a result of such direction by the appropriate agency of the Federal Government, the Contractor may request the United States to enter into such litigation to protect the interests of the United States.

16.4.2 Exemptions to the above Equal Opportunity conditions are Contracts and Subcontracts not exceeding $10,000, and Contracts and Subcontracts under which Work is performed outside the United States where no recruitment of workers within the United States is involved.

16.4.3 Unless otherwise provided, the above Equal Opportunity provisions are not required to be inserted in Sub-subcontracts except for Sub-subcontracts involving the performance of construction Work at the site of construction, in which case the provisions must be inserted in all such Sub-subcontracts.

16.5 CERTIFICATION OF NONSEGREGATED FACILITIES — (Applicable to Contracts and Subcontracts exceeding $10,000 which are not exempt from the provisions of Paragraph 16.4 "Equal Opportunity" of this Article 16.)

16.5.1 By entering into an agreement related to the Work described in the Contract Documents the Contractor or Subcontractor certifies that he does not maintain or provide for his employees any segregated facilities at any of his establishments, and that he does not permit his employees to perform their services at any location under his control where segregated facilities are maintained. The Contractor or Subcontractor further certifies

that he will not maintain or provide for his employees any segregated facilities at any of his establishments and that he will not permit his employees to perform their services at any location under his control where segregated facilities are maintained. The Contractor or Subcontractor agrees that a breach of this certification is a violation of Paragraph 16.4 "Equal Opportunity". As used herein, the term "segregated facilities" means any waiting rooms, work areas, rest rooms and wash rooms, restaurants and other eating areas, time clocks, locker rooms and other storage or dressing areas, parking lots, drinking fountains, recreation or entertainment areas, transportation, and housing facilities provided for employees on the basis of race, creed, color, or national origin, because of habit, local custom, or otherwise. The Contractor further agrees that (except where he has obtained identical certifications from proposed Subcontractors for specific time periods) he will obtain identical certifications from proposed Subcontractors prior to the award of Subcontracts exceeding $10,000 which are not exempt from the provisions of Paragraph 16.4 "Equal Opportunity"; that he will retain such certifications in his files; and that he will forward the following notice to such proposed Subcontractors (except where the proposed Subcontractors have submitted identical certifications for specific time periods):

"NOTICE TO PROSPECTIVE SUBCONTRACTORS OF REQUIREMENT FOR CERTIFICATIONS OF NONSEGREGATED FACILITIES"

A Certification of Nonsegregated Facilities, as required by the May 9, 1967, order (32 Federal Register 7439, May 19, 1967) on Elimination of Segregated Facilities, by the Secretary of Labor, must be submitted prior to the award of a Subcontract exceeding $10,000 which is not exempt from the provisions of Paragraph 16.4 "Equal Opportunity". The Certification may be submitted either for each Subcontract or for all Subcontracts during a period, i.e., quarterly, semiannually or annually.

16.5.2 The penalty for making false statements in Certifications required by Subparagraph 16.5.1 is prescribed in 18 USC 1001.

16.6 PREVAILING WAGES

16.6.1 All mechanics and laborers, including apprentices and trainees, employed or working directly upon the site of the Work shall be paid unconditionally, and not less often than once a week, and without subsequent deduction or rebate on any account [except such payroll deductions as are permitted by the Copeland Regulations (29 Code of Federal Regulations, Part 3)], the full amounts due at time of payment computed at wage rates not less than the aggregate of the basic hourly rates and the rates of payments, contributions, or costs for any fringe benefits contained in the wage determination decision of the Secretary of Labor **which is attached hereto and made a part hereof,** regardless of any contractual relationship which may be alleged to exist between the Contractor or Subcontractor and such laborers and mechanics, including apprentices and trainees, and the wage

4

AIA DOCUMENTS A201/SC • SUPPLEMENTARY CONDITIONS OF THE CONTRACT FOR CONSTRUCTION • FEDERAL EDITION • AIA® • AUGUST 1972 EDITION • THE AMERICAN INSTITUTE OF ARCHITECTS, 1735 NEW YORK AVE., N.W., WASHINGTON, D. C. 20006

216

determination decision shall be posted by the Contractor at the site of the Work in a prominent place where it can easily be seen by the workers.

16.6.2 The Contractor may discharge his obligation under Subparagraph 16.6.1 to workers in any classification for which the wage determination decision contains:

.1 Only a basic hourly rate of pay, by making payment not less than such basic hourly rate, except as otherwise provided in the Copeland Regulations (29 CFR, Part 3); or

.2 Both a basic hourly rate of pay and fringe benefit payments, by making payment in cash, by irrevocably making contributions pursuant to a fund, plan or program for and/or by assuming an enforceable commitment to bear the cost of bona fide fringe benefits contemplated by the Davis-Bacon Act, or by any combination thereof. These fringe benefit payments can be discharged only by making contributions to the same type or types of fringe benefits listed in the applicable determination. Contributions made, or costs assumed, on other than a weekly basis shall be considered as having been constructively made or assumed during a weekly period to the extent that they apply to such period. Where a fringe benefit is expressed in a wage determination in any manner other than as an hourly rate and the Contractor pays a cash equivalent or provides an alternative fringe benefit, he shall furnish information with his payrolls showing how he determined that the cost incurred to make the cash payment or to provide the alternative fringe benefit is equal to the cost of the wage determination fringe benefit. In the event of disagreement between or among the interested parties as to an equivalent of any fringe benefit, the Owner shall submit the question together with his recommendation through the appropriate Federal agency to the Secretary of Labor for final determination.

16.6.3 The assumption of an enforceable commitment to bear the cost of fringe benefits listed in the wage determination decision forming a part of the Contract may be considered as payment of wages only with the approval of the Secretary of Labor pursuant to a written request by the Contractor. The Secretary of Labor may require the Contractor to set aside assets, in a separate account, to meet his obligations under any unfunded plan or program.

16.6.4 The Owner shall require that any class of laborers or mechanics, including apprentices and trainees, which is not listed in the wage determination and which is to be employed under the Contract shall be classified or reclassified conformably to the wage determination and a report of the action taken shall be sent to the appropriate Federal agency. If the interested parties cannot agree on the proper classification or reclassification of a particular class of laborers or mechanics, including apprentices and trainees, to be used, the Owner shall submit the question

together with his recommendations through the appropriate Federal agency to the Secretary of Labor for final determination.

16.6.5 In the event it is found by the Owner that any laborer or mechanic, including apprentices and trainees, employed by the Contractor or any Subcontractor directly on the site of the Work has been or is being paid at a rate of wages less than the rate of wages required by Subparagraph 16.6.1, the Owner may (1) by written notice to the prime Contractor terminate his right to proceed with the Work, or such part of the Work as to which there has been a failure to pay said required wages, and (2) prosecute the Work to completion by Contract or otherwise, whereupon such Contractor and his sureties shall be liable to the Owner for any excess costs occasioned the Owner thereby.

16.7 CONTRACT WORK HOURS AND SAFETY STANDARDS ACT — OVERTIME COMPENSATION AND SAFETY STANDARDS
(40 United States Code 327-330)

16.7.1 The Contractor shall not require or permit any laborer or mechanic, including apprentices and trainees, in any work-week in which he is employed on any Work under this Contract to work in excess of 8 hours in any calendar day or in excess of 40 hours in such work-week on Work subject to the provisions of the Contract Work Hours and Safety Standards Act unless such laborer or mechanic, including apprentices and trainees, receives compensation at a rate not less than one and one-half times his basic rate of pay for all such hours worked in excess of 8 hours in any calendar day or in excess of 40 hours in such work-week, whichever is the greater number of overtime hours. The "basic rate of pay" as used in this provision shall be the amount paid per hour, exclusive of the Contractor's contribution or cost for fringe benefits, and any cash payment made in lieu of providing fringe benefits, or the basic hourly rate contained in the wage determination, whichever is greater.

16.7.2 In the event of any violation of the provisions of Subparagraph 16.7.1, the Contractor shall be liable to any affected employee for any amounts due, and to the United States for liquidated damages. Such liquidated damages shall be computed with respect to each individual laborer or mechanic, including apprentices and trainees, employed in violation of the provisions of Subparagraph 16.7.1 in the sum of $10 for each calendar day on which such employee was required or permitted to be employed on such Work in excess of 8 hours or in excess of the standard work-week of 40 hours without payment of the overtime wages required by Subparagraph 16.7.1.

16.7.3 The Contractor shall not require or permit any laborer or mechanic, including apprentices and trainees, employed in the performance of this contract to work in surroundings or under conditions which are unsanitary, hazardous, or dangerous to his health as determined under construction safety and health standards promulgated by the Secretary of Labor by regulation (29 CFR Part 1926, 36 FR 7340, April 17, 1971) pursuant to Section 107 of the Contract Work Hours and Safety Standards Act.

AIA DOCUMENTS A201/SC • SUPPLEMENTARY CONDITIONS OF THE CONTRACT FOR CONSTRUCTION • FEDERAL EDITION •
AIA® • AUGUST 1972 EDITION • THE AMERICAN INSTITUTE OF ARCHITECTS, 1735 NEW YORK AVE., N.W., WASHINGTON, D. C. 20006   **5**

217

16.8 APPRENTICES AND TRAINEES

16.8.1 Apprentices will be permitted to work as such only when they are registered, individually under a bona fide apprenticeship program registered with a State apprenticeship agency which is recognized by the Bureau of Apprenticeship and Training, U.S. Department of Labor; or, if no such recognized agency exists in a State, under a program registered with the Bureau of Apprenticeship and Training, U.S. Department of Labor. The allowable ratio of apprentices to journeymen in any craft classification shall not be greater than the ratio permitted to the Contractor as to his entire work force under the registered program. Any employee listed on a payroll at an apprentice wage rate who is not a trainee as defined in Subparagraph 16.8.2 or is not registered as above, shall be paid the wage rate determined by the Secretary of Labor for the classification of Work he actually performed. The Contractor or Subcontractor will be required to furnish to the Owner written evidence of the registration of his program and apprentices as well as of the appropriate ratios and wage rates for the area of construction prior to using any apprentices on the Contract Work.

16.8.2 Trainees will be permitted to work as such when they are bona fide trainees employed pursuant to a program approved by the U.S. Department of Labor, Manpower Administration, Bureau of Apprenticeship and Training and, where Subparagraph 16.8.3 is applicable, in accordance with the provisions of Paragraph 16.8.

16.8.3 On contracts in excess of $10,000 the employment of all laborers and mechanics, including apprentices and trainees, shall also be subject to the provisions of Paragraph 16.8.

16.8.4 The Contractor agrees:

.1 That he will make a diligent effort to hire for the performance of the Contract, a number of apprentices or trainees, or both, in each occupation, which bears to the average number of the journeymen in that occupation to be employed in the performance of the Contract the applicable ratio as determined by the Secretary of Labor;

.2 That he will assure that 25 percent of such apprentices or trainees in each occupation are in their first-year of training, where feasible. Feasibility here involves a consideration of (1) the availability of training opportunities for first-year apprentices; (2) the hazardous nature of the Work for beginning workers; and (3) excessive unemployment of apprentices in their second and subsequent years of training.

.3 That during the performance of the Contract he will, to the greatest extent possible, employ the number of apprentices or trainees necessary to meet currently the requirements of Subparagraph 16.8.4.1 and 16.8.4.2.

.4 To maintain records of employment by trade of the number of apprentices and trainees, apprentices and trainees by first-year of training, and of journeymen, and the wages paid and hours of work of such apprentices, trainees, and journeymen. The Contractor also agrees to make these records available for inspection upon request of the Department of Labor and/or the Department of Health, Education, and Welfare.

.5 That if he claims compliance based on the criterion in Subparagraph 16.8.5.1(2) to maintain records of employment, as described in Subparagraph 16.8.4.4 during the performance of this Contract in the same labor market area, and to make these records available for inspection upon request of the Department of Labor and the Department of Health, Education, and Welfare.

.6 To supply one copy of the written notices required in accordance with Subparagraph 16.8.5.1(3) at the request of Federal agency compliance officers. The Contractor also agrees to supply at 3-month intervals during performance of the Contract and after completion of contract performance, a statement describing steps taken toward making a diligent effort and containing a breakdown by craft, of hours worked and wages paid for first-year apprentices and trainees, other apprentices and trainees and journeymen. One copy of the statement will be sent to the Department of Health, Education, and Welfare and one to the Secretary of Labor.

.7 To insert in any Subcontract under this Contract the requirements contained in Paragraph 16.8 including the criteria for measuring diligent effort contained in Subparagraph 16.8.5. The term "Contractor" as used in such clauses in any Subcontract shall mean the Subcontractor.

.8 That the provisions of Paragraph 16.8.4 shall not apply with regard to any Contract if the Secretary of Health, Education, and Welfare finds it likely that making of the Contract with the clauses contained in Paragraph 16.8.4 will prejudice the national security.

16.8.5 CRITERIA FOR MEASURING DILIGENT EFFORT

A Contractor will be deemed to have made a "diligent effort" as required by Subparagraph 16.8.4 if during the performance of his Contract he accomplishes at least one of the following three objectives:

(1) The Contractor employs on this Project a number of apprentices and trainees by craft as required by the contract clauses at least equal to the ratios established in accordance with Subparagraph 16.8.6.

(2) The Contractor employs on all his public and private construction work combined in the labor market area of this Project, an average number of apprentices and trainees by craft as required by the contract clauses, at least equal to the ratios established in accordance with Subparagraph 16.8.6.

(3) .1 Before commencement of work on the Project, the Contractor if covered by a collective bargaining agreement will give

AIA DOCUMENTS A201/SC • SUPPLEMENTARY CONDITIONS OF THE CONTRACT FOR CONSTRUCTION • FEDERAL EDITION • AIA® • AUGUST 1972 EDITION • THE AMERICAN INSTITUTE OF ARCHITECTS, 1735 NEW YORK AVE., N.W., WASHINGTON, D. C. 20006 written notice to all joint apprenticeship committees; the local U.S. Employment Security Office; local chapter of the Urban League, or other local organizations concerned with minority employment; and the Bureau of Apprenticeship and Training Representative, U.S. Department of Labor, for the locality. The Contractor if not covered by a collective bargaining agreement will give written notice to all the groups stated above except joint apprenticeship committees; this Contractor also will notify all non-joint apprenticeship sponsors in the labor market area.

.2 The notice will include at least the Contractor's name and address, the job site address, value of contract, expected starting and completion dates, the estimated average number of employees in each occupation to be employed over the duration of the Contract, and a statement of his willingness to employ a number of apprentices and trainees at least equal to the ratios established in accordance with Paragraph 16.8.6.

.3 The Contractor must employ all qualified applicants referred to him through normal channels (such as the Employment Service, the Joint Apprenticeship Committees and, where applicable, minority organizations and apprentice outreach programs which have been delegated this function) at least up to the number of such apprentices and trainees required by the applicable provision of Subparagraph 16.8.6.

16.8.6 DETERMINATION OF RATIOS OF APPRENTICES OR TRAINEES TO JOURNEYMEN

The Secretary of Labor has determined that the applicable ratios of apprentices and trainees to journeymen in any occupation shall be as follows:

(1) In any occupation the applicable ratio of apprentices and trainees to journeymen shall be equal to the predominant ratio for the occupation in the area where the construction is to be undertaken, as set forth in collective bargaining agreements, or other employment agreements, and available through the Regional Manager for the Bureau of Apprenticeship and Training for the applicable area.

(2) For any occupation for which no such ratio is found, the ratio of apprentices and trainees to journeymen shall be determined by the the Contractor in accordance with the recommendations set forth in the standards of the National Joint Apprentice Committee for the occupation, which are filed with the U.S. Department of Labor's Bureau of Apprenticeship and Training.

(3) For any occupation for which no such recommendations are found, the ratio of apprentices and trainees to journeymen shall be at least one apprentice or trainee for every five journeymen.

16.8.7 VARIATIONS, TOLERANCES, AND EXEMPTIONS

Variations, tolerances and exemptions from any requirement of this part with respect to any Contract or Subcontract may be granted when such action is necessary and proper in the public interest, or to prevent injustice or undue hardship. A request for a variation, tolerance, or exemption may be made in writing by any interested person to the Department of Health, Education, and Welfare.

16.8.8 ENFORCEMENT

.1 The Department of Health, Education, and Welfare shall insure that the contract clauses required by Paragraph 16.8, are inserted in every Federally assisted construction contract subject thereto. The Department of Health, Education, and Welfare shall also promulgate regulations and procedures necessary to insure that contracts for the construction work subject to Paragraph 16.8, will contain the clauses required thereby.

.2 Enforcement activities, including the investigation of complaints of violations to assure compliance with the requirements of this Paragraph, shall be the primary duty of the Department of Health, Education, and Welfare. The Department of Labor will coordinate its efforts with the Department of Health, Education, and Welfare as may be necessary to assure consistent enforcement of the requirements of this Paragraph.

16.9 PAYROLLS AND BASIC RECORDS

16.9.1 The Contractor shall maintain payrolls and basic records relating thereto during the course of the Work and shall preserve them for a period of three years thereafter for all laborers and mechanics, including apprentices and trainees, working at the site of the Work. Such records shall contain the name and address of each employee, his correct classification, rate of pay (including rates of contributions for, or costs assumed to provide, fringe benefits), daily and weekly number of hours worked, deductions made and actual wages paid. Whenever the Contractor has obtained approval from the Secretary of Labor as provided in Subparagraph 16.6.3, he shall maintain records which show the commitment, its approval, written communication of the plan or program to the laborers or mechanics, including apprentices and trainees, affected, and the costs anticipated or incurred under the plan or program.

16.9.2 The Contractor shall submit weekly a copy of all payrolls to the Owner. The prime Contractor shall be responsible for the submission of copies of payrolls of all Subcontractors. Each such copy shall be accompanied by a statement signed by the Contractor indicating that the

AIA DOCUMENTS A201/SC • SUPPLEMENTARY CONDITIONS OF THE CONTRACT FOR CONSTRUCTION • FEDERAL EDITION •
AIA® • AUGUST 1972 EDITION • THE AMERICAN INSTITUTE OF ARCHITECTS, 1735 NEW YORK AVE., N.W., WASHINGTON, D. C. 20006   7

219

payrolls are correct and complete, that the wage rates contained therein are not less than those determined by the Secretary of Labor, and that the classifications set forth for each laborer or mechanic, including apprentices and trainees, conform with the Work he performed. Submission of the "Weekly Statement of Compliance" required under this Contract and the Copeland Regulations of the Secretary of Labor (29 CFR, Part 3) shall satisfy the requirement for submission of the above statement. The Contractor shall submit also a copy of any approval by the Secretary of Labor with respect to fringe benefits which is required by Subparagraph 16.6.3.

16.9.3 The Contractor shall make the records required under Subparagraphs 16.9.1 and 16.9.2 available for inspection by authorized representatives of the Owner, the State, the appropriate Federal agency and the U.S. Department of Labor, and shall permit such representatives to interview employees during working hours on the job.

16.10   COMPLIANCE WITH COPELAND REGULATIONS

16.10.1   The Contractor shall comply with the Copeland Regulations of the Secretary of Labor (29 CFR, Part 3) which are incorporated herein by reference. In addition, the Weekly Statement of Compliance required by these Regulations shall also contain a statement that the fringe benefits paid are equal to or greater than those set forth in the minimum wage decision.

16.11   WITHHOLDING OF FUNDS

16.11.1   The Owner may withhold or cause to be withheld from the prime Contractor so much of the accrued payments or advances as may be considered necessary (1) to pay the laborers and mechanics, including apprentices and trainees, employed by the Contractor or any Subcontractor on the Work the full amount of wages required by the Contract, and (2) to satisfy any liability of any Contractor for liquidated damages under Paragraph 16.7 hereof entitled "Contract Work Hours and Safety Standards Act — Overtime Compensation and Safety Standards (40 USC 327-330)".

16.11.2   If the Contractor or any Subcontractor fails to pay any laborer or mechanic, including apprentices and trainees, employed or working on the site of the Work, all or part of the wages required by the Contract, the Owner may, after written notice to the prime Contractor, take such action as may be necessary to cause suspension of any further payments or advances until such violations have ceased.

16.12   SUBCONTRACTS

16.12.1   The Contractor will insert in all Subcontracts Paragraphs 16.6 through 16.12 inclusive, respectively entitled "Prevailing Wages", "Contract Work Hours and Safety Standards Act — Overtime Compensation and Safety Standards (40 USC 327-330)", "Apprentices and Trainees", "Payrolls and Basic Records", "Compliance with Copeland Regulations", "Withholding of Funds", "Subcontracts" and "Contract Termination — Debarment", and shall further require all Subcontractors to in-

corporate physically these same Paragraphs in all Sub-subcontracts.

16.12.2   The term "Contractor" as used in such Paragraphs in any Subcontract shall be deemed to refer to the Subcontractor except when the phrase "prime Contractor" is used.

16.13   CONTRACT TERMINATION — DEBARMENT

16.13.1   A breach of Paragraphs 16.6 through 16.12 inclusive, respectively entitled "Prevailing Wages", "Contract Work Hours and Safety Standards Act — Overtime Compensation and Safety Standards (40 USC 327-330)", "Apprentices and Trainees", "Payrolls and Basic Records", "Compliance with Copeland Regulations", "Withholding of Funds" and "Subcontracts", may be grounds for termination of the Contract and for debarment as provided in 29 CFR 5.6.

16.14   USE AND OCCUPANCY OF PROJECT PRIOR TO ACCEPTANCE BY THE OWNER

16.14.1 The Contractor agrees to use and occupancy of a portion or unit of the Project before formal acceptance by the Owner under the following conditions:

.1   A Certificate of Substantial Completion shall be prepared and executed as provided in Subparagraph 9.7.1 of the accompanying General Conditions of the Contract for Construction, except that when, in the opinion of the Architect, the Contractor is chargeable with unwarranted delay in completing Work or other Contract requirements the signature of the Contractor will not be required. The Certificate of Substantial Completion shall be accompanied by a written endorsement of the Contractor's insurance carrier and surety permitting occupancy by the Owner during the remaining period of Project Work.

.2   Occupancy by the Owner shall not be construed by the Contractor as being an acceptance of that part of the Project to be occupied.

.3   The Contractor shall not be held responsible for any damage to the occupied part of the Project resulting from the Owner's occupancy.

.4   Occupancy by the Owner shall not be deemed to constitute a waiver of existing claims in behalf of the Owner or Contractor against each other.

.5   If the Project consists of more than one building, and one of the buildings is to be occupied, the Owner, prior to occupancy of the building, shall secure permanent property insurance on the building to be occupied and necessary permits which may be required for use and occupancy.

16.14.2 With the exception of Clause 16.14.1.5, use and occupancy by the Owner prior to Project acceptance does not relieve the Contractor of his responsibility to maintain all insurance and bonds required of the Contractor under the Contract until the Project is completed and accepted by the Owner.

8

AIA DOCUMENTS A201/SC • SUPPLEMENTARY CONDITIONS OF THE CONTRACT FOR CONSTRUCTION • FEDERAL EDITION • AIA® • AUGUST 1972 EDITION • THE AMERICAN INSTITUTE OF ARCHITECTS, 1735 NEW YORK AVE., N.W., WASHINGTON, D. C. 20006

220

16.15 ENUMERATION OF THE DRAWINGS, THE SPECIFICATIONS AND ADDENDA

16.15.1 Following are the Drawings, Specifications and Addenda which form a part of this Contract, as set forth in Subparagraph 1.1.1 of the accompanying General Conditions of the Contract for Construction.

(List below all Drawings, Specifications and Addenda, showing page or sheet numbers and dates where applicable.)

DRAWINGS

SPECIFICATIONS: **DATED** _____

ADDENDA:

AIA DOCUMENTS A201/SC • SUPPLEMENTARY CONDITIONS OF THE CONTRACT FOR CONSTRUCTION • FEDERAL EDITION •
AIA® • AUGUST 1972 EDITION • THE AMERICAN INSTITUTE OF ARCHITECTS, 1735 NEW YORK AVE., N.W., WASHINGTON, D. C. 20006   **9**

221

THE AMERICAN INSTITUTE OF ARCHITECTS

AIA Document A401

SUBCONTRACT

Standard Form of Agreement Between Contractor and Subcontractor

Use with the latest edition of the appropriate AIA Documents as follows:

A101, Owner-Contractor Agreement — Stipulated Sum
A107, Owner-Contractor Agreement — Short Form for Small Construction Contracts
A111, Owner-Contractor Agreement — Cost plus Fee
A201, General Conditions of the Contract for Construction.

*THIS DOCUMENT HAS IMPORTANT LEGAL CONSEQUENCES; CONSULTATION WITH
AN ATTORNEY IS ENCOURAGED WITH RESPECT TO ITS COMPLETION OR MODIFICATION*

This document has been approved and endorsed by the American Subcontractors Association
and the Associated Specialty Contractors, Inc.

AGREEMENT

made this   day of   in the year Nineteen
Hundred and

BETWEEN the Contractor:

and the Subcontractor:

Project:

Owner:

Architect:

The Contractor and Subcontractor agree as set forth below.

AIA DOCUMENT A401 • CONTRACTOR-SUBCONTRACTOR AGREEMENT • JANUARY 1972 EDITION • AIA®
©1972 • THE AMERICAN INSTITUTE OF ARCHITECTS, 1735 NEW YORK AVE., N.W., WASHINGTON, D.C. 20006

1

ARTICLE 1
THE CONTRACT DOCUMENTS

The Contract Documents for this Subcontract consist of this Agreement and any Exhibits attached hereto, the Agreement between the Owner and Contractor dated   the Conditions of the Contract between the Owner and Contractor (General, Supplementary and other Conditions), Drawings, Specifications, all Addenda issued prior to execution of the Agreement between the Owner and Contractor, and all Modifications issued subsequent thereto.

All of the above documents are a part of this Subcontract and shall be available for inspection by the Subcontractor upon his request.

The only Addenda and Modifications issued prior to the execution of this Subcontract and applicable to it are as follows:

ARTICLE 2
THE WORK

The Subcontractor shall furnish

(Insert above a precise description of the Work covered by this Subcontract and refer to numbers of Drawings and pages of Specifications including Addenda and accepted Alternates.)

ARTICLE 3
TIME OF COMMENCEMENT AND COMPLETION

(Here insert the specific provisions that are applicable to this Subcontract including any information pertaining to notice to proceed or other method of notification for commencement of Work, starting and completion dates, or duration, and any provisions for liquidated damages relating to failure to complete on time.)

Time is of the essence of this Subcontract.
No extension of time will be valid without the Contractor's written consent after claim made by the Subcontractor in accordance with Paragraph 11.4.

ARTICLE 4
THE CONTRACT SUM

The Contractor shall pay the Subcontractor in current funds for the performance of the Work, subject to additions and deductions by Change Order, the total sum of

ARTICLE 5
PROGRESS PAYMENTS

The Contractor shall pay the Subcontractor monthly progress payments in accordance with Paragraphs 12.3 through 12.6 inclusive of this Subcontract.

(Here insert details on unit prices, payment procedures and date of monthly applications, or other procedure if on other than a monthly basis, considera- tion of materials and equipment safely and suitably stored at the site or other location agreed upon in writing, and any provisions for limiting or reducing the amount retained after the Work reaches a certain stage of completion which should be consistent with the Contract Documents.)

Applications for monthly progress payments shall be in writing and in accordance with Subparagraph 11.2.1, shall state the estimated percentage of the Work in this Subcontract that has been satisfactorily completed and shall be submitted to the Contractor on or before the   day of each month.

ARTICLE 6
FINAL PAYMENT

Final payment shall be due when the Work described in this Subcontract is fully completed and performed in ac- cordance with the Contract Documents and is satisfactory to the Architect. Such payment shall be in accordance with Article 5 and with Paragraphs 12.3 through 12.6 inclusive of this Contract.

Before issuance of the final payment the Subcontractor, if required, shall submit evidence satisfactory to the Con- tractor that all payrolls, bills for materials and equipment, and all known indebtedness connected with the Sub- contractor's Work have been satisfied.

ARTICLE 7
PERFORMANCE AND LABOR
AND MATERIAL PAYMENT BONDS

(Here insert any requirement for the furnishing of bonds by the Subcontractor.)

ARTICLE 8
TEMPORARY FACILITIES AND SERVICES

Unless otherwise provided in this Subcontract, the Contractor shall furnish and make available at no cost to the Subcontractor the following temporary facilities and services:

ARTICLE 9
INSURANCE

Prior to starting work, the Subcontractor shall obtain the required insurance from a responsible insurer, and shall furnish satisfactory evidence to the Contractor that the Subcontractor has complied with the requirements of this Article 9. Similarly, the Contractor shall furnish to the Subcontractor satisfactory evidence of insurance required by the Contract Documents.

The Contractor and Subcontractor waive all rights against each other and against the Owner and all other Subcontractors for damages caused by fire or other perils to the extent covered by property insurance provided under the General Conditions, except such rights as they may have to the proceeds of such insurance.

(Here insert any insurance requirements and Subcontractor's responsibility for obtaining, maintaining and paying for necessary insurance, not less than limits as may be specified in the Contract Documents, shown below, or required by law. If applicable, this shall include fire insurance and extended coverage, public liability, property damage, employer's liability, and workmen's compensation insurance for the Subcontractor and his employees. The insertion should cover provisions for notice of cancellation, allocation of insurance proceeds, and other aspects of insurance.)

ARTICLE 10
WORKING CONDITIONS

(Here insert any applicable arrangements concerning working conditions and labor matters for the Project.)

ARTICLE 11
SUBCONTRACTOR'S RESPONSIBILITIES

11.1 The Subcontractor shall be bound to the Contractor by the terms of this Agreement and of the Contract Documents between the Owner and Contractor, and shall assume toward the Contractor all the obligations and responsibilities which the Contractor, by those Documents, assumes toward the Owner, and shall have the benefit of all rights, remedies and redress against the Contractor which the Contractor, by those Documents, has against the Owner, insofar as applicable to this Subcontract, provided that where any provision of the Contract Documents between the Owner and Contractor is inconsistent with any provision of this Agreement, this Agreement shall govern.

11.2 The Subcontractor shall submit to the Contractor applications for payment at such times as stipulated in Article 5 to enable the Contractor to apply for payment.

11.2.1 If payments are made on the valuation of Work done, the Subcontractor shall, before the first application, submit to the Contractor a schedule of values of the various parts of the Work aggregating the total sum of this Subcontract, made out in such detail as the Subcontractor and Contractor may agree upon, or as required by the Owner, and supported by such evidence as to its correctness as the Contractor may direct. This schedule, when approved by the Contractor, shall be used as a basis for Applications for Payment, unless it be found to be in error. In applying for payment, the Subcontractor shall submit a statement based upon this schedule.

11.2.2 If payments are made on account of materials or equipment not incorporated in the Work but delivered and suitably stored at the site, or at some other location agreed upon in writing, such payments shall be in accordance with the terms and conditions of the Contract Documents.

11.3 The Subcontractor shall pay for all materials, equipment and labor used in, or in connection with, the performance of this Subcontract through the period covered by previous payments received from the Contractor, and shall furnish satisfactory evidence, when requested by the Contractor, to verify compliance with the above requirements.

11.4 The Subcontractor shall make all claims promptly to the Contractor for additional work, extensions of time, and damage for delays or otherwise, in accordance with the Contract Documents.

11.5 In carrying out his Work the Subcontractor shall take necessary precautions to protect properly the finished work of other trades from damage caused by his operations.

11.6 The Subcontractor shall at all times keep the building and premises clean of debris arising out of the operations of this Subcontract. Unless otherwise provided, the Subcontractor shall not be held responsible for unclean conditions caused by other contractors or subcontractors.

11.7 The Subcontractor shall take all reasonable safety precautions with respect to his Work, shall comply with all safety measures initiated by the Contractor and with all applicable laws, ordinances, rules, regulations and orders of any public authority for the safety of persons or property in accordance with the requirements of the Contract Documents. The Subcontractor shall report within three days to the Contractor any injury to any of the Subcontractor's employees at the site.

11.8 The Subcontractor shall not assign this Subcontract without the written consent of the Contractor, nor subcontract the whole of this Subcontract without the written consent of the Contractor, nor further subcontract portions of this Subcontract without written notification to the Contractor when such notification is requested by the Contractor. The Subcontractor shall not assign any amounts due or to become due under this subcontract without written notice to the Contractor.

11.9 The Subcontractor warrants that all materials and equipment furnished and incorporated by him in the Project shall be new unless otherwise specified, and that all Work under this Subcontract shall be of good quality, free from faults and defects and in conformance with the Contract Documents. All Work not conforming to these standards may be considered defective. The warranty provided in this Paragraph 11.9 shall be in addition to and not in limitation of any other warranty or remedy required by law or by the Contract Documents.

11.10 The Subcontractor agrees that if he should neglect to prosecute the Work diligently and properly or fail to perform any provisions of this Subcontract, the Contractor, after three working days' written notice to the Subcontractor, may, without prejudice to any other remedy he may have, make good such deficiencies and may deduct the cost thereof from the payments then or thereafter due the Subcontractor, provided, however, that if such action is based upon faulty workmanship or materials and equipment, the Architect shall first have determined that the workmanship or materials and equipment are not in accordance with the Contract Documents.

11.11 The Subcontractor agrees that the Contractor's equipment will be available to the Subcontractor only at the Contractor's discretion and on mutually satisfactory terms.

11.12 The Subcontractor shall furnish periodic progress reports on the Work as mutually agreed, including information on the status of materials and equipment under this Subcontract which may be in the course of preparation or manufacture.

11.13 The Subcontractor shall make any and all changes in the Work from the Drawings and Specifications of the Contract Documents without invalidating this Subcontract when specifically ordered to do so in writing by the Contractor. The Subcontractor, prior to the commencement of such changed or revised work, shall submit promptly to the Contractor written copies of the cost or credit proposal for such revised Work in a manner consistent with the Contract Documents.

11.14 The Subcontractor shall cooperate with the Contractor and other subcontractors whose work might interfere with the Subcontractor's Work, and shall participate in the preparation of coordinated drawings in areas of congestion as required by the Contract Documents, specifically noting and advising the Contractor of any such interference.

11.15 The Subcontractor shall cooperate with the Contractor in scheduling and performing his Work to avoid conflict or interference with the work of others.

11.16 The Subcontractor shall promptly submit shop drawings and samples as required in order to perform his work efficiently, expeditiously and in a manner that will not cause delay in the progress of the Work of the Contractor or other Subcontractors.

11.17 The Subcontractor shall give all notices and comply with all laws, ordinances, rules, regulations and orders of any public authority bearing on the performance of the Work under this Subcontract. The Subcontractor shall secure and pay for all permits, fees and licenses necessary for the execution of the Work described in the Contract Documents as applicable to this Subcontract.

11.18 The Subcontractor shall comply with Federal, State and local tax laws, social security acts, unemployment compensation acts and workmen's compensation acts insofar as applicable to the performance of this Subcontract.

11.19 The Subcontractor agrees that all Work shall be done subject to the final approval of the Architect. The Architect's decisions in matters relating to artistic effect shall be final if consistent with the intent of the Contract Documents.

11.20 The Subcontractor shall indemnify and hold harmless the Contractor and all of his agents and employees from and against all claims, damages, losses and expenses including attorney's fees arising out of or resulting from the performance of the Subcontractor's Work under this Subcontract, provided that any such claim, damage, loss, or expense (a) is attributable to bodily injury, sickness, disease, or death, or to injury to or destruction of tangible property (other than the Work itself) including the loss of use resulting therefrom, and (b) is caused in whole or in part by any negligent act or omission of the Subcontractor or anyone directly or indirectly employed by him or anyone for whose acts he may be liable, regardless of whether it is caused in part by a party indemnified hereunder.

11.20.1 In any and all claims against the Contractor or any of his agents or employees by any employee of the Subcontractor, anyone directly or indirectly employed by him or anyone for whose acts he may be liable, the indemnification obligation under this Paragraph 11.20 shall not be limited in any way by any limitation on the amount or type of damages, compensation or benefits payable by or for the Subcontractor under workmen's compensation acts, disability benefit acts or other employee benefit acts.

11.20.2 The obligations of the Subcontractor under this Paragraph 11.20 shall not extend to the liability of the Architect, his agents or employees arising out of (1) the preparation or approval of maps, drawings, opinions, reports, surveys, Change Orders, designs or specifications, or (2) the giving of or the failure to give directions or instructions by the Architect, his agents or employees provided such giving or failure to give is the primary cause of the injury or damage.

ARTICLE 12
CONTRACTOR'S RESPONSIBILITIES

12.1 The Contractor shall be bound to the Subcontractor by the terms of this agreement and of the Contract Documents between the Owner and the Contractor and shall assume toward the Subcontractor all the obligations and responsibilities that the Owner, by those Documents, assumes toward the Contractor, and shall have the benefit of all rights, remedies and redress against the Subcontractor which the Owner, by those Documents, has against the Contractor, insofar as applicable to this Subcontract, provided that where any provision of the Contract Documents between the Owner and the Contractor is inconsistent with any provision in this Agreement, this Agreement shall govern.

12.2 The Contractor shall promptly notify the Subcontractor of all modifications to the Contract between the Owner and the Contractor which affect this Subcontract and which were issued or entered into subsequent to the execution of this Subcontract.

12.3 Unless otherwise provided in the Contract Documents, the Contractor shall pay the Subcontractor each progress payment and the final payment under this Subcontract within three (3) working days after he receives payment from the Owner. The amount of each progress payment to the Subcontractor shall be equal to the percentage of completion allowed to the Contractor for the Work of this Subcontractor applied to the Contract Sum of this Subcontract, plus the amount allowed for materials and equipment suitably stored by the Subcontractor, less the aggregate of previous payments to the Subcontractor and less the percentage retained as provided in this Subcontract.

12.4 The Contractor shall permit the Subcontractor to obtain directly from the Architect evidence of percentages of completion certified on his account.

12.5 Unless otherwise provided in the Contract Documents, if the Architect fails to issue a Certificate for Payment or the Contractor does not receive payment for any cause which is not the fault of the Subcontractor, the Contractor shall pay the Subcontractor, on demand, a progress payment computed as provided in Paragraph 12.3 or the final payment as provided in Article 6.

12.6 The Contractor agrees that if he fails to make payments to the Subcontractor as herein provided for any cause not the fault of the Subcontractor, within seven days from the time payment should be made as provided in Paragraphs 12.3 and 12.5, the Subcontractor may, upon seven days' additional written notice to the Contractor, stop his Work without prejudice to any other remedy he may have.

12.7 The Contractor shall make no demand for liquidated damages for delay in any sum in excess of such amount as may be specifically named in this Subcontract, and no liquidated damages shall be assessed against this Subcontractor for delays or causes attributed to other Subcontractors or arising outside the scope of this Subcontract.

12.8 The Contractor agrees that no claim for payment for services rendered or materials and equipment furnished by the Contractor to the Subcontractor shall be

AIA DOCUMENT A401 • CONTRACTOR-SUBCONTRACTOR AGREEMENT • JANUARY 1972 EDITION • AIA®
©1972 • THE AMERICAN INSTITUTE OF ARCHITECTS, 1735 NEW YORK AVE., N.W., WASHINGTON, D. C. 20006

valid without prior notice to the Subcontractor and unless written notice thereof is given by the Contractor during the first ten days of the calendar month following that in which the claim originated.

12.9 The Contractor shall not give instructions or orders directly to employees or workmen of the Subcontractor except to persons designated as authorized representatives of the Subcontractor.

12.10 The Contractor shall cooperate with the Subcontractor in scheduling and performing his Work to avoid conflicts or interference in the Subcontractor's Work.

12.11 The Contractor shall permit the Subcontractor to be present and to submit evidence in any arbitration proceeding involving his rights.

12.12 The Contractor shall permit the Subcontractor to exercise whatever rights the Contractor may have under the Contract Documents in the choice of arbitrators in any dispute, if the sole cause of the dispute is the Work, materials, equipment, rights or responsibilities of the Subcontractor; or if the dispute involves the Subcontractor and any other Subcontractor or Subcontractors jointly, the Contractor shall permit them to exercise such rights jointly.

ARTICLE 13
INTEREST

Any monies not paid when due to either party under this Subcontract shall bear interest at the legal rate in effect at the place of the Project.

ARTICLE 14
ARBITRATION

14.1 All claims, disputes and other matters in question arising out of, or relating to, this Subcontract, or the breach thereof, shall be decided by arbitration in the same manner and under the same procedure as provided in the Contract Documents with respect to disputes between the Owner and the Contractor except that a decision by the Architect shall not be a condition precedent to arbitration.

14.2 This Article shall not be deemed a limitation on any rights or remedies which the Subcontractor may have under any Federal or State mechanics' lien laws or under any applicable labor and material payment bonds unless such rights or remedies are expressly waived by him.

This Agreement executed the day and year first written above.

CONTRACTOR   SUBCONTRACTOR

THE AMERICAN INSTITUTE OF ARCHITECTS

AIA Document B141

Standard Form of Agreement Between Owner and Architect

1977 EDITION

*THIS DOCUMENT HAS IMPORTANT LEGAL CONSEQUENCES; CONSULTATION WITH
AN ATTORNEY IS ENCOURAGED WITH RESPECT TO ITS COMPLETION OR MODIFICATION*

AGREEMENT

made as of the   day of   in the year of Nineteen
Hundred and

BETWEEN the Owner:

and the Architect:

For the following Project:
(Include detailed description of Project location and scope.)

The Owner and the Architect agree as set forth below.

AIA DOCUMENT B141 • OWNER-ARCHITECT AGREEMENT • THIRTEENTH EDITION • JULY 1977 • AIA® • © 1977
THE AMERICAN INSTITUTE OF ARCHITECTS, 1735 NEW YORK AVENUE, N.W., WASHINGTON, D.C. 20006

B141-1977   1

TERMS AND CONDITIONS OF AGREEMENT BETWEEN OWNER AND ARCHITECT

ARTICLE 1
ARCHITECT'S SERVICES AND RESPONSIBILITIES

BASIC SERVICES

The Architect's Basic Services consist of the five phases described in Paragraphs 1.1 through 1.5 and include normal structural, mechanical and electrical engineering services and any other services included in Article 15 as part of Basic Services.

1.1 SCHEMATIC DESIGN PHASE

1.1.1 The Architect shall review the program furnished by the Owner to ascertain the requirements of the Project and shall review the understanding of such requirements with the Owner.

1.1.2 The Architect shall provide a preliminary evaluation of the program and the Project budget requirements, each in terms of the other, subject to the limitations set forth in Subparagraph 3.2.1.

1.1.3 The Architect shall review with the Owner alternative approaches to design and construction of the Project.

1.1.4 Based on the mutually agreed upon program and Project budget requirements, the Architect shall prepare, for approval by the Owner, Schematic Design Documents consisting of drawings and other documents illustrating the scale and relationship of Project components.

1.1.5 The Architect shall submit to the Owner a Statement of Probable Construction Cost based on current area, volume or other unit costs.

1.2 DESIGN DEVELOPMENT PHASE

1.2.1 Based on the approved Schematic Design Documents and any adjustments authorized by the Owner in the program or Project budget, the Architect shall prepare, for approval by the Owner, Design Development Documents consisting of drawings and other documents to fix and describe the size and character of the entire Project as to architectural, structural, mechanical and electrical systems, materials and such other elements as may be appropriate.

1.2.2 The Architect shall submit to the Owner a further Statement of Probable Construction Cost.

1.3 CONSTRUCTION DOCUMENTS PHASE

1.3.1 Based on the approved Design Development Documents and any further adjustments in the scope or quality of the Project or in the Project budget authorized by the Owner, the Architect shall prepare, for approval by the Owner, Construction Documents consisting of Drawings and Specifications setting forth in detail the requirements for the construction of the Project.

1.3.2 The Architect shall assist the Owner in the preparation of the necessary bidding information, bidding forms, the Conditions of the Contract, and the form of Agreement between the Owner and the Contractor.

1.3.3 The Architect shall advise the Owner of any adjust-

ments to previous Statements of Probable Construction Cost indicated by changes in requirements or general market conditions.

1.3.4 The Architect shall assist the Owner in connection with the Owner's responsibility for filing documents required for the approval of governmental authorities having jurisdiction over the Project.

1.4 BIDDING OR NEGOTIATION PHASE

1.4.1 The Architect, following the Owner's approval of the Construction Documents and of the latest Statement of Probable Construction Cost, shall assist the Owner in obtaining bids or negotiated proposals, and assist in awarding and preparing contracts for construction.

1.5 CONSTRUCTION PHASE—ADMINISTRATION OF THE CONSTRUCTION CONTRACT

1.5.1 The Construction Phase will commence with the award of the Contract for Construction and, together with the Architect's obligation to provide Basic Services under this Agreement, will terminate when final payment to the Contractor is due, or in the absence of a final Certificate for Payment or of such due date, sixty days after the Date of Substantial Completion of the Work, whichever occurs first.

1.5.2 Unless otherwise provided in this Agreement and incorporated in the Contract Documents, the Architect shall provide administration of the Contract for Construction as set forth below and in the edition of AIA Document A201, General Conditions of the Contract for Construction, current as of the date of this Agreement.

1.5.3 The Architect shall be a representative of the Owner during the Construction Phase, and shall advise and consult with the Owner. Instructions to the Contractor shall be forwarded through the Architect. The Architect shall have authority to act on behalf of the Owner only to the extent provided in the Contract Documents unless otherwise modified by written instrument in accordance with Subparagraph 1.5.16.

1.5.4 The Architect shall visit the site at intervals appropriate to the stage of construction or as otherwise agreed by the Architect in writing to become generally familiar with the progress and quality of the Work and to determine in general if the Work is proceeding in accordance with the Contract Documents. However, the Architect shall not be required to make exhaustive or continuous on-site inspections to check the quality or quantity of the Work. On the basis of such on-site observations as an architect, the Architect shall keep the Owner informed of the progress and quality of the Work, and shall endeavor to guard the Owner against defects and deficiencies in the Work of the Contractor.

1.5.5 The Architect shall not have control or charge of and shall not be responsible for construction means, methods, techniques, sequences or procedures, or for safety precautions and programs in connection with the Work, for the acts or omissions of the Contractor, Subcontractors or any other persons performing any of the Work, or for the failure of any of them to carry out the Work in accordance with the Contract Documents.

1.5.6 The Architect shall at all times have access to the Work wherever it is in preparation or progress.

1.5.7 The Architect shall determine the amounts owing to the Contractor based on observations at the site and on evaluations of the Contractor's Applications for Payment, and shall issue Certificates for Payment in such amounts, as provided in the Contract Documents.

1.5.8 The issuance of a Certificate for Payment shall constitute a representation by the Architect to the Owner, based on the Architect's observations at the site as provided in Subparagraph 1.5.4 and on the data comprising the Contractor's Application for Payment, that the Work has progressed to the point indicated; that, to the best of the Architect's knowledge, information and belief, the quality of the Work is in accordance with the Contract Documents (subject to an evaluation of the Work for conformance with the Contract Documents upon Substantial Completion, to the results of any subsequent tests required by or performed under the Contract Documents, to minor deviations from the Contract Documents correctable prior to completion, and to any specific qualifications stated in the Certificate for Payment); and that the Contractor is entitled to payment in the amount certified. However, the issuance of a Certificate for Payment shall not be a representation that the Architect has made any examination to ascertain how and for what purpose the Contractor has used the moneys paid on account of the Contract Sum.

1.5.9 The Architect shall be the interpreter of the requirements of the Contract Documents and the judge of the performance thereunder by both the Owner and Contractor. The Architect shall render interpretations necessary for the proper execution or progress of the Work with reasonable promptness on written request of either the Owner or the Contractor, and shall render written decisions, within a reasonable time, on all claims, disputes and other matters in question between the Owner and the Contractor relating to the execution or progress of the Work or the interpretation of the Contract Documents.

1.5.10 Interpretations and decisions of the Architect shall be consistent with the intent of and reasonably inferable from the Contract Documents and shall be in written or graphic form. In the capacity of interpreter and judge, the Architect shall endeavor to secure faithful performance by both the Owner and the Contractor, shall not show partiality to either, and shall not be liable for the result of any interpretation or decision rendered in good faith in such capacity.

1.5.11 The Architect's decisions in matters relating to artistic effect shall be final if consistent with the intent of the Contract Documents. The Architect's decisions on any other claims, disputes or other matters, including those in question between the Owner and the Contractor, shall be subject to arbitration as provided in this Agreement and in the Contract Documents.

1.5.12 The Architect shall have authority to reject Work which does not conform to the Contract Documents. Whenever, in the Architect's reasonable opinion, it is necessary or advisable for the implementation of the intent of the Contract Documents, the Architect will have authority to require special inspection or testing of the Work in accordance with the provisions of the Contract Documents, whether or not such Work be then fabricated, installed or completed.

1.5.13 The Architect shall review and approve or take other appropriate action upon the Contractor's submittals such as Shop Drawings, Product Data and Samples, but only for conformance with the design concept of the Work and with the information given in the Contract Documents. Such action shall be taken with reasonable promptness so as to cause no delay. The Architect's approval of a specific item shall not indicate approval of an assembly of which the item is a component.

1.5.14 The Architect shall prepare Change Orders for the Owner's approval and execution in accordance with the Contract Documents, and shall have authority to order minor changes in the Work not involving an adjustment in the Contract Sum or an extension of the Contract Time which are not inconsistent with the intent of the Contract Documents.

1.5.15 The Architect shall conduct inspections to determine the Dates of Substantial Completion and final completion, shall receive and forward to the Owner for the Owner's review written warranties and related documents required by the Contract Documents and assembled by the Contractor, and shall issue a final Certificate for Payment.

1.5.16 The extent of the duties, responsibilities and limitations of authority of the Architect as the Owner's representative during construction shall not be modified or extended without written consent of the Owner, the Contractor and the Architect.

1.6 PROJECT REPRESENTATION BEYOND BASIC SERVICES

1.6.1 If the Owner and Architect agree that more extensive representation at the site than is described in Paragraph 1.5 shall be provided, the Architect shall provide one or more Project Representatives to assist the Architect in carrying out such responsibilities at the site.

1.6.2 Such Project Representatives shall be selected, employed and directed by the Architect, and the Architect shall be compensated therefor as mutually agreed between the Owner and the Architect as set forth in an exhibit appended to this Agreement, which shall describe the duties, responsibilities and limitations of authority of such Project Representatives.

1.6.3 Through the observations by such Project Representatives, the Architect shall endeavor to provide further protection for the Owner against defects and deficiencies in the Work, but the furnishing of such project representation shall not modify the rights, responsibilities or obligations of the Architect as described in Paragraph 1.5.

1.7 ADDITIONAL SERVICES

The following Services are not included in Basic Services unless so identified in Article 15. They shall be provided if authorized or confirmed in writing by the Owner, and they shall be paid for by the Owner as provided in this Agreement, in addition to the compensation for Basic Services.

AIA DOCUMENT B141 • OWNER-ARCHITECT AGREEMENT • THIRTEENTH EDITION • JULY 1977 • AIA® • © 1977
THE AMERICAN INSTITUTE OF ARCHITECTS, 1735 NEW YORK AVENUE, N.W., WASHINGTON, D.C. 20006

1.7.1 Providing analyses of the Owner's needs, and programming the requirements of the Project.

1.7.2 Providing financial feasibility or other special studies.

1.7.3 Providing planning surveys, site evaluations, environmental studies or comparative studies of prospective sites, and preparing special surveys, studies and submissions required for approvals of governmental authorities or others having jurisdiction over the Project.

1.7.4 Providing services relative to future facilities, systems and equipment which are not intended to be constructed during the Construction Phase.

1.7.5 Providing services to investigate existing conditions or facilities or to make measured drawings thereof, or to verify the accuracy of drawings or other information furnished by the Owner.

1.7.6 Preparing documents of alternate, separate or sequential bids or providing extra services in connection with bidding, negotiation or construction prior to the completion of the Construction Documents Phase, when requested by the Owner.

1.7.7 Providing coordination of Work performed by separate contractors or by the Owner's own forces.

1.7.8 Providing services in connection with the work of a construction manager or separate consultants retained by the Owner.

1.7.9 Providing Detailed Estimates of Construction Cost, analyses of owning and operating costs, or detailed quantity surveys or inventories of material, equipment and labor.

1.7.10 Providing interior design and other similar services required for or in connection with the selection, procurement or installation of furniture, furnishings and related equipment.

1.7.11 Providing services for planning tenant or rental spaces.

1.7.12 Making revisions in Drawings, Specifications or other documents when such revisions are inconsistent with written approvals or instructions previously given, are required by the enactment or revision of codes, laws or regulations subsequent to the preparation of such documents or are due to other causes not solely within the control of the Architect.

1.7.13 Preparing Drawings, Specifications and supporting data and providing other services in connection with Change Orders to the extent that the adjustment in the Basic Compensation resulting from the adjusted Construction Cost is not commensurate with the services required of the Architect, provided such Change Orders are required by causes not solely within the control of the Architect.

1.7.14 Making investigations, surveys, valuations, inventories or detailed appraisals of existing facilities, and services required in connection with construction performed by the Owner.

1.7.15 Providing consultation concerning replacement of any Work damaged by fire or other cause during con-struction, and furnishing services as may be required in connection with the replacement of such Work.

1.7.16 Providing services made necessary by the default of the Contractor, or by major defects or deficiencies in the Work of the Contractor, or by failure of performance of either the Owner or Contractor under the Contract for Construction.

1.7.17 Preparing a set of reproducible record drawings showing significant changes in the Work made during construction based on marked-up prints, drawings and other data furnished by the Contractor to the Architect.

1.7.18 Providing extensive assistance in the utilization of any equipment or system such as initial start-up or testing, adjusting and balancing, preparation of operation and maintenance manuals, training personnel for operation and maintenance, and consultation during operation.

1.7.19 Providing services after issuance to the Owner of the final Certificate for Payment, or in the absence of a final Certificate for Payment, more than sixty days after the Date of Substantial Completion of the Work.

1.7.20 Preparing to serve or serving as an expert witness in connection with any public hearing, arbitration proceeding or legal proceeding.

1.7.21 Providing services of consultants for other than the normal architectural, structural, mechanical and electrical engineering services for the Project.

1.7.22 Providing any other services not otherwise included in this Agreement or not customarily furnished in accordance with generally accepted architectural practice.

1.8 TIME

1.8.1 The Architect shall perform Basic and Additional Services as expeditiously as is consistent with professional skill and care and the orderly progress of the Work. Upon request of the Owner, the Architect shall submit for the Owner's approval, a schedule for the performance of the Architect's services which shall be adjusted as required as the Project proceeds, and shall include allowances for periods of time required for the Owner's review and approval of submissions and for approvals of authorities having jurisdiction over the Project. This schedule, when approved by the Owner, shall not, except for reasonable cause, be exceeded by the Architect.

ARTICLE 2

THE OWNER'S RESPONSIBILITIES

2.1 The Owner shall provide full information regarding requirements for the Project including a program, which shall set forth the Owner's design objectives, constraints and criteria, including space requirements and relationships, flexibility and expandability, special equipment and systems and site requirements.

2.2 If the Owner provides a budget for the Project it shall include contingencies for bidding, changes in the Work during construction, and other costs which are the responsibility of the Owner, including those described in this Article 2 and in Subparagraph 3.1.2. The Owner shall, at the request of the Architect, provide a statement of funds available for the Project, and their source.

2.3 The Owner shall designate, when necessary, a representative authorized to act in the Owner's behalf with respect to the Project. The Owner or such authorized representative shall examine the documents submitted by the Architect and shall render decisions pertaining thereto promptly, to avoid unreasonable delay in the progress of the Architect's services.

2.4 The Owner shall furnish a legal description and a certified land survey of the site, giving, as applicable, grades and lines of streets, alleys, pavements and adjoining property; rights-of-way, restrictions, easements, encroachments, zoning, deed restrictions, boundaries and contours of the site; locations, dimensions and complete data pertaining to existing buildings, other improvements and trees; and full information concerning available service and utility lines both public and private, above and below grade, including inverts and depths.

2.5 The Owner shall furnish the services of soil engineers or other consultants when such services are deemed necessary by the Architect. Such services shall include test borings, test pits, soil bearing values, percolation tests, air and water pollution tests, ground corrosion and resistivity tests, including necessary operations for determining subsoil, air and water conditions, with reports and appropriate professional recommendations.

2.6 The Owner shall furnish structural, mechanical, chemical and other laboratory tests, inspections and reports as required by law or the Contract Documents.

2.7 The Owner shall furnish all legal, accounting and insurance counseling services as may be necessary at any time for the Project, including such auditing services as the Owner may require to verify the Contractor's Applications for Payment or to ascertain how or for what purposes the Contractor uses the moneys paid by or on behalf of the Owner.

2.8 The services, information, surveys and reports required by Paragraphs 2.4 through 2.7 inclusive shall be furnished at the Owner's expense, and the Architect shall be entitled to rely upon the accuracy and completeness thereof.

2.9 If the Owner observes or otherwise becomes aware of any fault or defect in the Project or nonconformance with the Contract Documents, prompt written notice thereof shall be given by the Owner to the Architect.

2.10 The Owner shall furnish required information and services and shall render approvals and decisions as expeditiously as necessary for the orderly progress of the Architect's services and of the Work.

ARTICLE 3

CONSTRUCTION COST

3.1 DEFINITION

3.1.1 The Construction Cost shall be the total cost or estimated cost to the Owner of all elements of the Project designed or specified by the Architect.

3.1.2 The Construction Cost shall include at current market rates, including a reasonable allowance for overhead and profit, the cost of labor and materials furnished by the Owner and any equipment which has been de-signed, specified, selected or specially provided for by the Architect.

3.1.3 Construction Cost does not include the compensation of the Architect and the Architect's consultants, the cost of the land, rights-of-way, or other costs which are the responsibility of the Owner as provided in Article 2.

3.2 RESPONSIBILITY FOR CONSTRUCTION COST

3.2.1 Evaluations of the Owner's Project budget, Statements of Probable Construction Cost and Detailed Estimates of Construction Cost, if any, prepared by the Architect, represent the Architect's best judgment as a design professional familiar with the construction industry. It is recognized, however, that neither the Architect nor the Owner has control over the cost of labor, materials or equipment, over the Contractor's methods of determining bid prices, or over competitive bidding, market or negotiating conditions. Accordingly, the Architect cannot and does not warrant or represent that bids or negotiated prices will not vary from the Project budget proposed, established or approved by the Owner, if any, or from any Statement of Probable Construction Cost or other cost estimate or evaluation prepared by the Architect.

3.2.2 No fixed limit of Construction Cost shall be established as a condition of this Agreement by the furnishing, proposal or establishment of a Project budget under Subparagraph 1.1.2 or Paragraph 2.2 or otherwise, unless such fixed limit has been agreed upon in writing and signed by the parties hereto. If such a fixed limit has been established, the Architect shall be permitted to include contingencies for design, bidding and price escalation, to determine what materials, equipment, component systems and types of construction are to be included in the Contract Documents, to make reasonable adjustments in the scope of the Project and to include in the Contract Documents alternate bids to adjust the Construction Cost to the fixed limit. Any such fixed limit shall be increased in the amount of any increase in the Contract Sum occurring after execution of the Contract for Construction.

3.2.3 If the Bidding or Negotiation Phase has not commenced within three months after the Architect submits the Construction Documents to the Owner, any Project budget or fixed limit of Construction Cost shall be adjusted to reflect any change in the general level of prices in the construction industry between the date of submission of the Construction Documents to the Owner and the date on which proposals are sought.

3.2.4 If a Project budget or fixed limit of Construction Cost (adjusted as provided in Subparagraph 3.2.3) is exceeded by the lowest bona fide bid or negotiated proposal, the Owner shall (1) give written approval of an increase in such fixed limit, (2) authorize rebidding or renegotiating of the Project within a reasonable time, (3) if the Project is abandoned, terminate in accordance with Paragraph 10.2, or (4) cooperate in revising the Project scope and quality as required to reduce the Construction Cost. In the case of (4), provided a fixed limit of Construction Cost has been established as a condition of this Agreement, the Architect, without additional charge, shall modify the Drawings and Specifications as necessary to comply with the fixed limit. The providing of such service shall be the limit of the Architect's responsibility arising from the establishment of such fixed limit, and having done so, the Architect shall be entitled to compensation for all services performed, in accordance with this Agreement, whether or not the Construction Phase is commenced.

ARTICLE 4

DIRECT PERSONNEL EXPENSE

4.1 Direct Personnel Expense is defined as the direct salaries of all the Architect's personnel engaged on the Project, and the portion of the cost of their mandatory and customary contributions and benefits related thereto, such as employment taxes and other statutory employee benefits, insurance, sick leave, holidays, vacations, pensions and similar contributions and benefits.

ARTICLE 5

REIMBURSABLE EXPENSES

5.1 Reimbursable Expenses are in addition to the Compensation for Basic and Additional Services and include actual expenditures made by the Architect and the Architect's employees and consultants in the interest of the Project for the expenses listed in the following Subparagraphs:

5.1.1 Expense of transportation in connection with the Project; living expenses in connection with out-of-town travel; long distance communications, and fees paid for securing approval of authorities having jurisdiction over the Project.

5.1.2 Expense of reproductions, postage and handling of Drawings, Specifications and other documents, excluding reproductions for the office use of the Architect and the Architect's consultants.

5.1.3 Expense of data processing and photographic production techniques when used in connection with Additional Services.

5.1.4 If authorized in advance by the Owner, expense of overtime work requiring higher than regular rates.

5.1.5 Expense of renderings, models and mock-ups requested by the Owner.

5.1.6 Expense of any additional insurance coverage or limits, including professional liability insurance, requested by the Owner in excess of that normally carried by the Architect and the Architect's consultants.

ARTICLE 6

PAYMENTS TO THE ARCHITECT

6.1 PAYMENTS ON ACCOUNT OF BASIC SERVICES

6.1.1 An initial payment as set forth in Paragraph 14.1 is the minimum payment under this Agreement.

6.1.2 Subsequent payments for Basic Services shall be made monthly and shall be in proportion to services performed within each Phase of services, on the basis set forth in Article 14.

6.1.3 If and to the extent that the Contract Time initially established in the Contract for Construction is exceeded

or extended through no fault of the Architect, compensation for any Basic Services required for such extended period of Administration of the Construction Contract shall be computed as set forth in Paragraph 14.4 for Additional Services.

6.1.4 When compensation is based on a percentage of Construction Cost, and any portions of the Project are deleted or otherwise not constructed, compensation for such portions of the Project shall be payable to the extent services are performed on such portions, in accordance with the schedule set forth in Subparagraph 14.2.2, based on (1) the lowest bona fide bid or negotiated proposal or, (2) if no such bid or proposal is received, the most recent Statement of Probable Construction Cost or Detailed Estimate of Construction Cost for such portions of the Project.

**6.2 PAYMENTS ON ACCOUNT OF
ADDITIONAL SERVICES**

6.2.1 Payments on account of the Architect's Additional Services as defined in Paragraph 1.7 and for Reimbursable Expenses as defined in Article 5 shall be made monthly upon presentation of the Architect's statement of services rendered or expenses incurred.

6.3 PAYMENTS WITHHELD

6.3.1 No deductions shall be made from the Architect's compensation on account of penalty, liquidated damages or other sums withheld from payments to contractors, or on account of the cost of changes in the Work other than those for which the Architect is held legally liable.

6.4 PROJECT SUSPENSION OR TERMINATION

6.4.1 If the Project is suspended or abandoned in whole or in part for more than three months, the Architect shall be compensated for all services performed prior to receipt of written notice from the Owner of such suspension or abandonment, together with Reimbursable Expenses then due and all Termination Expenses as defined in Paragraph 10.4. If the Project is resumed after being suspended for more than three months, the Architect's compensation shall be equitably adjusted.

ARTICLE 7

ARCHITECT'S ACCOUNTING RECORDS

7.1 Records of Reimbursable Expenses and expenses pertaining to Additional Services and services performed on the basis of a Multiple of Direct Personnel Expense shall be kept on the basis of generally accepted accounting principles and shall be available to the Owner or the Owner's authorized representative at mutually convenient times.

ARTICLE 8

OWNERSHIP AND USE OF DOCUMENTS

8.1 Drawings and Specifications as instruments of service are and shall remain the property of the Architect whether the Project for which they are made is executed or not. The Owner shall be permitted to retain copies, including reproducible copies, of Drawings and Specifications for information and reference in connection with the Owner's use and occupancy of the Project. The Drawings and Specifications shall not be used by the Owner on other projects, for additions to this Project, or for completion of this Project by others provided the Architect is not in default under this Agreement, except by agreement in writing and with appropriate compensation to the Architect.

8.2 Submission or distribution to meet official regulatory requirements or for other purposes in connection with the Project is not to be construed as publication in derogation of the Architect's rights.

ARTICLE 9

ARBITRATION

9.1 All claims, disputes and other matters in question between the parties to this Agreement, arising out of or relating to this Agreement or the breach thereof, shall be decided by arbitration in accordance with the Construction Industry Arbitration Rules of the American Arbitration Association then obtaining unless the parties mutually agree otherwise. No arbitration, arising out of or relating to this Agreement, shall include, by consolidation, joinder or in any other manner, any additional person not a party to this Agreement except by written consent containing a specific reference to this Agreement and signed by the Architect, the Owner, and any other person sought to be joined. Any consent to arbitration involving an additional person or persons shall not constitute consent to arbitration of any dispute not described therein or with any person not named or described therein. This Agreement to arbitrate and any agreement to arbitrate with an additional person or persons duly consented to by the parties to this Agreement shall be specifically enforceable under the prevailing arbitration law.

9.2 Notice of the demand for arbitration shall be filed in writing with the other party to this Agreement and with the American Arbitration Association. The demand shall be made within a reasonable time after the claim, dispute or other matter in question has arisen. In no event shall the demand for arbitration be made after the date when institution of legal or equitable proceedings based on such claim, dispute or other matter in question would be barred by the applicable statute of limitations.

9.3 The award rendered by the arbitrators shall be final, and judgment may be entered upon it in accordance with applicable law in any court having jurisdiction thereof.

ARTICLE 10

TERMINATION OF AGREEMENT

10.1 This Agreement may be terminated by either party upon seven days' written notice should the other party fail substantially to perform in accordance with its terms through no fault of the party initiating the termination.

10.2 This Agreement may be terminated by the Owner upon at least seven days' written notice to the Architect in the event that the Project is permanently abandoned.

10.3 In the event of termination not the fault of the Architect, the Architect shall be compensated for all services performed to termination date, together with Reimbursable Expenses then due and all Termination Expenses as defined in Paragraph 10.4.

10.4 Termination Expenses include expenses directly attributable to termination for which the Architect is not otherwise compensated, plus an amount computed as a percentage of the total Basic and Additional Compensation earned to the time of termination, as follows:

  .1 20 percent if termination occurs during the Schematic Design Phase; or

  .2 10 percent if termination occurs during the Design Development Phase; or

  .3 5 percent if termination occurs during any subsequent phase.

ARTICLE 11

MISCELLANEOUS PROVISIONS

11.1 Unless otherwise specified, this Agreement shall be governed by the law of the principal place of business of the Architect.

11.2 Terms in this Agreement shall have the same meaning as those in AIA Document A201, General Conditions of the Contract for Construction, current as of the date of this Agreement.

11.3 As between the parties to this Agreement: as to all acts or failures to act by either party to this Agreement, any applicable statute of limitations shall commence to run and any alleged cause of action shall be deemed to have accrued in any and all events not later than the relevant Date of Substantial Completion of the Work, and as to any acts or failures to act occurring after the relevant Date of Substantial Completion, not later than the date of issuance of the final Certificate for Payment.

11.4 The Owner and the Architect waive all rights against each other and against the contractors, consultants, agents and employees of the other for damages covered by any property insurance during construction as set forth in the edition of AIA Document A201, General Conditions, current as of the date of this Agreement. The Owner and the Architect each shall require appropriate similar waivers from their contractors, consultants and agents.

ARTICLE 12

SUCCESSORS AND ASSIGNS

12.1 The Owner and the Architect, respectively, bind themselves, their partners, successors, assigns and legal representatives to the other party to this Agreement and to the partners, successors, assigns and legal representatives of such other party with respect to all covenants of this Agreement. Neither the Owner nor the Architect shall assign, sublet or transfer any interest in this Agreement without the written consent of the other.

ARTICLE 13

EXTENT OF AGREEMENT

13.1 This Agreement represents the entire and integrated agreement between the Owner and the Architect and supersedes all prior negotiations, representations or agreements, either written or oral. This Agreement may be amended only by written instrument signed by both Owner and Architect.

ARTICLE 14

BASIS OF COMPENSATION

The Owner shall compensate the Architect for the Scope of Services provided, in accordance with Article 6, Payments to the Architect, and the other Terms and Conditions of this Agreement, as follows:

14.1   AN INITIAL PAYMENT of   dollars ($)

shall be made upon execution of this Agreement and credited to the Owner's account as follows:

14.2   BASIC COMPENSATION

14.2.1   FOR BASIC SERVICES, as described in Paragraphs 1.1 through 1.5, and any other services included in Article 15 as part of Basic Services, Basic Compensation shall be computed as follows:

(Here insert basis of compensation, including fixed amounts, multiples or percentages, and identify Phases to which particular methods of compensation apply, if necessary.)

14.2.2   Where compensation is based on a Stipulated Sum or Percentage of Construction Cost, payments for Basic Services shall be made as provided in Subparagraph 6.1.2, so that Basic Compensation for each Phase shall equal the following percentages of the total Basic Compensation payable:

(Include any additional Phases as appropriate.)

Schematic Design Phase:	percent (%)
Design Development Phase:	percent (%)
Construction Documents Phase:	percent (%)
Bidding or Negotiation Phase:	percent (%)
Construction Phase:	percent (%)

14.3   FOR PROJECT REPRESENTATION BEYOND BASIC SERVICES, as described in Paragraph 1.6, Compensation shall be computed separately in accordance with Subparagraph 1.6.2.

14.4　COMPENSATION FOR ADDITIONAL SERVICES

14.4.1　FOR ADDITIONAL SERVICES OF THE ARCHITECT, as described in Paragraph 1.7, and any other services included in Article 15 as part of Additional Services, but excluding Additional Services of consultants, Compensation shall be computed as follows:

(Here insert basis of compensation, including rates and/or multiples of Direct Personnel Expense for Principals and employees, and identify Principals and classify employees, if required. Identify specific services to which particular methods of compensation apply, if necessary.)

14.4.2　FOR ADDITIONAL SERVICES OF CONSULTANTS, including additional structural, mechanical and electrical engineering services and those provided under Subparagraph 1.7.21 or identified in Article 15 as part of Additional Services, a multiple of　　　　　　　　　　　(　　　　　　　) times the amounts billed to the Architect for such services.

(Identify specific types of consultants in Article 15, if required.)

14.5　FOR REIMBURSABLE EXPENSES, as described in Article 5, and any other items included in Article 15 as Reimbursable Expenses, a multiple of　　　　　　　　　　(　　　　　　　) times the amounts expended by the Architect, the Architect's employees and consultants in the interest of the Project.

14.6　Payments due the Architect and unpaid under this Agreement shall bear interest from the date payment is due at the rate entered below, or in the absence thereof, at the legal rate prevailing at the principal place of business of the Architect.

(Here insert any rate of interest agreed upon.)

(Usury laws and requirements under the Federal Truth in Lending Act, similar state and local consumer credit laws and other regulations at the Owner's and Architect's principal places of business, the location of the Project and elsewhere may affect the validity of this provision. Specific legal advice should be obtained with respect to deletion, modification, or other requirements such as written disclosures or waivers.)

14.7　The Owner and the Architect agree in accordance with the Terms and Conditions of this Agreement that:

14.7.1　IF THE SCOPE of the Project or of the Architect's Services is changed materially, the amounts of compensation shall be equitably adjusted.

14.7.2　IF THE SERVICES covered by this Agreement have not been completed within

(　) months of the date hereof, through no fault of the Architect, the amounts of compensation, rates and multiples set forth herein shall be equitably adjusted.

ARTICLE 15

OTHER CONDITIONS OR SERVICES

This Agreement entered into as of the day and year first written above.

OWNER   ARCHITECT

_____   _____

_____   _____

_____   _____

BY_____   BY_____

THE AMERICAN INSTITUTE OF ARCHITECTS

AIA Document B801

Standard Form of Agreement Between Owner and Construction Manager

Recommended for use with the current editions of standard AIA agreement forms and documents.

THIS DOCUMENT HAS IMPORTANT LEGAL CONSEQUENCES; CONSULTATION WITH
AN ATTORNEY IS ENCOURAGED WITH RESPECT TO ITS COMPLETION OR MODIFICATION

AGREEMENT

made this   day of   in the year of Nineteen
Hundred and

BETWEEN the Owner:

and the Construction Manager:

For services in connection with the following described Project:
(Include detailed description of Project location and scope)

The Architect for the Project is:

The Owner and the Construction Manager agree as set forth below.

I. THE CONSTRUCTION MANAGER, as agent of the Owner, shall provide services in accordance with the Terms and Conditions of this Agreement.

The Construction Manager accepts the relationship of trust and confidence established between him and the Owner by this Agreement. He covenants with the Owner to furnish his professional skill and judgment and to cooperate with the Architect in furthering the interests of the Owner. He agrees to furnish efficient business administration and superintendence and to use his professional efforts at all times in an expeditious and economical manner consistent with the interests of the Owner.

II. THE OWNER shall compensate the Construction Manager in accordance with the Terms and Conditions of this Agreement.

    A. *FOR BASIC SERVICES,* as described in Paragraph 1.1, compensation shall be computed as follows: **(select one)**

        1. On the basis of a PROFESSIONAL FEE PLUS EXPENSES.

            a. A Professional Fee of

                                                dollars ($),

        PLUS

            b. An amount computed as follows:

            Principals' time at the fixed rate of

                                            dollars ($) per hour.

            For the purposes of this Agreement, the Principals are:

            Employees' time (other than Principals) at the following multiples of the employees' Direct Personnel Expense as defined in Article 4.

                For those assigned to the Project in the office, a multiple of

                                              (   ).

                For those assigned to the Project at the construction site, a multiple of

                                            (   ).

            c. Services of professional consultants at a multiple of

                                            (   ) times
            the amount billed to the Construction Manager for such services.

OR 2. On the basis of a MULTIPLE OF DIRECT PERSONNEL EXPENSE.

     a. Principals' time at the fixed rate of

                                                       dollars (\$) per hour.
       For the purposes of this Agreement, the Principals are:

     b. Employees' time (other than Principals) at the following multiples of the employees' Direct Personnel Expense as defined in Article 4.

        For those assigned to the Project in the office, a multiple of

                                             (   ).

        For those assigned to the Project at the construction site, a multiple of

                                           (   ).

     c. Services of professional consultants at a multiple of   (   )
       times the amount billed to the Construction Manager for such services.

OR 3. On the basis of a FIXED FEE of

                                     dollars (\$).

OR 4. On the basis of a PERCENTAGE OF CONSTRUCTION COST, as defined in Article 3, of
                                          percent (%).

  B. *AN INITIAL PAYMENT of*
                                    dollars (\$).
  shall be made upon the execution of this Agreement and credited to the Owner's account.

  C. *FOR ADDITIONAL SERVICES,* as described in Paragraph 1.2, compensation shall be computed as follows:

    1. Principals' time at the fixed rate of
                                     dollars (\$) per hour.
      For the purposes of this Agreement, the Principals are:

    2. Employees' time (other than Principals) at the following multiples of the employees' Direct Personnel Expense as defined in Article 4.

       For those assigned to the Project in the office, a multiple of

                                       (   ).

       For those assigned to the Project at the construction site, a multiple of

                                     (   ).

    3. Services of professional consultants at a multiple of   (   )
      times the amount billed to the Construction Manager for such services.

D. *FOR REIMBURSABLE EXPENSES,* amounts expended as defined in Article 5.

III. THE PARTIES agree in accordance with the Terms and Conditions of this Agreement that:

A. *IF THE SCOPE* of the Project is changed materially, compensation for Basic Services shall be subject to renegotiation.

B. *IF THE SERVICES* covered by this Agreement have not been completed within
( ) months of the date hereof, the amounts of compensation, rates and multiples set forth in Paragraph II shall be subject to renegotiation.

ARTICLE 1

CONSTRUCTION MANAGER'S SERVICES

1.1 BASIC SERVICES

The Construction Manager's Basic Services consist of the two phases described below and any other services included in Article 13 as Basic Services.

DESIGN PHASE

1.1.1 *Consultation During Project Development:* Review conceptual designs during development. Advise on site use and improvements, selection of materials, building systems and equipment. Provide recommendations on relative construction feasibility, availability of materials and labor, time requirements for installation and construction, and factors related to cost including costs of alternative designs or materials, preliminary budgets, and possible economies.

1.1.2 *Scheduling:* Provide and periodically update a Project time schedule that coordinates and integrates the Architect's services with construction schedules.

1.1.3 *Project Budget:* Prepare a Project budget for the Owner's approval as soon as major Project requirements have been identified and update periodically. Prepare an estimate of construction cost based on a quantity survey of Drawings and Specifications at the end of the Schematic Design Phase for approval by the Owner. Update and refine this estimate for Owner's approval as the development of the Drawings and Specifications proceeds, and advise the Owner and the Architect if it appears that the Project budget will not be met and make recommendations for corrective action.

1.1.4 *Coordination of Contract Documents:* Review the Drawings and Specifications as they are being prepared, recommending alternative solutions whenever design details affect construction feasibility or schedules.

1.1.4.1 Verify that the requirements and assignment of responsibilities for safety precautions and programs, temporary Project facilities and for equipment, materials and services for common use of Contractors are included in the proposed Contract Documents.

1.1.4.2 Advise on the method to be used for selecting Contractors and awarding contracts. If separate contracts are to be awarded, review the Drawings and Specifications to (1) ascertain if areas of jurisdiction overlap, (2) verify that all Work has been included, and (3) allow for phased construction.

1.1.4.3 Investigate and recommend a schedule for purchase by the Owner of all materials and equipment requiring long lead time procurement, and coordinate the schedule with the early preparation of Contract Documents by the Architect. Expedite and coordinate delivery of these purchases.

1.1.5 *Labor:* Provide an analysis of the types and quantity of labor required for the Project and review the availability of appropriate categories of labor required for critical phases.

1.1.5.1 Determine applicable requirements for equal employment opportunity programs for inclusion in the proposed Contract Documents.

1.1.6 *Bidding:* Prepare pre-qualification criteria for bidders and develop Contractor interest in the Project. Establish bidding schedules and conduct pre-bid conferences to familiarize bidders with the bidding documents and management techniques and with any special systems, materials or methods.

1.1.6.1 Receive bids, prepare bid analyses and make recommendations to the Owner for award of contracts or rejection of bids.

1.1.7 *Contract Awards:* Conduct pre-award conferences with successful bidders. Assist the Owner in preparing Construction Contracts and advise the Owner on the acceptability of Subcontractors and material suppliers proposed by Contractors.

CONSTRUCTION PHASE

The Construction Phase will commence with the award of the first Construction Contract or purchase order and will terminate 30 days after the final Certificate for Payment is issued by the Architect.

1.1.8 *Project Control:* Coordinate the Work of the Contractors with the activities and responsibilities of the Owner and Architect to complete the Project in accordance with the Owner's objectives on cost, time and quality. Provide sufficient personnel at the Project site with authority to achieve these objectives.

1.1.8.1 Schedule and conduct pre-construction and progress meetings at which Contractors, Owner, Architect and Construction Manager can discuss jointly such matters as procedures, progress, problems and scheduling.

1.1.8.2 Provide a detailed schedule for the operations of Contractors on the Project, including realistic activity sequences and durations, allocation of labor and materials, processing of shop drawings and samples, and delivery of products requiring long lead time procurement; include the Owner's occupancy requirements showing portions of the Project having occupancy priority.

1.1.8.3 Provide regular monitoring of the schedule as construction progresses. Identify potential variances between scheduled and probable completion dates. Review schedule for Work not started or incomplete and recommend to the Owner and Contractor adjustments in the schedule to meet the probable completion date. Provide summary reports of each monitoring, and document all changes in schedule.

1.1.8.4 Recommend courses of action to the Owner when requirements of a contract are not being fulfilled.

1.1.9 *Cost Control:* Revise and refine the approved estimate of construction cost, incorporate approved changes as they occur, and develop cash flow reports and forecasts as needed.

1.1.9.1 Provide regular monitoring of the approved estimate of construction cost, showing actual costs for activities in process and estimates for uncompleted tasks. Identify variances between actual and budgeted or estimated costs, and advise the Owner and Architect whenever projected costs exceed budgets or estimates.

1.1.9.2 Arrange for the maintenance of cost accounting records on authorized Work performed under unit costs, actual costs for labor and materials, or other bases requiring accounting records.

1.1.9.3 Develop and implement a system for review and processing of Change Orders.

1.1.9.4 Recommend necessary or desirable changes to the Owner and the Architect, review requests for changes, submit recommendations to the Owner and the Architect, and assist in negotiating Change Orders.

1.1.9.5 Develop and implement a procedure for the review and processing of applications by Contractors for progress and final payments. Make recommendations to the Architect for certification to the Owner for payment.

1.1.10 *Permits and Fees:* Assist in obtaining all building permits and special permits for permanent improvements, excluding permits for inspection or temporary facilities required to be obtained directly by the various Contractors. Verify that the Owner has paid all applicable fees and assessments for permanent facilities. Assist in obtaining approvals from all the authorities having jurisdiction.

1.1.11 *Owner's Consultants:* If required, assist the Owner in selecting and retaining professional services of a surveyor, special consultants and testing laboratories. Coordinate these services.

1.1.12 *Inspection:* Inspect the work of Contractors to assure that the Work is being performed in accordance with the requirements of the Contract Documents. Endeavor to guard the Owner against defects and deficiencies in the Work. Require any Contractor to stop Work or any portion thereof, and require special inspection or testing of any Work not in accordance with the provisions of the Contract Documents whether or not such Work be then fabricated, installed or completed. Reject Work which does not conform to the requirements of the Contract Documents.

1.1.12.1 The Construction Manager shall not be responsible for construction means, methods, techniques, sequences and procedures employed by Contractors in performance of their contract, and he shall not be responsible for the failure of any Contractors to carry out the Work in accordance with the Contract Documents.

1.1.13 *Contract Performance:* Consult with the Architect and the Owner if any Contractor requests interpretations of the meaning and intent of the Drawings and Specifications, and assist in the resolution of any questions which may arise.

1.1.14 *Shop Drawings and Samples:* In collaboration with the Architect, establish and implement procedures for expediting the processing and approval of shop drawings and samples.

1.1.15 *Reports and Records:* Record the progress of the Project. Submit written progress reports to the Owner and the Architect including information on the Contractors and Work, the percentage of completion and the number and amounts of Change Orders. Keep a daily log available to the Owner and the Architect.

1.1.15.1 Maintain at the Project site, on a current basis: records of all Contracts; shop drawings; samples; purchases; materials; equipment; applicable handbooks; federal, commercial and technical standards and specifications; maintenance and operating manuals and instructions; and any other related documents and revisions which arise out of the Contract or the Work. Obtain data from Contractors and maintain a current set of record drawings, specifications and operating manuals. At the completion of the Project, deliver all such records to the Owner.

1.1.16 *Owner-Purchased Items:* Accept delivery and arrange storage, protection and security for all Owner-purchased materials, systems and equipment which are a part of the Work until such items are turned over to the Contractors.

1.1.17 *Substantial Completion:* Upon the Contractors' determination of Substantial Completion of the Work or designated portions thereof, prepare for the Architect a list of incomplete or unsatisfactory items and a schedule for their completion. After the Architect certifies the Date of Substantial Completion, supervise the correction and completion of Work.

1.1.18 *Start-Up:* With the Owner's maintenance personnel, direct the checkout of utilities, operational systems and equipment for readiness and assist in their initial start-up and testing.

1.1.19 *Final Completion:* Determine final completion and provide written notice to the Owner and Architect that the Work is ready for final inspection. Secure and transmit to the Architect required guarantees, affidavits, releases, bonds and waivers. Turn over to the Owner all keys, manuals, record drawings and maintenance stocks.

1.2 ADDITIONAL SERVICES

The following Additional Services shall be performed upon authorization in writing from the Owner and shall be paid for as hereinbefore provided.

1.2.1 Services related to investigation, appraisals or valuations of existing conditions, facilities or equipment, or verifying the accuracy of existing drawings or other Owner-furnished information.

1.2.2 Services related to Owner-furnished equipment, furniture and furnishings which are not a part of the Work.

1.2.3 Services for tenant or rental spaces.

1.2.4 Services related to construction performed by the Owner.

1.2.5 Consultation on replacement of Work damaged by

fire or other cause during construction, and furnishing services for the replacement of such Work.

1.2.6 Services made necessary by the default of a Contractor.

1.2.7 Preparing to serve or serving as an expert witness in connection with any public hearing, arbitration proceeding, or legal proceeding.

1.2.8 Finding housing for construction labor, and defining requirements for establishment and maintenance of base camps.

1.2.9 Obtaining or training maintenance personnel or negotiating maintenance service contracts.

1.2.10 Services related to Work items required by the Conditions of the Contract and the Specifications which are not provided by Contractors.

1.2.11 Inspections of and services related to the Project after completion of the services under this Agreement.

1.2.12 Providing any other service not otherwise included in this Agreement.

ARTICLE 2

THE OWNER'S RESPONSIBILITIES

2.1 The Owner shall provide full information regarding his requirements for the Project.

2.2 The Owner shall designate a representative who shall be fully acquainted with the scope of the Work, and has authority to render decisions promptly and furnish information expeditiously.

2.3 The Owner shall retain an Architect for design and to prepare construction documents for the Project. The Architect's services, duties and responsibilities are described in the Agreement between the Owner and the Architect, pertinent parts of which will be furnished to the Construction Manager and will not be modified without written notification to him.

2.4 The Owner shall furnish such legal, accounting and insurance counselling services as may be necessary for the Project, and such auditing services as he may require to ascertain how or for what purposes the Contractors have used the moneys paid to them under the Construction Contracts.

2.5 The Owner shall furnish the Construction Manager with a sufficient quantity of construction documents.

2.6 If the Owner becomes aware of any fault or defect in the Project or nonconformance with the Contract Documents, he shall give prompt written notice thereof to the Construction Manager.

2.7 The services, information, surveys and reports required by Paragraphs 2.3 through 2.5 inclusive shall be furnished at the Owner's expense, and the Construction Manager shall be entitled to rely upon the accuracy and completeness thereof.

ARTICLE 3

CONSTRUCTION COST

3.1 If the Construction Cost is to be used as the basis for determining the Construction Manager's compensation for Basic Services, it shall be the cost of all Work, including work items in the Conditions of the Contract and the Specifications and shall be determined as follows:

3.1.1 For completed construction, the total construction cost of all such Work;

3.1.2 For Work not constructed, (1) the sum of the lowest bona fide bids received from qualified bidders for any or all of such Work or (2) if the Work is not bid, the sum of the bona fide negotiated proposals submitted for any or all of such Work; or

3.1.3 For Work for which no such bids or proposals are received, the Construction Cost contained in the Construction Manager's latest Project construction budget approved by the Owner.

3.2 Construction Cost shall not include the compensation of the Construction Manager (except for costs of Work items in the Conditions of the Contract and in the Specifications), the Architect and consultants, the cost of the land, rights-of-way, or other costs which are the responsibility of the Owner as provided in Paragraphs 2.3 through 2.5 inclusive.

3.3 The cost of labor, materials and equipment furnished by the Owner shall be included in the Construction Cost at current market rates, including a reasonable allowance for overhead and profit.

3.4 Cost estimates prepared by the Construction Manager represent his best judgment as a professional familiar with the construction industry. It is recognized, however, that neither the Construction Manager nor the Owner has any control over the cost of labor, materials or equipment, over Contractors' methods of determining bid prices, or other competitive bidding or market conditions.

3.5 No fixed limit of Construction Cost shall be deemed to have been established unless it is in writing and signed by the parties hereto. When a fixed limit of Construction Cost is established in writing as a condition of this Agreement, the Construction Manager shall advise what materials, equipment, component systems and types of construction should be included in the Contract Documents, and shall suggest reasonable adjustments in the scope of the Project to bring it within the fixed limit.

3.5.1. If responsive bids are not received as scheduled, any fixed limit of Construction Cost established as a condition of this Agreement shall be adjusted to reflect any change in the general level of prices occurring between the originally scheduled date and the date on which bids are received.

3.5.2 When the fixed limit of Construction Cost is exceeded, the Owner shall (1) give written approval of an increase in such fixed limit, (2) authorize rebidding within a reasonable time, or (3) cooperate in revising the scope and quality of the Work as required to reduce the Construction Cost. In the case of (3) the Construction Manager, without additional compensation, shall cooperate with the Architect as necessary to bring the Construction Cost within the fixed limit.

ARTICLE 4

DIRECT PERSONNEL EXPENSE

Direct Personnel Expense is defined as the salaries of professional, technical and clerical employees engaged on the Project by the Construction Manager, and the cost of their mandatory and customary benefits such as statutory employee benefits, insurance, sick leave, holidays, vacations, pensions and similar benefits.

ARTICLE 5

REIMBURSABLE EXPENSES

5.1 Reimbursable Expenses are in addition to the compensation for Basic and Additional Services and include actual expenditures made by the Construction Manager, his employees, or his professional consultants in the interest of the Project:

5.1.1 Long distance calls and telegrams, and fees paid for securing approval of authorities having jurisdiction over the Project.

5.1.2 Handling, shipping, mailing and reproduction of Project related materials.

5.1.3 Transportation and living when traveling in connection with the Project, relocation costs, and overtime work requiring higher than regular rates if authorized in advance by the Owner.

5.1.4 Electronic data processing service and rental of electronic data processing equipment when used in connection with Additional Services.

5.1.5 Premiums for insurance required in Article 10.

5.1.6 Providing construction support activities such as Work items included in the Conditions of the Contract and in the Specifications unless they are provided by the Contractors.

ARTICLE 6

PAYMENTS TO THE CONSTRUCTION MANAGER

6.1 Payments shall be made monthly upon presentation of the Construction Manager's statement of services as follows:

6.1.1 An initial payment as set forth in Paragraph IIB is the minimum payment under this Agreement.

6.1.2 When compensation is computed as described in Paragraphs IIA1a, IIA3, or IIA4, subsequent payments for Basic Services shall be made in proportion to services performed. The compensation at the completion of each Phase shall equal the following percentages of the total compensation for Basic Services:

Design Phase 20%
Construction Phase 100%

6.1.3 Payments for Reimbursable Expenses shall be made upon presentation of the Construction Manager's statement.

6.2 No deductions shall be made from the Construction Manager's compensation on account of penalty, liquidated damages, or other sums withheld from payments to Contractors.

6.3 If the Project is suspended for more than three months or abandoned in whole or in part, the Construction Manager shall be paid his compensation for services performed prior to receipt of written notice from the Owner of such suspension or abandonment, together with Reimbursable Expenses then due, and all termination expenses as defined in Paragraph 7.2 resulting from such suspension or abandonment. If the Project is resumed after being suspended for more than three months, the Construction Manager's compensation shall be subject to renegotiation.

6.4 If construction of the Project has started and is delayed by reason of strikes, or other circumstance not due to the fault of the Construction Manager, the Owner shall reimburse the Construction Manager for the costs of his Project-site staff as provided for by this Agreement. The Construction Manager shall reduce the size of his Project-site staff after a 60-day delay, or sooner if feasible, for the remainder of the delay period as directed by the Owner and, during the period, the Owner shall reimburse the Construction Manager for the direct personnel expense of such staff plus any relocation costs. Upon the termination of the delay, the Construction Manager shall restore his Project-site staff to its former size, subject to the approval of the Owner.

6.5 If the Project time schedule established in Subparagraph 1.1.2 is exceeded by more than thirty days through no fault of the Construction Manager, compensation for Basic Services performed by Principals, employees and professional consultants required beyond the thirtieth day to complete the services under this Agreement shall be as set forth in Paragraph IIC.

6.6 Payments due the Construction Manager which are unpaid for more than 60 days from date of billing shall bear interest at the legal rate of interest applicable at the construction site.

ARTICLE 7

TERMINATION OF AGREEMENT

7.1 This Agreement may be terminated by either party upon seven days' written notice should the other party fail substantially to perform in accordance with its terms through no fault of the party initiating termination. In the event of termination due to the fault of others than the Construction Manager, the Construction Manager shall be paid his compensation plus Reimbursable Expenses for services performed to termination date and all termination expenses.

7.2 Termination expenses are defined as Reimbursable Expenses directly attributable to termination, plus an

AIA DOCUMENT B801 • OWNER-CONSTRUCTION MANAGER AGREEMENT • DECEMBER 1973 EDITION • AIA®
© 1973 • THE AMERICAN INSTITUTE OF ARCHITECTS, 1735 NEW YORK AVE., N.W., WASHINGTON, D. C. 20006

amount computed as a percentage of the total compensation earned to the time of termination, as follows:

- 20 percent if termination occurs during the Design Phase; or
- 10 percent if termination occurs during the Construction Phase.

ARTICLE 8

SUCCESSORS AND ASSIGNS

The Owner and the Construction Manager each binds himself, his partners, successors, assigns and legal representatives to the other party to this Agreement and to the partners, successors, assigns and legal representatives of such other party with respect to all covenants of this Agreement. Neither the Owner nor the Construction Manager shall assign, sublet or transfer his interest in this Agreement without the written consent of the other.

ARTICLE 9

ARBITRATION

9.1 All claims, disputes and other matters in question between the parties to this Agreement, arising out of, or relating to this Agreement or the breach thereof, shall be decided by arbitration in accordance with the Construction Industry Arbitration Rules of the American Arbitration Association then obtaining unless the parties mutually agree otherwise. No arbitration, arising out of, or relating to this Agreement, shall include, by consolidation, joinder or in any other manner, any additional party not a party to this Agreement except by written consent containing a specific reference to this Agreement and signed by all the parties hereto. Any consent to arbitration involving an additional party or parties shall not constitute consent to arbitration of any dispute not described therein or with any party not named or described therein. This Agreement to arbitrate and any agreement to arbitrate with an additional party or parties duly consented to by the parties hereto shall be specifically enforceable under the prevailing arbitration law.

9.2 Notice of the demand for arbitration shall be filed in writing with the other party to this Agreement and with the American Arbitration Association. The demand shall be made within a reasonable time after the claim, dispute or other matter in question has arisen. In no event shall the demand for arbitration be made after the date when institution of legal or equitable proceedings based on such claim, dispute or other matter in question would be barred by the applicable statute of limitations.

9.3 The award rendered by the arbitrators shall be final, and judgment may be entered upon it in accordance with applicable law in any court having jurisdiction thereof.

ARTICLE 10

INSURANCE

The Construction Manager shall purchase and maintain insurance to protect himself from claims under workmen's compensation acts; claims for damages because of bodily injury including personal injury, sickness or disease, or death of any of his employees or of any person other than his employees; and from claims for damages because of injury to or destruction of tangible property including loss of use resulting therefrom; and from claims arising out of the performance of professional services caused by any errors, omissions or negligent acts for which he is legally liable.

ARTICLE 11

EXTENT OF AGREEMENT

11.1 This Agreement represents the entire and integrated agreement between the Owner and the Construction Manager and supersedes all prior negotiations, representations or agreements, either written or oral. This Agreement shall not be superseded by provisions of contracts for construction and may be amended only by written instrument signed by both Owner and Construction Manager.

11.2 Nothing contained herein shall be deemed to create any contractual relationship between the Construction Manager and the Architect or any of the Contractors, Subcontractors or material suppliers on the Project; nor shall anything contained herein be deemed to give any third party any claim or right of action against the Owner or the Construction Manager which does not otherwise exist without regard to this Agreement.

ARTICLE 12

GOVERNING LAW

Unless otherwise specified, this Agreement shall be governed by the law in effect at the location of the Project.

ARTICLE 13

OTHER CONDITIONS OR SERVICES

This Agreement executed the day and year first written above.

OWNER   CONSTRUCTION MANAGER

_____   _____

THE AMERICAN INSTITUTE OF ARCHITECTS

<u>COMPARISON OF THE 1970 AND 1976 EDITIONS, AIA DOCUMENT A201</u>

General Conditions of the Contract for Construction

*THIS DOCUMENT HAS IMPORTANT LEGAL CONSEQUENCES; CONSULTATION
WITH AN ATTORNEY IS ENCOURAGED WITH RESPECT TO ITS MODIFICATION*

SEPTEMBER 27, 1976 (REVISED TO CONFORM TO FINAL PRINTING OF A201)

This document has been published by The American Institute of Architects
for the information of users of the AIA General Conditions. Additions to
the 1970 Edition of A201 are indicated by <u>underlined wording</u>. Deletions
from the 1970 Edition are indicated by wording which has been stricken.

This document IS NOT TO BE USED AS A CONTRACT DOCUMENT. The 1976 Edition has
been published in a form suitable for use as a contract document, available
from AIA and its distributors. Because of the numerous changes in the document,
it is recommended that practitioners carefully coordinate their Supplementary
Conditions, Specifications, and Instructions to Bidders with this document.

OWNER-ARCHITECT AGREEMENTS for projects which will be constructed under
the terms of the 1976 Edition of A201 must also be coordinated with this
document to insure that there are no variations between them. This can be
accomplished through the ammendment of such Owner-Architect Agreements or
through Supplementary Conditions to this document, or both.

Requests for additional information on, additional copies of, or interpretations of this document should be directed to the Department of Professional
Practice at AIA Headquarters. Published commentary and cross-reference to
the 1970 Edition and 1976 Edition are currently available. In addition, new
editions of all AIA Documents relating to the General Conditions will be
published during 1976, including Chapter 13 of the Architect's Handbook of
Professional Practice, which describes the General Conditions in detail.

The AIA is indebted to the members of the Documents Board for their inten-
sive efforts over more than two years, especially the A201 Task Force chaired
by Dean F. Hilfinger, FAIA. The cooperation of, and the contributions made
by, the following national organizations is greatly appreciated:

ACEC	American Consulting Engineers Council
ASA	American Subcontractors Association
ASC	Associated Specialty Contractors, Inc.
CSI	Construction Specifications Institute
NSPE	National Society of Professional Engineers

This 1976 Edition of A201 has been endorsed and approved by The Associated General Contractors of America (AGC), whose assistance is especially appreciated.

AIA DOCUMENT A201 • GENERAL CONDITIONS OF THE CONTRACT FOR CONSTRUCTION • THIRTEENTH EDITION • AUGUST 1976
EDITION • AIA®•©1976 • THE AMERICAN INSTITUTE OF ARCHITECTS, 1735 NEW YORK AVENUE, N.W., WASHINGTON, D.C. 20006   1

GENERAL CONDITIONS OF THE CONTRACT FOR CONSTRUCTION

ARTICLE 1

CONTRACT DOCUMENTS

1.1 DEFINITIONS

1.1.1 THE CONTRACT DOCUMENTS

The Contract Documents consist of the Agreement, the Conditions of the Contract (General, Supplementary and other Conditions), the Drawings, the Specifications, all Addenda issued prior to execution of the Contract, and all Modifications thereto. A Modification is (1) a written amendment to the Contract signed by both parties, (2) a Change Order, (3) a written interpretation issued by the Architect pursuant to Subparagraph 1.2.5, or (4) a written order for a minor change in the Work issued by the Architect pursuant to Paragraph 12.3. A Modification may be made only after execution of the Contract.

```
New Language in 1.1.1 replaces
"including the bidding documents"
in former 1.1.2                    ➜
```

1.1.2 THE CONTRACT

The Contract Documents form the Contract. The Contract represents the entire and integrated agreement between the parties hereto and supersedes all prior negotiations, representations, or agreements, either written or oral, including the bidding documents. The Contract may be amended or modified only by a Modification as defined in Subparagraph 1.1.1.

```
New language in 1.1.2 replaces
former 2.1.2 and 5.1.3.           ➜
```

1.1.3 THE WORK

The term Work includes all labor necessary to produce the construction required by the Contract Documents, and all materials and equipment incorporated or to be incorporated in such construction.

1.1.4 THE PROJECT

The Project is the total construction designed by the Architect of which the Work performed under the Contract Documents may be the whole or a part.

ARTICLE I

CONTRACT DOCUMENTS

I.I DEFINITIONS

I.I.I THE CONTRACT DOCUMENTS

The Contract Documents consist of the Owner-Contractor Agreement, the Conditions of the Contract (General, Supplementary and other Conditions), the Drawings, the Specifications, and all Addenda issued prior to and all Modifications issued after execution of the Contract . and all Modifications thereto. A Modification is (I) a written amendment to the Contract signed by both parties, (2) a Change Order, (3) a written interpretation issued by the Architect pursuant to Subparagraph 1.2.5 2.2.8, or (4) a written order for a minor change in the Work issued by the Architect pursuant to Paragraph 12.3. A Modification may be made only after execution of the Contract. The Contract Documents do not include Bidding Documents such as the Advertisement or Invitation to Bid, the Instructions to Bidders, sample forms, the Contractor's Bid or portions of Addenda relating to any of these, or any other documents, unless specifically enumerated in the Owner-Contractor Agreement.

I.I.2 THE CONTRACT

The Contract Documents form the Contract for Construction. This Contract represents the entire and integrated agreement between the parties hereto and supersedes all prior negotiations, representations, or agreements, either written or oral . including the bidding documents. The Contract may be amended or modified only by a Modification as defined in Subparagraph I.I.I. Nothing contained in The Contract Documents shall not be construed to create any contractual relationship of any kind between the Architect and the Contractor, but the Architect shall be entitled to performance of obligations intended for his benefit, and to enforcement thereof. Nothing contained in the Contract Documents shall create any contractual relationship between the Owner or the Architect and any Subcontractor or Sub-subcontractor.

I.I.3 THE WORK

The term Work comprises the completed construction required by the Contract Documents and includes all labor necessary to produce the such construction , required by the Contract Documents, and all materials and equipment incorporated or to be incorporated in such construction.

I.I.4 THE PROJECT

The Project is the total construction designed by the Architect of which the Work performed under the Contract Documents may be the whole or a part.

1.2 **EXECUTION, CORRELATION, INTENT AND INTERPRETATIONS**

1.2 EXECUTION, CORRELATION AND INTENT ~~AND INTERPRETATIONS~~

1.2.1 The Contract Documents shall be signed in not less than triplicate by the Owner and Contractor. If either the Owner or the Contractor or both do not sign the Conditions of the Contract, Drawings, Specifications, or any of the other Contract Documents, the Architect shall identify them.

1.2.1 The Contract Documents shall be signed in not less than triplicate by the Owner and Contractor. If either the Owner or the Contractor or both do not sign the Conditions of the Contract, Drawings, Specifications, or any of the other Contract Documents, the Architect shall identify ~~them~~ such Documents.

1.2.2 By executing the Contract, the Contractor represents that he has visited the site, familiarized himself with the local conditions under which the Work is to be performed, and correlated his observations with the requirements of the Contract Documents.

1.2.2 By executing the Contract, the Contractor represents that he has visited the site, familiarized himself with the local conditions under which the Work is to be performed, and correlated his observations with the requirements of the Contract Documents.

1.2.3 The Contract Documents are complementary, and what is required by any one shall be as binding as if required by all. The intention of the Documents is to include all labor, materials, equipment and other items as provided in Subparagraph 4.4.1 necessary for the proper execution and completion of the Work. It is not intended that Work not covered under any heading, section, branch, class or trade of the Specifications shall be supplied unless it is required elsewhere in the Contract Documents or is reasonably inferable therefrom as being necessary to produce the intended results. Words which have well-known technical or trade meanings are used herein in accordance with such recognized meanings.

1.2.3 The ~~intention~~ intent of the Contract Documents is to include all ~~labor, materials, equipment and other items as provided in Subparagraph 4.4.1~~ items necessary for the proper execution and completion of the Work. The Contract Documents are complementary, and what is required by any one shall be as binding as if required by all. ~~It is not intended that~~ Work not covered ~~under any heading, section, branch, class or trade of the Specifications shall be supplied unless it is required elsewhere~~ in the Contract Documents will not be required unless it is consistent therewith and ~~or~~ is reasonably inferable therefrom as being necessary to produce the intended results. Words and abbreviations which have well-known technical or trade meanings are used ~~herein~~ in the Contract Documents in accordance with such recognized meanings.

1.2.4 The organization of the Specifications into divisions, sections and articles, and the arrangement of Drawings shall not control the Contractor in dividing the Work among Subcontractors or in establishing the extent of Work to be performed by any trade.

1.2.4 The organization of the Specifications into divisions, sections and articles, and the arrangement of Drawings shall not control the Contractor in dividing the Work among Subcontractors or in establishing the extent of Work to be performed by any trade.

1.2.5 Written interpretations necessary for the proper execution or progress of the Work, in the form of drawings or otherwise, will be issued with reasonable promptness by the Architect and in accordance with any schedule agreed upon. Either party to the Contract may make written request to the Architect for such interpretations. Such interpretations shall be consistent with and reasonably inferable from the Contract Documents, and may be effected by Field Order.

Former 1.2.5 becomes new 2.2.8

1.3 **COPIES FURNISHED AND OWNERSHIP**

1.3 ~~COPIES FURNISHED AND~~ OWNERSHIP AND USE OF DOCUMENTS

1.3.1 Unless otherwise provided in the Contract Documents, the Contractor will be furnished, free of charge, all copies of Drawings and Specifications reasonably necessary for the execution of the Work.

Former 1.3.1 is new 3.2.5

1.3.2 All Drawings, Specifications and copies thereof furnished by the Architect are and shall remain his property. They are not to be used on any other project, and, with the exception of one contract set for each party to the Contract, are to be returned to the Architect on request at the completion of the Work.

1.3.1 All Drawings, Specifications and copies thereof furnished by the Architect are and shall remain his property. They are to be used only with respect to this Project and are not to be used on any other project . , and, With the exception of one contract set for each party to the Contract, such documents are to be returned or suitably accounted for to the Architect on request at the completion of the Work. Submission or distribution to meet official regulatory requirements or for other purposes in connection with the Project is not to be construed as publication in derogation of the Architect's common law copyright or other reserved rights.

ARTICLE 2

ARCHITECT

2.1 DEFINITION

2.1.1 The Architect is the person or organization licensed to practice architecture and identified as such in the Agreement and is referred to throughout the Contract Documents as if singular in number and masculine in gender. The term Architect means the Architect or his authorized representative.

ARTICLE 2

ARCHITECT

2.1 DEFINITION

2.1.1 The Architect is the person or organization lawfully licensed to practice architecture, and or an entity lawfully practicing architecture identified as such in the Owner-Contractor Agreement , and is referred to throughout the Contract Documents as if singular in number and masculine in gender. The term Architect means the Architect or his authorized representative.

2.1.2 Nothing contained in the Contract Documents shall create any contractual relationship between the Architect and the Contractor.

Former 2.1.2 included in new 1.1.2

2.2 ADMINISTRATION OF THE CONTRACT

2.2.1 The Architect will provide general Administration of the Construction Contract, including performance of the functions hereinafter described.

2.2 ADMINISTRATION OF THE CONTRACT

2.2.1 The Architect will provide general administration of the construction Contract , including performance of the functions as hereinafter described.

2.2.2 The Architect will be the Owner's representative during construction and until final payment. The Architect will have authority to act on behalf of the Owner to the extent provided in the Contract Documents, unless otherwise modified by written instrument which will be shown to the Contractor. The Architect will advise and consult with the Owner, and all of the Owner's instructions to the Contractor shall be issued through the Architect.

2.2.2 The Architect will be the Owner's representative during construction and until final payment is due. The Architect will advise and consult with the Owner. The Owner's instructions to the Contractor shall be forwarded through the Architect. The Architect will have authority to act on behalf of the Owner only to the extent provided in the Contract Documents, unless otherwise modified by written instrument which will be shown to the Contractor in accordance with Subparagraph 2.2.18. The Architect will advise and consult with the Owner, and all of the Owner's instructions to the Contractor shall be issued through the Architect.

2.2.3 The Architect shall at all times have access to the Work wherever it is in preparation and progress. The Contractor shall provide facilities for such access so the Architect may perform his functions under the Contract Documents.

Former 2.2.3 is new 2.2.5

1970 Edition

2.2.4 The Architect will make periodic visits to the site to familiarize himself generally with the progress and quality of the Work and to determine in general if the Work is proceeding in accordance with the Contract Documents. On the basis of his on-site observations as an architect, he will keep the Owner informed of the progress of the Work, and will endeavor to guard the Owner against defects and deficiencies in the Work of the Contractor. The Architect will not be required to make exhaustive or continuous on-site inspections to check the quality or quantity of the Work. The Architect will not be responsible for construction means, methods, techniques, sequences or procedures, or for safety precautions and programs in connection with the Work, and he will not be responsible for the Contractor's failure to carry out the Work in accordance with the Contract Documents.

New 2.2.4 replaces last sentence of former 2.2.4 and former 2.2.18 ———▶

New 2.2.5 is former 2.2.3 ———▶

2.2.5 Based on such observations and the Contractor's Applications for Payment, the Architect will determine the amounts owing to the Contractor and will issue Certificates for Payment in such amounts, as provided in Paragraph 9.4.

2.2.6 The Architect will be, in the first instance, the interpreter of the requirements of the Contract Documents and the judge of the performance thereunder by both the Owner and Contractor. The Architect will, within a reasonable time, render such interpretations as he may deem necessary for the proper execution or progress of the Work.

New 2.2.8 replaces former 1.2.5 ———▶

1976 Edition

2.2.3 The Architect will ~~make periodic visits to~~ visit the site <u>at intervals appropriate to the stage of construction</u> to familiarize himself generally with the progress and quality of the Work and to determine in general if the Work is proceeding in accordance with the Contract Documents. <u>However, the Architect will not be required to make exhaustive or continuous on-site inspections to check the quality or quantity of the Work.</u> On the basis of his on-site observations as an architect, he will keep the Owner informed of the progress of the Work, and will endeavor to guard the Owner against defects and deficiencies in the Work of the Contractor. ~~The Architect will not be required to make exhaustive or continuous on-site inspections to check the quality or quantity of the Work.~~

2.2.4 The Architect will not be responsible for <u>and will not have control or charge of</u> construction means, methods, techniques, sequences or procedures, or for safety precautions and programs in connection with the Work, and he will not be responsible for the Contractor's failure to carry out the Work in accordance with the Contract Documents. The Architect will not be responsible for <u>or have control or charge over</u> the acts or omissions of the Contractor, ~~any~~ Subcontractors, <u>or any of their agents or employees,</u> or any other persons performing any of the Work.

2.2.5 The Architect shall at all times have access to the Work wherever it is in preparation and progress. The Contractor shall provide facilities for such access so the Architect may perform his functions under the Contract Documents.

2.2.6 Based on ~~such~~ <u>the Architect's</u> observations <u>and an evaluation of</u> the Contractor's Applications for Payment, the Architect will determine the amounts owing to the Contractor and will issue Certificates for Payment in such amounts, as provided in Paragraph 9.4.

2.2.7 The Architect will be ~~, in the first instance,~~ the interpreter of the requirements of the Contract Documents and the judge of the performance thereunder by both the Owner and Contractor. ~~The Architect will, within a reasonable time, render such interpretations as he may deem necessary for the proper execution or progress of the Work.~~

2.2.8 ~~Written~~ The Architect will render interpretations necessary for the proper execution or progress of the Work, ~~in the form of drawings or otherwise, will be issued~~ with reasonable promptness ~~by the Architect~~ and in accordance with any ~~schedule~~ <u>time limit</u> agreed upon. Either party to the Contract may make written request to the Architect for such interpretations. ~~Such interpretations shall be consistent with and reasonably inferable from the Contract Documents, and may be effected by Field Order.~~

2.2.7 Claims, disputes and other matters in question between the Contractor and the Owner relating to the execution or progress of the Work or the interpretation of the Contract Documents shall be referred initially to the Architect for decision which he will render in writing within a reasonable time.

2.2.8 All interpretations and decisions of the Architect shall be consistent with the intent of the Contract Documents. In his capacity as interpreter and judge, he will exercise his best efforts to insure faithful performance by both the Owner and the Contractor and will not show partiality to either.

2.2.9 The Architect's decisions in matters relating to artistic effect will be final if consistent with the intent of the Contract Documents.

2.2.10 Any claim, dispute or other matter that has been referred to the Architect, except those relating to artistic effect as provided in Subparagraph 2.2.9 and except any which have been waived by the making or acceptance of final payment as provided in Subparagraphs 9.7.5 and 9.7.6, shall be subject to arbitration upon the written demand of either party. However, no demand for arbitration of any such claim, dispute or other matter may be made until the earlier of:

2.2.10.1 The date on which the Architect has rendered his written decision, or

.2 the tenth day after the parties have presented their evidence to the Architect or have been given a reasonable opportunity to do so, if the Architect has not rendered his written decision by that date.

2.2.11 If a decision of the Architect is made in writing and states that it is final but subject to appeal, no demand for arbitration of a claim, dispute or other matter covered by such decision may be made later than thirty days after the date on which the party making the demand received the decision. The failure to demand arbitration within said thirty days' period will result in the Architect's decision becoming final and binding upon the Owner and the Contractor. If the Architect renders a decision after arbitration proceedings have been initiated, such decision may be entered as evidence but will not supersede any arbitration proceedings unless the decision is acceptable to the parties concerned.

2.2.9 Claims, disputes and other matters in question between the Contractor and the Owner relating to the execution or progress of the Work or the interpretation of the Contract Documents shall be referred initially to the Architect for decision which he will render in writing within a reasonable time.

2.2.10 All interpretations and decisions of the Architect shall be consistent with the intent of and reasonably inferable from the Contract Documents and will be in writing or in the form of drawings. In his capacity as interpreter and judge, he will exercise his best efforts to insure endeavor to secure faithful performance by both the Owner and the Contractor , and will not show partiality to either , and will not be liable for the result of any interpretation or decision rendered in good faith in such capacity.

2.2.11 The Architect's decisions in matters relating to artistic effect will be final if consistent with the intent of the Contract Documents.

2.2.12 Any claim, dispute or other matter in question between the Contractor and the Owner that has been referred to the Architect, except those relating to artistic effect as provided in Subparagraph 2.2.9 2.2.11 and except any those which have been waived by the making or acceptance of final payment as provided in Subparagraphs 9.7.5 9.9.4 and 9.7.6 9.9.5, shall be subject to arbitration upon the written demand of either party. However, no demand for arbitration of any such claim, dispute or other matter may be made until the earlier of (1) the date on which the Architect has rendered his a written decision, or (2) the tenth day after the parties have presented their evidence to the Architect or have been given a reasonable opportunity to do so, if the Architect has not rendered his written decision by that date. If When such a written decision of the Architect is made in writing and states (1) that it the decision is final but subject to appeal, no, and (2) that any demand for arbitration of a claim, dispute or other matter covered by such decision may must be made later than within thirty days after the date on which the party making the demand receives the written decision , The failure to demand arbitration within said thirty days' period will result in the Architect's decision becoming final and binding upon the Owner and the Contractor. If the Architect renders a decision after arbitration proceedings have been initiated, such decision may be entered as evidence but will not supersede any arbitration proceedings unless the decision is acceptable to the all parties concerned.

2.2.12 The Architect will have authority to reject Work which does not conform to the Contract Documents. Whenever, in his reasonable opinion, he considers it necessary or advisable to insure the proper implementation of the intent of the Contract Documents, he will have authority to require special inspection or testing of the Work in accordance with Subparagraph 7.8.2 whether or not such Work be then fabricated, installed or completed. However, neither the Architect's authority to act under this Subparagraph 2.2.12, nor any decision made by him in good faith either to exercise or not to exercise such authority, shall give rise to any duty or responsibility of the Architect to the Contractor, any Subcontractor, any of their agents or employees, or any other person performing any of the Work.

2.2.13 The Architect will review Shop Drawings and Samples as provided in Subparagraphs 4.13.1 through 4.13.8 inclusive.

New 2.2.14 replaces former 2.2.13 and 4.13.5   ⟶

2.2.14 The Architect will prepare Change Orders in accordance with Article 12, and will have authority to order minor changes in the Work as provided in Subparagraph 12.3.1.

2.2.15 The Architect will conduct inspections to determine the dates of Substantial Completion and final completion, will receive and review written guarantees and related documents required by the Contract and assembled by the Contractor and will issue a final Certificate for Payment.

2.2.16 If the Owner and Architect agree, the Architect will provide one or more Full-Time Project Representatives to assist the Architect in carrying out his responsibilities at the site. The duties, responsibilities and limitations of authority of any such Project Representative shall be as set forth in an exhibit to be incorporated in the Contract Documents.

2.2.17 The duties, responsibilities and limitations of authority of the Architect as the Owner's representative during construction as set forth in Articles 1 through 14 inclusive of these General Conditions will not be modified or extended without written consent of the Owner, the Contractor and the Architect.

2.2.13 The Architect will have authority to reject Work which does not conform to the Contract Documents. Whenever, in his ~~reasonable~~ opinion, he considers it necessary or advisable ~~to insure~~ for the ~~proper~~ implementation of the intent of the Contract Documents, he will have authority to require special inspection or testing of the Work in accordance with Subparagraph ~~7.8.2~~ 7.7.2 whether or not such Work be then fabricated, installed or completed. However, neither the Architect's authority to act under this Subparagraph ~~2.2.14~~ 2.2.13, nor any decision made by him in good faith either to exercise or not to exercise such authority, shall give rise to any duty or responsibility of the Architect to the Contractor, any Subcontractor, any of their agents or employees, or any other person performing any of the Work.

2.2.14 The Architect will review and approve or take other appropriate action upon Contractor's submittals such as Shop Drawings , Product Data and Samples , ~~with reasonable promptness so as to cause no delay,~~ but only for conformance with the design concept of the ~~Project~~ Work and with the information given in the Contract Documents. Such action shall be taken with reasonable promptness so as to cause no delay. The Architect's approval of a ~~separate~~ specific item shall not indicate approval of an assembly ~~in~~ of which the item ~~functions.~~ is a component.

2.2.15 The Architect will prepare Change Orders in accordance with Article 12, and will have authority to order minor changes in the Work as provided in Subparagraph ~~12.3.1.~~ 12.4.1.

2.2.16 The Architect will conduct inspections to determine the dates of Substantial Completion and final completion, will receive and forward to the Owner for the Owner's review written ~~guarantees~~ warranties and related documents required by the Contract and assembled by the Contractor, and will issue a final Certificate for Payment upon compliance with Paragraph 9.9.

2.2.17 If the Owner and Architect agree, the Architect will provide one or more ~~Full-Time~~ Project Representatives to assist the Architect in carrying out his responsibilities at the site. The duties, responsibilities and limitations of authority of any such Project Representative shall be as set forth in an exhibit to be incorporated in the Contract Documents.

2.2.18 The duties, responsibilities and limitations of authority of the Architect as the Owner's representative during construction as set forth in ~~Articles 1 through 14 inclusive of these General Conditions~~ the Contract Documents will not be modified or extended without written consent of the Owner, the Contractor and the Architect.

2.2.18 The Architect will not be responsible for the acts or omissions of the Contractor, any Subcontractors, or any of their agents or employees, or any other persons performing any of the Work.

← Former 2.2.18 included in new 2.2.4

2.2.19 In case of the termination of the employment of the Architect, the Owner shall appoint an architect against whom the Contractor makes no reasonable objection whose status under the Contract Documents shall be that of the former architect. Any dispute in connection with such appointment shall be subject to arbitration.

2.2.19 In case of the termination of the employment of the Architect, the Owner shall appoint an architect against whom the Contractor makes no reasonable objection whose status under the Contract Documents shall be that of the former architect. Any dispute in connection with such appointment shall be subject to arbitration.

ARTICLE 3

OWNER

3.1 DEFINITION

3.1.1 The Owner is the person or organization identified as such in the Agreement and is referred to throughout the Contract Documents as if singular in number and masculine in gender. The term Owner means the Owner or his authorized representative.

3.2 INFORMATION AND SERVICES REQUIRED OF THE OWNER

ARTICLE 3

OWNER

3.1 DEFINITION

3.1.1 The Owner is the person or ~~organization~~ entity identified as such in the Owner-Contractor Agreement and is referred to throughout the Contract Documents as if singular in number and masculine in gender. The term Owner means the Owner or his authorized representative.

3.2 INFORMATION AND SERVICES REQUIRED OF THE OWNER

New Subparagraph ⟶

3.2.1 The Owner shall, at the request of the Contractor, at the time of execution of the Owner-Contractor Agreement, furnish to the Contractor reasonable evidence that he has made financial arrangements to fulfill his obligations under the Contract. Unless such reasonable evidence is furnished, the Contractor is not required to execute the Owner-Contractor Agreement or to commence the Work.

3.2.1 The Owner shall furnish all surveys describing the physical characteristics, legal limits and utility locations for the site of the Project.

3.2.2 The Owner shall furnish all surveys describing the physical characteristics, legal ~~limits~~ limitations and utility locations for the site of the Project, and a legal description of the site.

3.2.2 The Owner shall secure and pay for easements for permanent structures or permanent changes in existing facilities.

3.2.3 Except as provided in Subparagraph 4.7.1, the Owner shall secure and pay for necessary approvals, easements, assessments and charges required for the construction, use or occupancy of permanent structures or for permanent changes in existing facilities.

3.2.3 Information or services under the Owner's control shall be furnished by the Owner with reasonable promptness to avoid delay in the orderly progress of the Work.

3.2.4 Information or services under the Owner's control shall be furnished by the Owner with reasonable promptness to avoid delay in the orderly progress of the Work.

New 3.2.5 is former 1.3.1 ⟶

3.2.5 Unless otherwise provided in the Contract Documents, the Contractor will be furnished, free of charge, all copies of Drawings and Specifications reasonably necessary for the execution of the Work.

3.2.4 The Owner shall issue all instructions to the Contractor through the Architect.

3.2.6 The Owner shall ~~issue~~ forward all instructions to the Contractor through the Architect.

3.2.5 The foregoing are in addition to other duties and responsibilities of the Owner enumerated herein and especially those in respect to Payment and Insurance in Articles 9 and 11 respectively.

3.3 OWNER'S RIGHT TO STOP THE WORK
3.3.1 If the Contractor fails to correct defective Work or persistently fails to supply materials or equipment in accordance with the Contract Documents, the Owner may order the Contractor to stop the Work, or any portion thereof, until the cause for such order has been eliminated.

3.4 OWNER'S RIGHT TO CARRY OUT THE WORK
3.4.1 If the Contractor defaults or neglects to carry out the Work in accordance with the Contract Documents or fails to perform any provision of the Contract, the Owner may, after seven days' written notice to the Contractor and without prejudice to any other remedy he may have, make good such deficiencies. In such case an appropriate Change Order shall be issued deducting from the payments then or thereafter due the Contractor the cost of correcting such deficiencies, including the cost of the Architect's additional services made necessary by such default, neglect or failure. The Architect must approve both such action and the amount charged to the Contractor. If the payments then or thereafter due the Contractor are not sufficient to cover such amount, the Contractor shall pay the difference to the Owner.

ARTICLE 4
CONTRACTOR
4.1 DEFINITION
4.1.1 The Contractor is the person or organization identified as such in the Agreement and is referred to throughout the Contract Documents as if singular in number and masculine in gender. The term Contractor means the Contractor or his authorized representative.

3.2.7 The foregoing are in addition to other duties and responsibilities of the Owner enumerated herein and especially those in respect to Work by Owner or by Separate Contractors, Payments and Completion, and Insurance in Articles 6, 9 and 11 respectively.

3.3 OWNER'S RIGHT TO STOP THE WORK
3.3.1 If the Contractor fails to correct defective Work as required by Paragraph 13.2 or persistently fails to supply materials or equipment carry out the Work in accordance with the Contract Documents, the Owner, by a written order signed personally or by an agent specifically so empowered by the Owner in writing, may order the Contractor to stop the Work, or any portion thereof, until the cause for such order has been eliminated ; however, this right of the Owner to stop the Work shall not give rise to any duty on the part of the Owner to exercise this right for the benefit of the Contractor or any other person or entity, except to the extent required by Subparagraph 6.1.3.

3.4 OWNER'S RIGHT TO CARRY OUT THE WORK
3.4.1 If the Contractor defaults or neglects to carry out the Work in accordance with the Contract Documents or and fails within seven days after receipt of written notice from the Owner to perform any provision of the Contract commence and continue correction of such default or neglect with diligence and promptness, the Owner may, after seven days following receipt by the Contractor of an additional written notice to the Contractor and without prejudice to any other remedy he may have, make good such deficiencies. In such case an appropriate Change Order shall be issued deducting from the payments then or thereafter due the Contractor the cost of correcting such deficiencies, including the cost of compensation for the Architect's additional services made necessary by such default, neglect or failure. The Architect must approve both Such action by the Owner and the amount charged to the Contractor are both subject to the prior approval of the Architect. If the payments then or thereafter due the Contractor are not sufficient to cover such amount, the Contractor shall pay the difference to the Owner.

ARTICLE 4
CONTRACTOR
4.1 DEFINITION
4.1.1 The Contractor is the person or organization entity identified as such in the Owner-Contractor Agreement and is referred to throughout the Contract Documents as if singular in number and masculine in gender. The term Contractor means the Contractor or his authorized representative.

4.2   REVIEW OF CONTRACT DOCUMENTS

4.2.1 The Contractor shall carefully study and compare the Contract Documents and shall at once report to the Architect any error, inconsistency or omission he may discover. The Contractor shall not be liable to the Owner or the Architect for any damage resulting from any such errors, inconsistencies or omissions in the Contract Documents. The Contractor shall do no Work without Drawings, Specifications or Modifications.

4.2   REVIEW OF CONTRACT DOCUMENTS

4.2.1   The Contractor shall carefully study and compare the Contract Documents and shall at once report to the Architect any error, inconsistency or omission he may discover. The Contractor shall not be liable to the Owner or the Architect for any damage resulting from any such errors, inconsistencies or omissions in the Contract Documents. The Contractor shall ~~do~~ perform no portion of the Work at any time without ~~Drawings, Specifications or Modifications~~ Contract Documents or, where required, approved Shop Drawings, Product Data or Samples for such portion of the Work.

4.3   SUPERVISION AND CONSTRUCTION PROCEDURES

4.3.1 The Contractor shall supervise and direct the Work, using his best skill and attention. He shall be solely responsible for all construction means, methods, techniques, sequences and procedures and for coordinating all portions of the Work under the Contract.

4.3   SUPERVISION AND CONSTRUCTION PROCEDURES

4.3.1   The Contractor shall supervise and direct the Work, using his best skill and attention. He shall be solely responsible for all construction means, methods, techniques, sequences and procedures and for coordinating all portions of the Work under the Contract.

New 4.3.2 replaces former 4.10.1   ⟶

4.3.2   The Contractor shall be responsible to the Owner for the acts and omissions of ~~all~~ his employees, ~~and all~~ Subcontractors and their agents and employees , and ~~all~~ other persons performing any of the Work under a contract with the Contractor.

New 4.3.3 replaces former 7.8.5   ⟶

4.3.3   ~~Neither the observations of the Architect in his Administration of the Construction Contract, nor inspections, tests or approvals by persons other than~~ The Contractor shall ~~relieve~~ not be relieved ~~the Contractor~~ from his obligations to perform the Work in accordance with the Contract Documents either by the activities or duties of the Architect in his administration of the Contract, or by inspections, tests or approvals required or performed under Paragraph 7.7 by persons other than the Contractor.

4.4   LABOR AND MATERIALS

4.4.1 Unless otherwise specifically noted, the Contractor shall provide and pay for all labor, materials, equipment, tools, construction equipment and machinery, water, heat, utilities, transportation, and other facilities and services necessary for the proper execution and completion of the Work.

4.4   LABOR AND MATERIALS

4.4.1   Unless otherwise ~~specifically noted~~ provided in the Contract Documents, the Contractor shall provide and pay for all labor, materials, equipment, tools, construction equipment and machinery, water, heat, utilities, transportation, and other facilities and services necessary for the proper execution and completion of the Work , whether temporary or permanent and whether or not incorporated or to be incorporated in the Work.

4.4.2 The Contractor shall at all times enforce strict discipline and good order among his employees and shall not employ on the Work any unfit person or anyone not skilled in the task assigned to him.

4.4.2   The Contractor shall at all times enforce strict discipline and good order among his employees and shall not employ on the Work any unfit person or anyone not skilled in the task assigned to him.

4.5 WARRANTY

4.5 WARRANTY

4.5.1 The Contractor warrants to the Owner and the Architect that all materials and equipment furnished under this Contract will be new unless otherwise specified, and that all Work will be of good quality, free from faults and defects and in conformance with the Contract Documents. All Work not so conforming to these standards may be considered defective. If required by the Architect, the Contractor shall furnish satisfactory evidence as to the kind and quality of materials and equipment.

4.5.1 The Contractor warrants to the Owner and the Architect that all materials and equipment furnished under this Contract will be new unless otherwise specified, and that all Work will be of good quality, free from faults and defects and in conformance with the Contract Documents. All Work not ~~so~~ conforming to these ~~standards~~ requirements, including substitutions not properly approved and authorized, may be considered defective. If required by the Architect, the Contractor shall furnish satisfactory evidence as to the kind and quality of materials and equipment. This warranty is not limited by the provisions of Paragraph 13.2.

4.6 TAXES

4.6 TAXES

4.6.1 The Contractor shall pay all sales, consumer, use and other similar taxes required by law.

4.6.1 The Contractor shall pay all sales, consumer, use and other similar taxes ~~required by law.~~ for the Work or portions thereof provided by the Contractor which are legally enacted at the time bids are received, whether or not yet effective.

4.7 PERMITS, FEES AND NOTICES

4.7 PERMITS, FEES AND NOTICES

4.7.1 The Contractor shall secure and pay for all permits, governmental fees and licenses necessary for the proper execution and completion of the Work, which are applicable at the time the bids are received. It is not the responsibility of the Contractor to make certain that the Drawings and Specifications are in accordance with applicable laws, statutes, building codes and regulations.

4.7.1 Unless otherwise provided in the Contract Documents, the Contractor shall secure and pay for the building permit and for all other permits and governmental fees , ~~and~~ licenses and inspections necessary for the proper execution and completion of the Work which are customarily secured after execution of the Contract and which are ~~applicable~~ legally required at the time the bids are received.

Last sentence of former 4.7.1 included in new 4.7.3

4.7.2 The Contractor shall give all notices and comply with all laws, ordinances, rules, regulations and orders of any public authority bearing on the performance of the Work. If the Contractor observes that any of the Contract Documents are at variance therewith in any respect, he shall promptly notify the Architect in writing, and any necessary changes shall be adjusted by appropriate Modification. If the Contractor performs any Work knowing it to be contrary to such laws, ordinances, rules and regulations, and without such notice to the Architect, he shall assume full responsibility therefor and shall bear all costs attributable thereto.

4.7.2 The Contractor shall give all notices and comply with all laws, ordinances, rules, regulations and lawful orders of any public authority bearing on the performance of the Work.

Second sentence of former 4.7.2 included in new 4.7.3

Last sentence of former 4.7.2 is new 4.7.4

New 4.7.3 replaces portions of former 4.7.1 and 4.7.2

4.7.3 It is not the responsibility of the Contractor to make certain that the ~~Drawings and Specifications~~ Contract Documents are in accordance with applicable laws, statutes, building codes and regulations. If the Contractor observes that any of the Contract Documents are at variance therewith in any respect, he shall promptly notify the Architect in writing, and any necessary changes shall be ~~adjusted~~ accomplished by appropriate Modification.

1970 Edition	1976 Edition

1970 Edition

New 4.7.4 is last sentence of
former 4.7.2 ⟶

4.8   CASH ALLOWANCES

4.8.1 The Contractor shall include in the Contract Sum all allowances stated in the Contract Documents. These allowances shall cover the net cost of the materials and equipment delivered and unloaded at the site, and all applicable taxes. The Contractor's handling costs on the site, labor, installation costs, overhead, profit and other expenses contemplated for the original allowance shall be included in the Contract Sum and not in the allowance. The Contractor shall cause the Work covered by these allowances to be performed for such amounts and by such persons as the Architect may direct, but he will not be required to employ persons against whom he makes a reasonable objection. If the cost, when determined, is more than or less than the allowance, the Contract Sum shall be adjusted accordingly by Change Order which will include additional handling costs on the site, labor, installation costs, overhead, profit and other expenses resulting to the Contractor from any increase over the original allowance.

New 4.8.2 replaces portions of
former 4.8.1 ⟶

4.9   SUPERINTENDENT

4.9.1 The Contractor shall employ a competent superintendent and necessary assistants who shall be in attendance at the Project site during the progress of the Work. The superintendent shall be satisfactory to the Architect, and shall not be changed except with the consent of the Architect, unless the superintendent proves to be unsatisfactory to the Contractor and ceases to be in his employ. The superintendent shall represent the Contractor and all communications given to the superintendent shall be as binding as if given to the Contractor. Important communications will be confirmed in writing. Other communications will be so confirmed on written request in each case.

1976 Edition

4.7.4   If the Contractor performs any Work knowing it to be contrary to such laws, ordinances, rules and regulations, and without such notice to the Architect, he shall assume full responsibility therefor and shall bear all costs attributable thereto.

4.8   ~~CASH~~ ALLOWANCES

4.8.1   The Contractor shall include in the Contract Sum all allowances stated in the Contract Documents. Items covered by these allowances shall be supplied for such amounts and by such persons as the Owner may direct, but the Contractor will not be required to employ persons against whom he makes a reasonable objection.

4.8.2   Unless otherwise provided in the Contract Documents:

.1   these allowances shall cover the ~~net~~ cost to the Contractor, less any applicable trade discount, of the materials and equipment required by the allowance delivered ~~and unloaded~~ at the site, and all applicable taxes;

.2   the Contractor's ~~handling~~ costs for unloading and handling on the site, labor, installation costs, overhead, profit and other expenses contemplated for the original allowance shall be included in the Contract Sum and not in the allowance ; ~~The Contractor shall cause the Work covered by these allowances to be performed for such amounts and by such persons as the Architect may direct, but he will not be required to employ persons against whom he makes a reasonable objection.~~

.3   ~~If~~ whenever the cost ~~, when determined,~~ is more than or less than the allowance, the Contract Sum shall be adjusted accordingly by Change Order , the amount of which will ~~include additional~~ recognize changes, if any, in handling costs on the site, labor, installation costs, overhead, profit and other expenses . ~~resulting to the Contractor from any increase over the original allowance.~~

4.9   SUPERINTENDENT

4.9.1   The Contractor shall employ a competent superintendent and necessary assistants who shall be in attendance at the Project site during the progress of the Work. ~~The superintendent shall be satisfactory to the Architect, and shall not be changed except with the consent of the Architect, unless the superintendent proves to be unsatisfactory to the Contractor and ceases to be in his employ.~~ The superintendent shall represent the Contractor and all communications given to the superintendent shall be as binding as if given to the Contractor. Important communications ~~will~~ shall be confirmed in writing. Other communications ~~will~~ shall be so confirmed on written request in each case.

-12-

4.10 RESPONSIBILITY FOR THOSE PERFORMING THE WORK

4.10.1 The Contractor shall be responsible to the Owner for the acts and omissions of all his employees and all Subcontractors, their agents and employees, and all other persons performing any of the Work under a contract with the Contractor.

4.11 PROGRESS SCHEDULE

4.11.1 The Contractor, immediately after being awarded the Contract, shall prepare and submit for the Architect's approval an estimated progress schedule for the Work. The progress schedule shall be related to the entire Project to the extent required by the Contract Documents. This schedule shall indicate the dates for the starting and completion of the various stages of construction and shall be revised as required by the conditions of the Work, subject to the Architect's approval.

4.12 DRAWINGS AND SPECIFICATIONS AT THE SITE

4.12.1 The Contractor shall maintain at the site for the Owner one copy of all Drawings, Specifications, Addenda, approved Shop Drawings, Change Orders and other Modifications, in good order and marked to record all changes made during construction. These shall be available to the Architect. The Drawings, marked to record all changes made during construction, shall be delivered to him for the Owner upon completion of the Work.

4.13 SHOP DRAWINGS AND SAMPLES

4.13.1 Shop Drawings are drawings, diagrams, illustrations, schedules, performance charts, brochures and other data which are prepared by the Contractor or any Subcontractor, manufacturer, supplier or distributor, and which illustrate some portion of the Work.

New Subparagraph   →

4.13.2 Samples are physical examples furnished by the Contractor to illustrate materials, equipment or workmanship, and to establish standards by which the Work will be judged.

← Former 4.10.1 becomes new 4.3.2

4.10   PROGRESS SCHEDULE

4.10.1 The Contractor, immediately after being awarded the Contract, shall prepare and submit for the Owner's and Architect's ~~approval~~ information an estimated progress schedule for the Work. The progress schedule shall be related to the entire Project to the extent required by the Contract Documents , and shall provide for expeditious and practicable execution of the Work. ~~This schedule shall indicate the dates for the starting and completion of the various stages of construction and shall be revised as required by the conditions of the Work, subject to the Architect's approval.~~

4.11   ~~DRAWINGS AND SPECIFICATIONS~~ DOCUMENTS AND SAMPLES AT THE SITE

4.11.1 The Contractor shall maintain at the site for the Owner one record copy of all Drawings, Specifications, Addenda, ~~approved Shop Drawings,~~ Change Orders and other Modifications, in good order and marked currently to record all changes made during construction , and approved Shop Drawings, Product Data and Samples. These shall be available to the Architect ~~. The Drawings, marked to record all changes made during construction~~ and shall be delivered to him for the Owner upon completion of the Work.

4.12   SHOP DRAWINGS , PRODUCT DATA AND SAMPLES

4.12.1 Shop Drawings are drawings, diagrams, ~~illustrations,~~ schedules ~~, performance charts, brochures~~ and other data ~~which are~~ specially prepared for the Work by the Contractor or any Subcontractor, manufacturer, supplier or distributor ~~, and which~~ to illustrate some portion of the Work.

4.12.2 Product Data are illustrations, standard schedules, performance charts, instructions, brochures, diagrams and other information furnished by the Contractor to illustrate a material, product or system for some portion of the Work.

4.12.3 Samples are physical examples ~~furnished by the Contractor to~~ which illustrate materials, equipment or workmanship ~~,~~ and ~~to~~ establish standards by which the Work will be judged.

4.13.3 The Contractor shall review, stamp with his approval and submit, with reasonable promptness and in orderly sequence so as to cause no delay in the Work or in the work of any other contractor, all Shop Drawings and Samples required by the Contract Documents or subsequently by the Architect as covered by Modifications. Shop Drawings and Samples shall be properly identified as specified, or as the Architect may require. At the time of submission the Contractor shall inform the Architect in writing of any deviation in the Shop Drawings or Samples from the requirements of the Contract Documents.

4.13.4 By approving and submitting Shop Drawings and Samples, the Contractor thereby represents that he has determined and verified all field measurements, field construction criteria, materials, catalog numbers and similar data, or will do so, and that he has checked and coordinated each Shop Drawing and Sample with the requirements of the Work and of the Contract Documents.

4.13.5 The Architect will review and approve Shop Drawings and Samples with reasonable promptness so as to cause no delay, but only for conformance with the design concept of the Project and with the information given in the Contract Documents. The Architect's approval of a separate item shall not indicate approval of an assembly in which the item functions.

New 4.12.6 replaces former 4.13.7 ➤

4.13.6 The Contractor shall make any corrections required by the Architect and shall resubmit the required number of corrected copies of Shop Drawings or new Samples until approved. The Contractor shall direct specific attention in writing or on resubmitted Shop Drawings to revisions other than the corrections requested by the Architect on previous submissions.

4.12.4 The Contractor shall review, ~~stamp with his approval~~ approve and submit, with reasonable promptness and in ~~orderly~~ such sequence ~~so~~ as to cause no delay in the Work or in the work of the Owner or any ~~other~~ separate contractor, all Shop Drawings , Product Data and Samples required by the Contract Documents . ~~or subsequently by the Architect as covered by Modifications. Shop Drawings and Samples shall be properly identified as specified, or as the Architect may require. At the time of submission the Contractor shall inform the Architect in writing of any deviation in the Shop Drawings or Samples from the requirements of the Contract Documents.~~

4.12.5 By approving and submitting Shop Drawings , Product Data and Samples, the Contractor ~~thereby~~ represents that he has determined and verified all materials, field measurements, and field construction criteria ~~, materials, catalog numbers and similar data,~~ related thereto, or will do so, and that he has checked and coordinated ~~each Shop Drawing and Sample~~ the information contained within such submittals with the requirements of the Work and of the Contract Documents.

◀ Former 4.13.5 included in new 2.2.14

4.12.6 ~~The Architect's approval of Shop Drawings and Samples~~ The Contractor shall not ~~relieve~~ be relieved ~~the Contractor~~ of responsibility for any deviation from the requirements of the Contract Documents by the Architect's approval of Shop Drawings, Product Data or Samples under Subparagraph 2.2.14 unless the Contractor has specifically informed the Architect in writing of such deviation at the time of submission and the Architect has given written approval to the specific deviation . ~~nor shall~~ The ~~Architect's approval relieve the~~ Contractor shall not be relieved from responsibility for errors or omissions in the Shop Drawings, Product Data or Samples by the Architect's approval thereof.

4.12.7 ~~The Contractor shall make any corrections required by the Architect and shall resubmit the required number of corrected copies of Shop Drawings or new Samples until approved.~~ The Contractor shall direct specific attention, in writing or on resubmitted Shop Drawings , Product Data or Samples, to revisions other than ~~the corrections~~ those requested by the Architect on previous ~~submissions.~~ submittals.

4.13.7 The Architect's approval of Shop Drawings or Samples shall not relieve the Contractor of responsibility for any deviation from the requirements of the Contract Documents unless the Contractor has informed the Architect in writing of such deviation at the time of submission and the Architect has given written approval to the specific deviation, nor shall the Architect's approval relieve the Contractor from responsibility for errors or omissions in the Shop Drawings or Samples.

Former 4.13.7 becomes new 4.12.6

4.13.8 No portion of the Work requiring a Shop Drawing or Sample submission shall be commenced until the submission has been approved by the Architect. All such portions of the Work shall be in accordance with approved Shop Drawings and Samples.

4.12.8 No portion of the Work requiring submission of a Shop Drawing , Product Data or Sample submission shall be commenced until the submission submittal has been approved by the Architect as provided in Subparagraph 2.2.14. All such portions of the Work shall be in accordance with approved Shop Drawings and Samples. submittals.

4.14 USE OF SITE
4.14.1 The Contractor shall confine operations at the site to areas permitted by law, ordinances, permits and the Contract Documents and shall not unreasonably encumber the site with any materials or equipment.

4.13 USE OF SITE

4.13.1 The Contractor shall confine operations at the site to areas permitted by law, ordinances, permits and the Contract Documents and shall not unreasonably encumber the site with any materials or equipment.

4.15 CUTTING AND PATCHING OF WORK
4.15.1 The Contractor shall do all cutting, fitting or patching of his Work that may be required to make its several parts fit together properly, and shall not endanger any Work by cutting, excavating or otherwise altering the Work or any part of it.

4.14 CUTTING AND PATCHING OF WORK

4.14.1 The Contractor shall do be responsible for all cutting, fitting or patching of his Work that may be required to complete the Work or to make its several parts fit together properly . and shall not endanger any Work by cutting, excavating or otherwise altering the Work or any part of it.

New 4.14.2 replaces former 6.3.1 →

4.14.2 The Contractor shall be responsible for any cutting, fitting and patching that may be required to complete his Work except as otherwise specifically provided in the Contract Documents. The Contractor shall not damage or endanger any portion of the Work or the work of the Owner or any other separate contractors by cutting, excavating patching or otherwise altering any work , or by excavation. and The Contractor shall not cut or otherwise alter the work of the Owner or any other separate contractor except with the written consent of the Architect Owner and of such separate contractor. The Contractor shall not unreasonably withhold from the Owner or any separate contractor his consent to cutting or otherwise altering the Work.

4.16 CLEANING UP
4.16.1 The Contractor at all times shall keep the premises free from accumulation of waste materials or rubbish caused by his operations. At the completion of the Work he shall remove all his waste materials and rubbish from and about the Project as well as all his tools, construction equipment, machinery and surplus materials, and shall clean all glass surfaces and leave the Work "broom-clean" or its equivalent, except as otherwise specified.

4.15 CLEANING UP

4.15.1 The Contractor at all times shall keep the premises free from accumulation of waste materials or rubbish caused by his operations. At the completion of the Work he shall remove all his waste materials and rubbish from and about the Project as well as all his tools, construction equipment, machinery and surplus materials . and shall clean all glass surfaces and leave the Work "broom-clean" or its equivalent, except as otherwise specified.

4.16.2 If the Contractor fails to clean up, the Owner may do so and the cost thereof shall be charged to the Contractor as provided in Paragraph 3.4.

4.15.2 If the Contractor fails to clean up ~~at the completion of the Work~~, the Owner may do so ~~as provided in Paragraph 3.4~~ and the cost thereof shall be charged to the Contractor . ~~as provided in Paragraph 3.4~~

4.16   COMMUNICATIONS

4.16.1   The Contractor shall forward all communications to the Owner through the Architect.

4.17   COMMUNICATIONS
4.17.1 The Contractor shall forward all communications to the Owner through the Architect.

4.17   ROYALTIES AND PATENTS

4.17.1   The Contractor shall pay all royalties and license fees. He shall defend all suits or claims for infringement of any patent rights and shall save the Owner harmless from loss on account thereof, except that the Owner shall be responsible for all such loss when a particular design, process or the product of a particular manufacturer or manufacturers is specified, but if the Contractor has reason to believe that the design, process or product specified is an infringement of a patent, he shall be responsible for such loss unless he promptly gives such information to the Architect.

New 4.17.1 is former 7.7.1   ⟶

4.18   INDEMNIFICATION
4.18.1 The Contractor shall indemnify and hold harmless the Owner and the Architect and their agents and employees from and against all claims, damages, losses and expenses including attorneys' fees arising out of or resulting from the performance of the Work, provided that any such claim, damage, loss or expense (1) is attributable to bodily injury, sickness, disease or death, or to injury to or destruction of tangible property (other than the Work itself) including the loss of use resulting therefrom, and (2) is caused in whole or in part by any negligent act or omission of the Contractor, any Subcontractor, anyone directly or indirectly employed by any of them or anyone for whose acts any of them may be liable, regardless of whether or not it is caused in part by a party indemnified hereunder.

4.18   INDEMNIFICATION

4.18.1   To the fullest extent permitted by law, the Contractor shall indemnify and hold harmless the Owner and the Architect and their agents and employees from and against all claims, damages, losses and expenses, including but not limited to attorneys' fees, arising out of or resulting from the performance of the Work, provided that any such claim, damage, loss or expense (1) is attributable to bodily injury, sickness, disease or death, or to injury to or destruction of tangible property (other than the Work itself) including the loss of use resulting therefrom, and (2) is caused in whole or in part by any negligent act or omission of the Contractor, any Subcontractor, anyone directly or indirectly employed by any of them or, anyone for whose acts any of them may be liable, regardless of whether or not it is caused in part by a party indemnified hereunder. Such obligation shall not be construed to negate, abridge, or otherwise reduce any other right or obligation of indemnity which would otherwise exist as to any party or person described in this Paragraph 4.18.

4.18.2 In any and all claims against the Owner or the Architect or any of their agents or employees by any employee of the Contractor, any Subcontractor, anyone directly or indirectly employed by any of them or anyone for whose acts any of them may be liable, the indemnification obligation under this Paragraph 4.18 shall not be limited in any way by any limitation on the amount or type of damages, compensation or benefits payable by or for the Contractor or any Subcontractor under workmen's compensation acts, disability benefit acts or other employee benefit acts.

4.18.2   In any and all claims against the Owner or the Architect or any of their agents or employees by any employee of the Contractor, any Subcontractor, anyone directly or indirectly employed by any of them or anyone for whose acts any of them may be liable, the indemnification obligation under this Paragraph 4.18 shall not be limited in any way by any limitation on the amount or type of damages, compensation or benefits payable by or for the Contractor or any Subcontractor under workers' or workmen's compensation acts, disability benefit acts or other employee benefit acts.

| |

1970 Edition

4.18.3 The obligations of the Contractor under this Paragraph 4.18 shall not extend to the liability of the Architect, his agents or employees arising out of (1) the preparation or approval of maps, drawings, opinions, reports, surveys, Change Orders, designs or specifications, or (2) the giving of or the failure to give directions or instructions by the Architect, his agents or employees provided such giving or failure to give is the primary cause of the injury or damage.

ARTICLE 5

SUBCONTRACTORS

5.1 DEFINITION

5.1.1 A Subcontractor is a person or organization who has a direct contract with the Contractor to perform any of the Work at the site. The term Subcontractor is referred to throughout the Contract Documents as if singular in number and masculine in gender and means a Subcontractor or his authorized representative.

5.1.2 A Sub-subcontractor is a person or organization who has a direct or indirect contract with a Subcontractor to perform any of the Work at the site. The term Sub-subcontractor is referred to throughout the Contract Documents as if singular in number and masculine in gender and means a Sub-subcontractor or an authorized representative thereof.

5.1.3 Nothing contained in the Contract Documents shall create any contractual relation between the Owner or the Architect and any Subcontractor or Sub-subcontractor.

5.2 AWARD OF SUBCONTRACTS AND OTHER CONTRACTS FOR PORTIONS OF THE WORK

5.2.1 Unless otherwise specified in the Contract Documents or in the Instructions to Bidders, the Contractor, as soon as practicable after the award of the Contract, shall furnish to the Architect in writing for acceptance by the Owner and the Architect a list of the names of the Subcontractors proposed for the principal portions of the Work. The Architect shall promptly notify the Contractor in writing if either the Owner or the Architect, after due investigation, has reasonable objection to any Subcontractor on such list and does not accept him. Failure of the Owner or Architect to make objection promptly to any Subcontractor on the list shall constitute acceptance of such Subcontractor.

1976 Edition

4.18.3 The obligations of the Contractor under this Paragraph 4.18 shall not extend to the liability of the Architect, his agents or employees, arising out of (1) the preparation or approval of maps, drawings, opinions, reports, surveys, change orders, designs or specifications, or (2) the giving of or the failure to give directions or instructions by the Architect, his agents or employees providing such giving or failure to give is the primary cause of the injury or damage.

ARTICLE 5

SUBCONTRACTORS

5.1 DEFINITION

5.1.1 A Subcontractor is a person or ~~organization~~ entity who has a direct contract with the Contractor to perform any of the Work at the site. The term Subcontractor is referred to throughout the Contract Documents as if singular in number and masculine in gender and means a Subcontractor or his authorized representative. The term Subcontractor does not include any separate contractor or his subcontractors.

5.1.2 A Sub-subcontractor is a person or ~~organization~~ entity who has a direct or indirect contract with a Subcontractor to perform any of the Work at the site. The term Sub-subcontractor is referred to throughout the Contract Documents as if singular in number and masculine in gender and means a Sub-subcontractor or an authorized representative thereof.

◄ Former 5.1.3 included in new 1.1.2

5.2 AWARD OF SUBCONTRACTS AND OTHER CONTRACTS FOR PORTIONS OF THE WORK

5.2.1 Unless otherwise ~~specified in~~ required by the Contract Documents or ~~in~~ the ~~Instructions to Bidders~~ Bidding Documents, the Contractor, as soon as practicable after the award of the Contract, shall furnish to the Owner and the Architect in writing ~~for acceptance by the Owner and the Architect a list of~~ the names of the ~~Subcontractors~~ persons or entities (including those who are to furnish materials or equipment fabricated to a special design) proposed for each of the principal portions of the Work. The Architect ~~shall~~ will promptly ~~notify~~ reply to the Contractor in writing stating whether or not ~~if either~~ the Owner or the Architect, after due investigation, has reasonable objection to any ~~Subcontractor~~ such proposed person or entity. ~~on such list and does not accept him.~~ Failure of the Owner or Architect to ~~make objection promptly to any Subcontractor on the list~~ reply promptly shall constitute ~~acceptance of such Subcontractor~~ notice of no reasonable objection.

-17-

1970 Edition	1976 Edition

5.2.2 The Contractor shall not contract with any Subcontractor or any person or organization (including those who are to furnish materials or equipment fabricated to a special design) proposed for portions of the Work designated in the Contract Documents or in the Instructions to Bidders or, if none is so designated, with any Subcontractor proposed for the principal portions of the Work who has been rejected by the Owner and the Architect. The Contractor will not be required to contract with any Subcontractor or person or organization against whom he has a reasonable objection.

5.2.2 The Contractor shall not contract with any ~~Subcontractor or any~~ such proposed person or entity ~~organization (including those who are to furnish materials or equipment fabricated to a special design) proposed for portions of the Work designated in the Contract Documents or in the Instructions to Bidders or, if none is so designated with any Subcontractor proposed for the principal portions of the Work who has been rejected by~~ to whom the Owner ~~and~~ or the Architect has made reasonable objection under the provisions of Subparagraph 5.2.1. The Contractor ~~will~~ shall not be required to contract with ~~any Subcontractor or person or organization against~~ anyone to whom he has a reasonable objection.

5.2.3 If the Owner or Architect refuses to accept any Subcontractor or person or organization on a list submitted by the Contractor in response to the requirements of the Contract Documents or the Instructions to Bidders, the Contractor shall submit an acceptable substitute and the Contract Sum shall be increased or decreased by the difference in cost occasioned by such substitution and an appropriate Change Order shall be issued; however, no increase in the Contract Sum shall be allowed for any such substitution unless the Contractor has acted promptly and responsively. in submitting for acceptance any list or lists of names as required by the Contract Documents or the Instructions to Bidders.

5.2.3 If the Owner or the Architect ~~refuses to accept~~ has reasonable objection to any such proposed ~~Subcontractor or~~ person or entity, ~~organization on a list submitted by the Contractor in response to the requirements of the Contract Documents or the Instructions to Bidders~~ an ~~acceptable~~ a substitute to whom the Owner or the Architect has no reasonable objection, and the Contract Sum shall be increased or decreased by the difference in cost occasioned by such substitution and an appropriate Change Order shall be issued; however, no increase in the Contract Sum shall be allowed for any such substitution unless the Contractor has acted promptly and responsively in submitting ~~for acceptance any list or lists of~~ names as required by ~~the Contract Documents or the Instructions to Bidders~~ Subparagraph 5.2.1.

5.2.4 If the Owner or the Architect requires a change of any proposed Subcontractor or person or organization previously accepted by them, the Contract Sum shall be increased or decreased by the difference in cost occasioned by such change and an appropriate Change Order shall be issued.

Former 5.2.4 Deleted

5.2.5 The Contractor shall not make any substitution for any Subcontractor or person or organization who has been accepted by the Owner and the Architect, unless the substitution is acceptable to the Owner and the Architect.

5.2.4 The Contractor shall ~~not~~ make ~~any~~ no substitution for any Subcontractor ~~or~~ , person or entity previously selected if ~~organization who has been accepted by~~ the Owner ~~and the~~ or Architect ~~unless the~~ makes reasonable objection to such substitution . ~~is acceptable to the Owner and the Architect.~~

5.3   SUBCONTRACTUAL RELATIONS

5.3.1 All work performed for the Contractor by a Subcontractor shall be pursuant to an appropriate agreement between the Contractor and the Subcontractor (and where appropriate between Subcontractors and Subsubcontractors) which shall contain provisions that:

  .1 preserve and protect the rights of the Owner and the Architect under the Contract with respect to the Work to be performed under the subcontract so that the subcontracting thereof will not prejudice such rights;

  .2 require that such Work be performed in accordance with the requirements of the Contract Documents;

5.3   SUBCONTRACTUAL RELATIONS

5.3.1 By an appropriate agreement, written where legally required for validity, the Contractor shall require each Subcontractor, to the extent of the Work to be performed by the Subcontractor, to be bound to the Contractor by the terms of the Contract Documents, and to assume toward the Contractor all the obligations and responsibilitites which the Contractor, by these Documents, assumes toward the Owner and the Architect. Said agreement shall preserve and protect the rights of the Owner and the Architect under the Contract Documents with respect to the Work to be performed ~~under the Subcontract~~ by the Subcontractor so that the subcontracting thereof will not prejudice such rights, and shall allow to the Subcontractor, unless specifically provided otherwise in the Contractor-Subcontractor agreement, the benefit of all rights, remedies and redress against the Contractor that the Contractor, by these Documents, has against the Owner. Where appropriate, the Contractor shall require each Subcontractor to enter into similar agreements with his Sub-subcontractors. The Contractor shall make available to each proposed Subcontractor, prior to the execution of the Subcontract, copies of the Contract Documents to which the Subcontractor will be bound by this Paragraph 5.3, and identify to the Subcontractor any terms and conditions of the proposed Subcontract which may be at variance with the Contract Documents. Each Subcontractor shall similarly make copies of such Documents available to his Sub-subcontractors.

.3 require submission to the Contractor of applications for payment under each subcontract to which the Contractor is a party, in reasonable time to enable the Contractor to apply for payment in accordance with Article 9;

.4 require that all claims for additional costs, extensions of time, damages for delays or otherwise with respect to subcontracted portions of the Work shall be submitted to the Contractor (via any Subcontractor or Sub-subcontractor where appropriate) in sufficient time so that the Contractor may comply in the manner provided in the Contract Documents for like claims by the Contractor upon the Owner;

.5 waive all rights the contracting parties may have against one another for damages caused by fire or other perils covered by the property insurance described in Paragraph 11.3, except such rights as they may have to the proceeds of such insurance held by the Owner as trustee under Paragraph 11.3; and

.6 obligate each Subcontractor specifically to consent to the provisions of this Paragraph 5.3.

5.4   PAYMENTS TO SUBCONTRACTORS

5.4.1 The Contractor shall pay each Subcontractor, upon receipt of payment from the Owner, an amount equal to the percentage of completion allowed to the Contractor on account of such Subcontractor's Work, less the percentage retained from payments to the Contractor. The Contractor shall also require each Subcontractor to make similar payments to his subcontractors.

5.4.2 If the Architect fails to issue a Certificate for Payment for any cause which is the fault of the Contractor and not the fault of a particular Subcontractor, the Contractor shall pay that Subcontractor on demand, made at any time after the Certificate for Payment should otherwise have been issued, for his Work to the extent completed, less the retained percentage.

5.4.3 The Contractor shall pay each Subcontractor a just share of any insurance moneys received by the Contractor under Article 11, and he shall require each Subcontractor to make similar payments to his subcontractors.

Former 5.3.1.3 and 5.3.1.4 Deleted

Former 5.3.1.5 included in new 11.3.6

Former 5.3.1.6 Deleted

Former 5.4.1 becomes new 9.5.2

Former 5.4.2 Deleted

Former 5.4.3 included in new 11.3.3

5.4.4 The Architect may, on request and at his discretion, furnish to any Subcontractor, if practicable, information regarding percentages of completion certified to the Contractor on account of Work done by such Subcontractors.

5.4.5 Neither the Owner nor the Architect shall have any obligation to pay or to see to the payment of any moneys to any Subcontractor except as may otherwise be required by law.

← Former 5.4.4 becomes new 9.5.3

← Former 5.4.5 is new 9.5.4

ARTICLE 6

SEPARATE CONTRACTS

6.1 OWNER'S RIGHT TO AWARD SEPARATE CONTRACTS

6.1.1 The Owner reserves the right to award other contracts in connection with other portions of the Project under these or similar Conditions of the Contract.

6.1.2 When separate contracts are awarded for different portions of the Project, "the Contractor" in the contract documents in each case shall be the contractor who signs each separate contract.

New Subparagraph →

6.2 MUTUAL RESPONSIBILITY OF CONTRACTORS

6.2.1 The Contractor shall afford other contractors reasonable opportunity for the introduction and storage of their materials and equipment and the execution of their work, and shall properly connect and coordinate his Work with theirs.

6.2.2 If any part of the Contractor's Work depends for proper execution or results upon the work of any other separate contractor, the Contractor shall inspect and promptly report to the Architect any apparent discrepancies or defects in such work that render it unsuitable for such proper execution and results. Failure of the Contractor so to inspect and report shall constitute an acceptance of the other contractor's work as fit and proper to receive his Work, except as to defects which may develop in the other separate contractor's work after the execution of the Contractor's Work.

ARTICLE 6

WORK BY OWNER OR BY SEPARATE ~~CONTRACTS~~ CONTRACTORS

6.1 OWNER'S RIGHT TO <u>PERFORM WORK AND TO</u> AWARD SEPARATE CONTRACTS

6.1.1 The Owner reserves the right to <u>perform work related to the Project with his own forces, and</u> to award ~~other~~ separate contracts in connection with other portions of the Project <u>or other work on the site</u> under these or similar Conditions of the Contract. <u>If the Contractor claims that delay or additional cost is involved because of such action by the Owner, he shall make such claim as provided elsewhere in the Contract Documents.</u>

6.1.2 When separate contracts are awarded for different portions of the Project <u>or other work</u> on the site, the <u>term</u> Contractor in the Contract Documents in each case shall ~~be~~ <u>mean</u> the Contractor who ~~signs~~ <u>executes</u> each separate ~~contract~~ <u>Owner-Contractor Agreement.</u>

6.1.3 <u>The Owner will provide for the coordination of the work of his own forces and of each separate contractor with the Work of the Contractor, who shall cooperate therewith as provided in Paragraph 6.2.</u>

6.2 MUTUAL RESPONSIBILITY ~~OF CONTRACTORS~~

6.2.1 The Contractor shall afford ~~other~~ <u>the Owner and separate</u> contractors reasonable opportunity for the introduction and storage of their materials and equipment and the execution of their work, and shall ~~properly~~ connect and coordinate his Work with theirs <u>as required by the Contract Documents.</u>

6.2.2 If any part of the Contractor's Work depends for proper execution or results upon the work of <u>the Owner or</u> any ~~other~~ separate contractor, the Contractor shall <u>,</u> ~~inspect and~~ <u>prior to proceeding with the Work,</u> promptly report to the Architect any apparent discrepancies or defects in such <u>other</u> work that render it unsuitable for such proper execution and results. Failure of the Contractor so to ~~inspect and~~ report shall constitute an acceptance of the ~~other~~ <u>Owner's or</u> separate contractors' work as fit and proper to receive his Work, except as to defects which may ~~develop~~ <u>subsequently become</u> apparent in ~~the other separate contractor's~~ <u>such</u> work <u>by others.</u> ~~after the execution of the Contractor's Work.~~

New 6.2.3 is former 6.3.2 ➞

6.2.3 Any costs caused by defective or ill-timed work shall be borne by the party responsible therefor.

New Subparagraph ➞

6.2.4 Should the Contractor wrongfully cause damage to the work or property of the Owner, or to other work on the site, the Contractor shall promptly remedy such damage as provided in Subparagraph 10.2.5.

6.2.3 Should the Contractor cause damage to the work or property of any separate contractor on the Project, the Contractor shall, upon due notice, settle with such other contractor by agreement or arbitration, if he will so settle. If such separate contractor sues the Owner or initiates an arbitration proceeding on account of any damage alleged to have been so sustained, the Owner shall notify the Contractor who shall defend such proceedings at the Owner's expense, and if any judgment or award against the Owner arises therefrom the Contractor shall pay or satisfy it and shall reimburse the Owner for all attorneys' fees and court or arbitration costs which the Owner has incurred.

6.2.5 Should the Contractor wrongfully cause damage to the work or property of any separate contractor ~~on the Project~~, the Contractor shall upon due notice promptly attempt to settle with such other contractor by agreement, or otherwise to resolve the dispute. ~~arbitration, if he will so settle.~~ If such separate contractor sues ~~the Owner~~ or initiates an arbitration proceeding against the Owner on account of any damage alleged to have been ~~so sustained,~~ caused by the Contractor, the Owner shall notify the Contractor who shall defend such proceedings at the Owner's expense, and if any judgment or award against the Owner arises therefrom the Contractor shall pay or satisfy it and shall reimburse the Owner for all attorneys' fees and court or arbitration costs which the Owner has incurred.

6.3 CUTTING AND PATCHING UNDER SEPARATE CONTRACTS

6.3.1 The Contractor shall be responsible for any cutting, fitting and patching that may be required to complete his Work except as otherwise specifically provided in the Contract Documents. The Contractor shall not endanger any work of any other contractors by cutting, excavating or otherwise altering any work and shall not cut or alter the work of any other contractor except with the written consent of the Architect.

⬅ Former 6.3.1 becomes new 4.14.2

6.3.2 Any costs caused by defective or ill-timed work shall be borne by the party responsible therefor.

⬅ Former 6.3.2 is new 6.2.3

6.4 OWNER'S RIGHT TO CLEAN UP

6.4.1 If a dispute arises between the separate contractors as to their responsibility for cleaning up as required by Paragraph 4.16, the Owner may clean up and charge the cost thereof to the several contractors as the Architect shall determine to be just.

6.3 OWNER'S RIGHT TO CLEAN UP

6.3.1 If a dispute arises between the Contractor and separate contractors as to their responsibility for cleaning up as required by Paragraph ~~4.16~~ 4.15, the Owner may clean up and charge the cost thereof to the ~~several~~ contractors responsible therefor as the Architect shall determine to be just.

ARTICLE 7

MISCELLANEOUS PROVISIONS

7.1 GOVERNING LAW

7.1.1 The Contract shall be governed by the law of the place where the Project is located.

ARTICLE 7

MISCELLANEOUS PROVISIONS

7.1 GOVERNING LAW

7.1.1 The Contract shall be governed by the law of the place where the Project is located.

7.2 SUCCESSORS AND ASSIGNS

7.2.1 The Owner and the Contractor each binds himself, his partners, successors, assigns and legal representatives to the other party hereto and to the partners, successors, assigns and legal representatives of such other party in respect to all covenants, agreements and obligations contained in the Contract Documents. Neither party to the Contract shall assign the Contract or sublet it as a whole without the written consent of the other, nor shall the Contractor assign any moneys due or to become due to him hereunder, without the previous written consent of the Owner.

7.3 WRITTEN NOTICE

7.3.1 Written notice shall be deemed to have been duly served if delivered in person to the individual or member of the firm or to an officer of the corporation for whom it was intended, or if delivered at or sent by registered or certified mail to the last business address known to him who gives the notice.

7.4 CLAIMS FOR DAMAGES

7.4.1 Should either party to the Contract suffer injury or damage to person or property because of any act or omission of the other party or of any of his employees, agents or others for whose acts he is legally liable, claim shall be made in writing to such other party within a reasonable time after the first observance of such injury or damage.

7.5 PERFORMANCE BOND AND LABOR AND MATERIAL PAYMENT BOND

7.5.1 The Owner shall have the right to require the Contractor to furnish bonds covering the faithful performance of the Contract and the payment of all obligations arising thereunder if and as required in the Instructions to Bidders or elsewhere in the Contract Documents.

7.6 RIGHTS AND REMEDIES

7.6.1 The duties and obligations imposed by the Contract Documents and the rights and remedies available thereunder shall be in addition to and not a limitation of any duties, obligations, rights and remedies otherwise imposed or available by law.

New Subparagraph   ━━━▶

7.2   SUCCESSORS AND ASSIGNS

7.2.1   The Owner and the Contractor each binds himself, his partners, successors, assigns and legal representatives to the other party hereto and to the partners, successors, assigns and legal representatives of such other party in respect to all covenants, agreements and obligations contained in the Contract Documents. Neither party to the Contract shall assign the Contract or sublet it as a whole without the written consent of the other, nor shall the Contractor assign any moneys due or to become due to him hereunder, without the previous written consent of the Owner.

7.3   WRITTEN NOTICE

7.3.1   Written notice shall be deemed to have been duly served if delivered in person to the individual or member of the firm or entity or to an officer of the corporation for whom it was intended, or if delivered at or sent by registered or certified mail to the last business address known to him who gives the notice.

7.4   CLAIMS FOR DAMAGES

7.4.1   Should either party to the Contract suffer injury or damage to person or property because of any act or omission of the other party or of any of his employees, agents or others for whose acts he is legally liable, claim shall be made in writing to such other party within a reasonable time after the first observance of such injury or damage.

7.5   PERFORMANCE BOND AND LABOR AND MATERIAL PAYMENT BOND

7.5.1   The Owner shall have the right to require the Contractor to furnish bonds covering the faithful performance of the Contract and the payment of all obligations arising thereunder if and as required in the ~~Instructions to Bidders~~ Bidding Documents or ~~elsewhere~~ in the Contract Documents.

7.6   RIGHTS AND REMEDIES

7.6.1   The duties and obligations imposed by the Contract Documents and the rights and remedies available thereunder shall be in addition to and not a limitation of any duties, obligations, rights and remedies otherwise imposed or available by law.

7.6.2   No action or failure to act by the Owner, Architect or Contractor shall constitute a waiver of any right or duty afforded any of them under the Contract, nor shall any such action or failure to act constitute an approval of or acquiescence in any breach thereunder, except as may be specifically agreed in writing.

-22-

7.7   ROYALTIES AND PATENTS

7.7.1 The Contractor shall pay all royalties and license fees. He shall defend all suits or claims for infringement of any patent rights and shall save the Owner harmless from loss on account thereof, except that the Owner shall be responsible for all such loss when a particular design, process or the product of a particular manufacturer or manufacturers is specified, but if the Contractor has reason to believe that the design, process or product specified is an infringement of a patent, he shall be responsible for such loss unless he promptly gives such information to the Architect.

Former 7.7.1 is new 4.17.1

7.8   TESTS

7.8.1 If the Contract Documents, laws, ordinances, rules, regulations or orders of any public authority having jurisdiction require any Work to be inspected, tested or approved, the Contractor shall give the Architect timely notice of its readiness and of the date arranged so the Architect may observe such inspection, testing or approval. The Contractor shall bear all costs of such inspections, tests and approvals unless otherwise provided.

7.7   TESTS

7.7.1 If the Contract Documents, laws, ordinances, rules, regulations or orders of any public authority having jurisdiction require any portion of the Work to be inspected, tested or approved, the Contractor shall give the Architect timely notice of its readiness ~~and of the date arranged~~ so the Architect may observe such inspection, testing or approval. The Contractor shall bear all costs of such inspections, tests ~~and~~ or approvals conducted by public authorities. Unless otherwise provided, the Owner shall bear all costs of other inspections, tests or approvals.

7.8.2 If after the commencement of the Work the Architect determines that any Work requires special inspection, testing, or approval which Subparagraph 7.8.1 does not include, he will, upon written authorization from the Owner, instruct the Contractor to order such special inspection, testing or approval, and the Contractor shall give notice as in Subparagraph 7.8.1. If such special inspection or testing reveals a failure of the Work to comply (1) with the requirements of the Contract Documents or (2), with respect to the performance of the Work, with laws, ordinances, rules, regulations or orders of any public authority having jurisdiction, the Contractor shall bear all costs thereof, including the Architect's additional services made necessary by such failure; otherwise the Owner shall bear such costs, and an appropriate Change Order shall be issued.

7.7.2 If ~~after the commencement of the Work~~ the Architect determines that any Work requires special inspection, testing, or approval which Subparagraph ~~7.8.1~~ 7.7.1 does not include, he will, upon written authorization from the Owner, instruct the Contractor to order such special inspection, testing or approval, and the Contractor shall give notice as provided in Subparagraph ~~7.8.1~~ 7.7.1. If such special inspection or testing reveals a failure of the Work to comply ~~(1)~~ with the requirements of the Contract Documents~~, or (2) with respect to the performance of the Work, with laws, ordinances, rules, regulations or orders of any public authority having jurisdiction~~, the Contractor shall bear all costs thereof, including compensation for the Architect's additional services made necessary by such failure; otherwise the Owner shall bear such costs, and an appropriate Change Order shall be issued.

7.8.3 Required certificates of inspection, testing or approval shall be secured by the Contractor and promptly delivered by him to the Architect.

7.7.3 Required certificates of inspection, testing or approval shall be secured by the Contractor and promptly delivered by him to the Architect.

7.8.4 If the Architect wishes to observe the inspections, tests or approvals required by this Paragraph 7.8, he will do so promptly and, where practicable, at the source of supply.

7.7.4 If the Architect ~~wishes~~ is to observe the inspections, tests or approvals required by ~~this Paragraph 7.8~~ the Contract Documents, he will do so promptly and, where practicable, at the source of supply.

7.8.5 Neither the observations of the Architect in his Administration of the Construction Contract, nor inspections, tests or approvals by persons other than the Contractor shall relieve the Contractor from his obligations to perform the Work in accordance with the Contract Documents.

Former 7.8.5 becomes new 4.3.3

7.9 INTEREST

7.9.1 Any moneys not paid when due to either party under this Contract shall bear interest at the legal rate in force at the place of the Project.

7.8 INTEREST

7.8.1 ~~Any moneys not paid when due to either party~~ Payments due and unpaid under ~~this~~ the Contract Documents shall bear interest from the date payment is due at such rate as the parties may agree upon in writing or, in the absence thereof, at the legal rate ~~in force~~ prevailing at the place of the Project.

7.10 ARBITRATION

7.10.1 All claims, disputes and other matters in question arising out of, or relating to, this Contract or the breach thereof, except as set forth in Subparagraph 2.2.9 with respect to the Architect's decisions on matters relating to artistic effect, and except for claims which have been waived by the making or acceptance of final payment as provided by Subparagraphs 9.7.5 and 9.7.6, shall be decided by arbitration in accordance with the Construction Industry Arbitration Rules of the American Arbitration Association then obtaining unless the parties mutually agree otherwise. This agreement to arbitrate shall be specifically enforceable under the prevailing arbitration law. The award rendered by the arbitrators shall be final, and judgment may be entered upon it in accordance with applicable law in any court having jurisdiction thereof.

7.9 ARBITRATION

7.9.1 All claims, disputes and other matters in question between the Contractor and the Owner arising out of, or relating to, ~~this~~ the Contract Documents or the breach thereof, except as ~~set forth~~ provided in Subparagraph ~~2.2.9~~ 2.2.11 with respect to the Architect's decisions on matters relating to artistic effect, and except for claims which have been waived by the making or acceptance of final payment as provided by Subparagraphs ~~9.7.4~~ 9.9.4 and ~~9.7.5,~~ 9.9.5, shall be decided by arbitration in accordance with the Construction Industry Arbitration Rules of the American Arbitration Association then obtaining unless the parties mutually agree otherwise. No arbitration arising out of or relating to the Contract Documents shall include, by consolidation, joinder or in any other manner, the Architect, his employees or consultants except by written consent containing a specific reference to the Owner-Contractor Agreement and signed by the Architect, the Owner, the Contractor and any other person sought to be joined. No arbitration shall include by consolidation, joinder or in any other manner, parties other than the Owner, the Contractor and any other persons substantially involved in a common question of fact or law, whose presence is required if complete relief is to be accorded in the arbitration. No person other than the Owner or Contractor shall be included as an original third party or additional third party to an arbitration whose interest or responsibility is insubstantial. Any consent to arbitration involving an additional person or persons shall not constitute consent to arbitration of any dispute not described therein or with any person not named or described therein. ~~This~~ The foregoing agreement to arbitrate and any other agreement to arbitrate with an additional person or perons duly consented to by the parties to the Owner-Contractor Agreement shall be specifically enforceable under the prevailing arbitration law. The award rendered by the arbitrators shall be final, and judgment may be entered upon it in accordance with applicable law in any court having jurisdiction thereof.

7.10.2 Notice of the demand for arbitration shall be filed in writing with the other party to the Contract and with the American Arbitration Association, and a copy shall be filed with the Architect. The demand for arbitration shall be made within the time limits specified in Subparagraphs 2.2.10 and 2.2.11 where applicable, and in all other cases within a reasonable time after the claim, dispute or other matter in question has arisen, and in no event shall it be made after the date when institution of legal or equitable proceedings based on such claim, dispute or other matter in question would be barred by the applicable statute of limitations.

7.10.3 The Contractor shall carry on the Work and maintain the progress schedule during any arbitration proceedings, unless otherwise agreed by him and the Owner in writing.

7.9.2 Notice of the demand for arbitration shall be filed in writing with the other party to the ~~Contract~~ Owner-Contractor Agreement and with the American Arbitration Association, and a copy shall be filed with the Architect. The demand for arbitration shall be made within the time limits specified in ~~Subparagraphs 2.2.10 and 2.2.11~~ Subparagraph 2.2.12 where applicable, and in all other cases within a reasonable time after the claim, dispute or other matter in question has arisen, and in no event shall it be made after the date when institution of legal or equitable proceedings based on such claim, dispute or other matter in question would be barred by the applicable statute of limitations.

7.9.3 Unless otherwise agreed in writing, the Contractor shall carry on the Work and maintain ~~the~~ its progress ~~schedule~~ during any arbitration proceedings, ~~unless otherwise agreed by him and the Owner in writing.~~ and the Owner shall continue to make payments to the Contractor in accordance with the Contract Documents.

ARTICLE 8

TIME

8.1 DEFINITIONS
8.1.1 The Contract Time is the period of time alloted in the Contract Documents for completion of the Work.

8.1.2 The date of commencement of the Work is the date established in a notice to proceed. If there is no notice to proceed, it shall be the date of the Agreement or such other date as may be established therein.

8.1.3 The Date of Substantial Completion of the Work or designated portion thereof is the Date certified by the Architect when construction is sufficiently complete, in accordance with the Contract Documents, so the Owner may occupy the Work or designated portion thereof for the use for which it is intended.

8.1.4 The term day as used in the Contract Documents shall mean calendar day.

8.2 PROGRESS AND COMPLETION
8.2.1 All time limits stated in the Contract Documents are of the essence of the Contract

8.2.2 The Contractor shall begin the Work on the date of commencement as defined in Subparagraph 8.1.2. He shall carry the Work forward expeditiously with adequate forces and shall complete it within the Contract Time.

ARTICLE 8

TIME

8.1 DEFINITIONS

8.1.1 Unless otherwise provided, the Contract Time is the period of time allotted in the Contract Documents for Substantial Completion of the Work as defined in Subparagraph 8.1.3, including authorized ~~extensions~~ adjustments thereto.

8.1.2 The date of commencement of the Work is the date established in a notice to proceed. If there is no notice to proceed, it shall be the date of the Owner-Contractor Agreement or such other date as may be established therein.

8.1.3 The Date of Substantial Completion of the Work or designated portion thereof is the Date certified by the Architect when construction is sufficiently complete, in accordance with the Contract Documents, so the Owner ~~may~~ can occupy or utilize the Work or designated portion thereof for the use for which it is intended.

8.1.4 The term day as used in the Contract Documents shall mean calendar day unless otherwise specifically designated.

8.2 PROGRESS AND COMPLETION

8.2.1 All time limits stated in the Contract Documents are of the essence of the Contract.

8.2.2 The Contractor shall begin the Work on the date of commencement as defined in Subparagraph 8.1.2. He shall carry the Work forward expeditiously with adequate forces and shall ~~complete it~~ achieve Substantial Completion within the Contract Time.

8.2.3 If a date or time of completion is included in the Contract, it shall be the Date of Substantial Completion as defined in Subparagraph 8.1.3, including authorized extensions thereto, unless otherwise provided.	← Former 8.2.3 included in new 8.1.1

8.3 DELAYS AND EXTENSIONS OF TIME

8.3.1 If the Contractor is delayed at any time in the progress of the Work by any act or neglect of the Owner or the Architect, or by any employee of either, or by any separate contractor employed by the Owner, or by changes ordered in the Work, or by labor disputes, fire, unusual delay in transportation, unavoidable casualties or any causes beyond the Contractor's control, or by delay authorized by the Owner pending arbitration, or by any cause which the Architect determines may justify the delay, then the Contract Time shall be extended by Change Order for such reasonable time as the Architect may determine.

8.3.2 All claims for extension of time shall be made in writing to the Architect no more than twenty days after the occurrence of the delay; otherwise they shall be waived. In the case of a continuing cause of delay only one claim is necessary.

8.3.3 If no schedule or agreement is made stating the dates upon which written interpretations as set forth in Subparagraph 1.2.5 shall be furnished, then no claim for delay shall be allowed on account of failure to furnish such interpretations until fifteen days after demand is made for them, and not then unless such claim is reasonable.

8.3.4 This Paragraph 8.3 does not exclude the recovery of damages for delay by either party under other provisions of the Contract Documents.

8.3 DELAYS AND EXTENSIONS OF TIME

8.3.1 If the Contractor is delayed at any time in the progress of the Work by any act or neglect of the Owner or the Architect, or by any employee of either, or by any separate contractor employed by the Owner, or by changes ordered in the Work, or by labor disputes, fire, unusual delay in transportation, adverse weather conditions not reasonably anticipatable, unavoidable casualties , or any causes beyond the Contractor's control, or by delay authorized by the Owner pending arbitration, or by any other cause which the Architect determines may justify the delay, then the Contract Time shall be extended by Change Order for such reasonable time as the Architect may determine.

8.3.2 ~~All claims~~ Any claim for extension of time shall be made in writing to the Architect not more than twenty days after the ~~occurrence~~ commencement of the delay; otherwise ~~they~~ it shall be waived. In the case of a continuing delay only one claim is necessary. The Contractor shall provide an estimate of the probable effect of such delay on the progress of the Work.

8.3.3 If no ~~schedule or~~ agreement is made stating the dates upon which ~~written~~ interpretations as ~~set forth~~ provided in Subparagraph ~~1.2.5~~ 2.2.8 shall be furnished, then no claim for delay shall be allowed on account of failure to furnish such interpretations until fifteen days after ~~demand~~ written request is made for them, and not then unless such claim is reasonable.

8.3.4 This Paragraph 8.3 does not exclude the recovery of damages for delay by either party under other provisions of the Contract Documents.

ARTICLE 9

PAYMENTS AND COMPLETION

9.1 CONTRACT SUM

9.1.1 The Contract Sum is stated in the Agreement and is the total amount payable by the Owner to the Contractor for the performance of the Work under the Contract Documents.

ARTICLE 9

PAYMENTS AND COMPLETION

9.1 CONTRACT SUM

9.1.1 The Contract Sum is stated in the Owner-Contractor Agreement and , including authorized adjustments thereto, is the total amount payable by the Owner to the Contractor for the performance of the Work under the Contract Documents.

9.2 SCHEDULE OF VALUES

9.2.1 Before the first Application for Payment, the Contractor shall submit to the Architect a schedule of values of the various portions of the Work, including quantities if required by the Architect, aggregating the total Contract Sum, divided so as to facilitate payments to Subcontractors in accordance with Paragraph 5.4, prepared in such form as specified or as the Architect and the Contractor may agree upon, and supported by such data to substantiate its correctness as the Architect may require. Each item in the schedule of values shall include its proper share of overhead and profit. This schedule, when approved by the Architect, shall be used only as a basis for the Contractor's Applications for Payment.

9.3 PROGRESS PAYMENTS

9.3.1 At least ten days before each progress payment falls due, the Contractor shall submit to the Architect an itemized Application for Payment, supported by such data substantiating the Contractor's right to payment as the Owner or the Architect may require.

9.3.2 If payments are to be made on account of materials or equipment not incorporated in the Work but delivered and suitably stored at the site, or at some other location agreed upon in writing, such payments shall be conditioned upon submission by the Contractor of bills of sale or such other procedures satisfactory to the Owner to establish the Owner's title to such materials or equipment or otherwise protect the Owner's interest including applicable insurance and transportation to the site.

9.3.3 The Contractor warrants and guarantees that title to all Work, materials and equipment covered by an Application for Payment, whether incorporated in the Project or not, will pass to the Owner upon the receipt of such payment by the Contractor, free and clear of all liens, claims, security interests or encumbrances, hereinafter referred to in this Article 9 as "liens"; and that no Work, materials or equipment covered by an Application for Payment will have been acquired by the Contractor; or by any other person performing the Work at the site or furnishing materials and equipment for the Project, subject to an agreement under which an interest therein or an encumbrance thereon is retained by the seller or otherwise imposed by the Contractor or such other person.

9.2   SCHEDULE OF VALUES

9.2.I   Before the first Application for Payment, the Contractor shall submit to the Architect a schedule of values ~~of~~ allocated to the various portions of the Work, ~~including quantities if required by the Architect, aggregating the total Contract Sum, divided so as to facilitate payments to Subcontractors in accordance with Paragraph 5.4,~~ prepared in such form ~~as specified or as the Architect and the Contractor may agree upon,~~ and supported by such data to substantiate its ~~correctness~~ accuracy as the Architect may require. ~~Each item in the schedule of values shall include its proper share of overhead and profit.~~ This schedule, ~~when approved~~ unless objected to by the Architect, shall be used only as a basis for the Contractor's Applications for Payment.

9.3   ~~PROGRESS PAYMENTS~~ APPLICATIONS FOR PAYMENT

9.3.I   At least ten days before ~~each progress payment falls due~~ the date for each progress payment established in the Owner-Contractor Agreement, the Contractor shall submit to the Architect an itemized Application for Payment, notarized if required, supported by such data substantiating the Contractor's right to payment as the Owner or the Architect may require , and reflecting retainage, if any, as provided elsewhere in the Contract Documents.

9.3.2   ~~If~~ Unless otherwise provided in the Contract Documents, payments ~~are to~~ will be made on account of materials or equipment not incorporated in the Work but delivered and suitably stored at the site ~~or~~ and, if approved in advance by the Owner, payments may similarly be made for materials or equipment suitably stored at some other location agreed upon in writing . ~~, such~~ Payments for materials or equipment stored on or off the site shall be conditioned upon submission by the Contractor of bills of sale or such other procedures satisfactory to the Owner to establish the Owner's title to such materials or equipment or otherwise protect the Owner's interest, including applicable insurance and transportation to the site for those materials and equipment stored off the site.

9.3.3   The Contractor warrants ~~and guarantees~~ that title to all Work, materials and equipment covered by an Application for Payment ~~, whether incorporated in the Project or not,~~ will pass to the Owner either by incorporation in the construction or upon the receipt of ~~such~~ payment by the Contractor, whichever occurs first, free and clear of all liens, claims, security interests or encumbrances, hereinafter referred to in this Article 9 as "liens"; and that no Work, materials or equipment covered by an Application for Payment will have been acquired by the Contractor, or by any other person performing ~~the~~ Work at the site or furnishing materials and equipment for the Project, subject to an agreement under which an interest therein or an encumbrance thereon is retained by the seller or otherwise imposed by the Contractor or such other person.

9.4 CERTIFICATES FOR PAYMENT

9.4.1 If the Contractor has made Application for Payment as above, the Architect will, with reasonable promptness but not more than seven days after the receipt of the Application, issue a Certificate for Payment to the Owner, with a copy to the Contractor, for such amount as he determines to be properly due, or state in writing his reasons for withholding a Certificate as provided in Subparagraph 9.5.1.

9.4.2 The issuance of a Certificate for Payment will constitute a representation by the Architect to the Owner, based on his observations at the site as provided in Subparagraph 2.2.4 and the data comprising the Application for Payment, that the Work has progressed to the point indicated; that, to the best of his knowledge, information and belief, the quality of the Work is in accordance with the Contract Documents (subject to an evaluation of the Work for conformance with the Contract Documents upon Substantial Completion, to the results of any subsequent tests required by the Contract Documents, to minor deviations from the Contract Documents correctable prior to completion, and to any specific qualifications stated in his Certificate); and that the Contractor is entitled to payment in the amount certified. In addition, the Architect's final Certificate for Payment will constitute a further representation that the conditions precedent to the Contractor's being entitled to final payment as set forth in Subparagraph 9.7.2 have been fulfilled. However, by issuing a Certificate for Payment, the Architect shall not thereby be deemed to represent that he has made exhaustive or continuous on-site inspections to check the quality or quantity of the Work or that he has reviewed the construction means, methods, techniques, sequences or procedures, or that he has made any examination to ascertain how or for what purpose the Contractor has used the moneys previously paid on account of the Contract Sum.

9.4.3 After the Architect has issued a Certificate for Payment, the Owner shall make payment in the manner provided in the Agreement.

New 9.5.2 replaces former 5.4.1 ⟶

9.4   CERTIFICATES FOR PAYMENT

9.4.1   ~~If the Contractor has made Application for Payment as above,~~ The Architect will, ~~with reasonable promptness but not more than~~ within seven days after the receipt of the ___Contractor's___ Application for Payment, ___either___ issue a Certificate for Payment to the Owner, with a copy to the Contractor, for such amount as ~~he~~ ___the Architect___ determines ~~to be~~ ___is___ properly due, or ~~state~~ ___notify the Contractor___ in writing his reasons for withholding a Certificate as provided in Subparagraph ~~9.5.1~~ ___9.6.1.___

9.4.2   The issuance of a Certificate for Payment will constitute a representation by the Architect to the Owner, based on his observations at the site as provided in Subparagraph ~~2.2.4~~ ___2.2.3___ and the data comprising the Application for Payment, that the Work has progressed to the point indicated; that, to the best of his knowledge, information and belief, the quality of the Work is in accordance with the Contract Documents (subject to an evaluation of the Work for conformance with the Contract Documents upon Substantial Completion, to the results of any subsequent tests required by ___or performed under___ the Contract Documents, to minor deviations from the Contract Documents correctable prior to completion, and to any specific qualifications stated in his Certificate); and that the Contractor is entitled to payment in the amount certified. ~~In addition, the Architect's final Certificate for Payment will constitute a further representation that the conditions precedent to the Contractor's being entitled to final payment as set forth in Subparagraph 9.7.2 have been fulfilled.~~ However, by issuing a Certificate for Payment, the Architect shall not thereby be deemed to represent that he has made exhaustive or continuous on-site inspections to check the quality or quantity of the Work or that he has reviewed the construction means, methods, techniques, sequences or procedures, or that he has made any examination to ascertain how or for what purpose the Contractor has used the moneys previously paid on account of the Contract Sum.

9.5   PROGRESS PAYMENTS

9.5.1   After the Architect has issued a Certificate for Payment, the Owner shall make payment in the manner ___and within the time___ provided in the ~~Agreement~~ ___Contract___ Documents.

9.5.2   The Contractor shall ___promptly___ pay each Subcontractor, upon receipt of payment from the Owner, ~~an amount equal to the percentage of completion allowed~~ ___out of the amount paid___ to the Contractor on account of such Subcontractor's Work, ___the amount to which said Subcontractor is entitled,___ ~~less~~ ___reflecting___ the percentage ___actually___ retained ___, if any,___ from payments to the Contractor ___on account of such Subcontractor's Work.___ The Contractor shall ___, by an appropriate___ agreement with each Subcontractor, ~~also~~ require each Subcontractor to make ~~similar~~ payments to his ___Sub-___subcontractors ___in similar manner.___

-28-

New 9.5.3 replaces former 5.4.4 ⟶

9.5.3 The Architect may, on request and at his discretion, furnish to any Subcontractor, if practicable, information regarding the percentages of completion ~~certified to~~ or the amounts applied for by the Contractor and the action taken thereon by the Architect on account of Work done by such ~~Subcontractors~~ Subcontractor.

9.5.4 Neither the Owner nor the Architect shall have any obligation to pay or to see to the payment of any moneys to any Subcontractor except as may otherwise be required by law.

New 9.5.4 is former 5.4.5 ⟶

9.4.4 No certificate for a progress payment, nor any progress payment, nor any partial or entire use or occupancy of the Project by the Owner, shall constitute an acceptance of any Work not in accordance with the Contract Documents.

9.5.5 No Certificate for a progress payment, nor any progress payment, nor any partial or entire use or occupancy of the Project by the Owner, shall constitute an acceptance of any Work not in accordance with the Contract Documents.

9.5 PAYMENTS WITHHELD

9.6 PAYMENTS WITHHELD

9.5.1 The Architect may decline to approve an Application for Payment and may withhold his Certificate in whole or in part, to the extent necessary reasonably to protect the Owner, if in his opinion he is unable to make representations to the Owner as provided in Subparagraph 9.4.2. The Architect may also decline to approve any Applications for Payment or, because of subsequently discovered evidence or subsequent inspections, he may nullify the whole or any part of any Certificate for Payment previously issued, to such extent as may be necessary in his opinion to protect the Owner from loss because of:

9.6.1 The Architect may decline to ~~approve an Application for Payment~~ certify payment and may withhold his Certificate in whole or in part, to the extent necessary reasonably to protect the Owner, if in his opinion he is unable to make representations to the Owner as provided in Subparagraph 9.4.2. If the Architect is unable to make representations to the Owner as provided in Subparagraph 9.4.2 and to certify payment in the amount of the Appplication, he will notify the Contractor as provided in Subparagraph 9.4.1. If the Contractor and the Architect cannot agree on a revised amount, the Architect will promptly issue a Certificate for Payment for the amount for which he is able to make such representations to the Owner. The Architect may also decline to ~~approve any Applications for Payment~~ certify payment or, because of subsequently discovered evidence or subsequent ~~inspections~~ observations, he may nullify the whole or any part of any Certificate for Payment previously issued, to such extent as may be necessary in his opinion to protect the Owner from loss because of:

.1 defective work not remedied,
.2 third party claims filed or reasonable evidence indicating probable filing of such claims,
.3 failure of the Contractor to make payments properly to Subcontractors or for labor, materials or equipment,
.4 reasonable doubt that the Work can be completed for the unpaid balance of the Contract Sum,
.5 damage to another contractor,
.6 reasonable indication that the Work will not be completed within the Contract Time, or
.7 unsatisfactory prosecution of the Work by the Contractor.

.1 defective work not remedied,
.2 third party claims filed or reasonable evidence indicating probable filing of such claims,
.3 failure of the Contractor to make payments properly to Subcontractors or for labor, materials or equipment,
.4 reasonable ~~doubt~~ evidence that the Work ~~can~~ cannot be completed for the unpaid balance of the Contract Sum,
.5 damage to the Owner or another contractor,
.6 reasonable ~~indication~~ evidence that the Work will not be completed within the Contract Time, or
.7 ~~unsatisfactory prosecution of~~ persistent failure to carry out the Work ~~by the Contractor.~~ in accordance with the Contract Documents.

9.5.2 When the above grounds in Subparagraph 9.5.1 are removed, payment shall be made for amounts withheld because of them.

9.6 FAILURE OF PAYMENT

9.6.1 If the Architect should fail to issue any Certificate for Payment, through no fault of the Contractor, within seven days after receipt of the Contractor's Application for Payment, or if the Owner should fail to pay the Contractor within seven days after the date of payment established in the Agreement any amount certified by the Architect or awarded by arbitration, then the Contractor may, upon seven additional days' written notice to the Owner and the Architect, stop the Work until payment of the amount owing has been received.

9.7 SUBSTANTIAL COMPLETION AND FINAL PAYMENT

9.7.1 When the Contractor determines that the Work or a designated portion thereof acceptable to the Owner is substantially complete, the Contractor shall prepare for submission to the Architect a list of items to be completed or corrected. The failure to include any items on such list does not alter the responsibility of the Contractor to complete all Work in accordance with the Contract Documents. When the Architect on the basis of an inspection determines that the Work is substantially complete, he will then prepare a Certificate of Substantial Completion which shall establish the Date of Substantial Completion, shall state the responsibilities of the Owner and the Contractor for maintenance, heat, utilities, and insurance, and shall fix the time within which the Contractor shall complete the items listed therein. The Certificate of Substantial Completion shall be submitted to the Owner and the Contractor for their written acceptance of the responsibilities assigned to them in such Certificate.

New Subparagraph ➤

9.6.2 When the above grounds in Subparagraph ~~9.5.1~~ 9.6.1 are removed, payment shall be made for amounts withheld because of them.

9.7 FAILURE OF PAYMENT

9.7.1 If the Architect ~~should fail to~~ does not issue a Certificate for Payment, through no fault of the Contractor, within seven days after receipt of the Contractor's Application for Payment, or if the Owner ~~should fail to~~ does not pay the Contractor within seven days after the date ~~of payment~~ established in the ~~Agreement~~ Contract Documents any amount certified by the Architect or awarded by arbitration, then the Contractor may, upon seven additional days' written notice to the Owner and the Architect, stop the Work until payment of the amount owing has been received. The Contract Sum shall be increased by the amount of the Contractor's reasonable costs of shut-down, delay and start-up, which shall be effected by appropriate Change Order in accordance with Paragraph 12.3.

9.8 SUBSTANTIAL COMPLETION ~~AND FINAL PAYMENT~~

9.8.1 When the Contractor ~~determines~~ considers that the Work, or a designated portion thereof which is acceptable to the Owner, is substantially complete as defined in Subparagraph 8.1.3, the Contractor shall prepare for submission to the Architect a list of items to be completed or corrected. The failure to include any items on such list does not alter the responsibility of the Contractor to complete all Work in accordance with the Contract Documents. When the Architect on the basis of an inspection determines that the Work or designated portion thereof is substantially complete, he will then prepare a Certificate of Substantial Completion which shall establish the Date of Substantial Completion, shall state the responsibilities of the Owner and the Contractor for security, maintenance, heat, utilities, damage to the Work, and insurance, and shall fix the time within which the Contractor shall complete the items listed therein. Warranties required by the Contract Documents shall commence on the Date of Substantial Completion of the Work or designated portion thereof unless otherwise provided in the Certificate of Substantial Completion. The Certificate of Substantial Completion shall be submitted to the Owner and the Contractor for their written acceptance of the responsibilities assigned to them in such Certificate.

9.8.2 Upon Substantial Completion of the Work or designated portion thereof and upon application by the Contractor and certification by the Architect, the Owner shall make payment, reflecting adjustment in retainage, if any, for such Work or portion thereof, as provided in the Contract Documents.

9.9   FINAL COMPLETION AND FINAL PAYMENT

9.7.2 Upon receipt of written notice that the Work is ready for final inspection and acceptance and upon receipt of a final Application for Payment, the Architect will promptly make such inspection and, when he finds the Work acceptable under the Contract Documents and the Contract fully performed, he will promptly issue a final Certificate for Payment stating that to the best of his knowledge, information and belief, and on the basis of his observations and inspections, the Work has been completed in accordance with the terms and conditions of the Contract Documents and that the entire balance found to be due the Contractor, and noted in said final Certificate, is due and payable.

9.9.1 Upon receipt of written notice that the Work is ready for final inspection and acceptance and upon receipt of a final Application for Payment, the Architect will promptly make such inspection and, when he finds the Work acceptable under the Contract Documents and the Contract fully performed, he will promptly issue a final Certificate for Payment stating that to the best of his knowledge, information and belief, and on the basis of his observations and inspections, the Work has been completed in accordance with the terms and conditions of the Contract Documents and that the entire balance found to be due the Contractor, and noted in said final Certificate, is due and payable. The Architect's final Certificate for Payment will constitute a further representation that the conditions precedent to the Contractor's being entitled to final payment as set forth in Subparagraph ~~9.7.2~~ 9.9.2 have been fulfilled.

9.7.3 Neither the final payment nor the remaining retained percentage shall become due until the Contractor submits to the Architect (1) an Affidavit that all payrolls, bills for materials and equipment, and other indebtedness connected with the Work for which the Owner or his property might in any way be responsible, have been paid or otherwise satisfied, (2) consent of surety, if any, to final payment and (3), if required by the Owner, other data establishing payment or satisfaction of all such obligations, such as receipts, releases and waivers of liens arising out of the Contract, to the extent and in such form as may be designated by the Owner. If any Subcontractor refuses to furnish a release or waiver required by the Owner, the Contractor may furnish a bond satisfactory to the Owner to indemnify him against any such lien. If any such lien remains unsatisfied after all payments are made, the Contractor shall refund to the Owner all moneys that the latter may be compelled to pay in discharging such lien, including all costs and reasonable attorneys' fees.

9.9.2 Neither the final payment nor the remaining retained percentage shall become due until the Contractor submits to the Architect (1) an affidavit that all payrolls, bills for materials and equipment, and other indebtedness connected with the Work for which the Owner or his property might in any way be responsible, have been paid or otherwise satisfied, (2) consent of surety, if any, to final payment and (3), if required by the Owner, other data establishing payment or satisfaction of all such obligations, such as receipts, releases and waivers of liens arising out of the Contract, to the extent and in such form as may be designated by the Owner. If any Subcontractor refuses to furnish a release or waiver required by the Owner, the Contractor may furnish a bond satisfactory to the Owner to indemnify him against any such lien. If any such lien remains unsatisfied after all payments are made, the Contractor shall refund to the Owner all moneys that the latter may be compelled to pay in discharging such lien, including all costs and reasonable attorneys' fees.

9.7.4 If after Substantial Completion of the Work final completion thereof is materially delayed through no fault of the Contractor, and the Architect so confirms, the Owner shall, upon certification by the Architect, and without terminating the Contract, make payment of the balance due for that portion of the Work fully completed and accepted. If the remaining balance for Work not fully completed or corrected is less than the retainage stipulated in the Agreement, and if bonds have been furnished as required in Subparagraph 7.5.1, the written consent of the surety to the payment of the balance due for that portion of the Work fully completed and accepted shall be submitted by the Contractor to the Architect prior to certification of such payment. Such payment shall be made under the terms and conditions governing final payment, except that it shall not constitute a waiver of claims.

9.9.3 If, after Substantial Completion of the Work, final completion thereof is materially delayed through no fault of the Contractor or by the issuance of Change Orders affecting final completion, and the Architect so confirms, the Owner shall, upon application by the Contractor and certification by the Architect, and without terminating the Contract, make payment of the balance due for that portion of the Work fully completed and accepted. If the remaining balance for Work not fully completed or corrected is less than the retainage stipulated in the ~~Agreement~~ Contract Documents, and if bonds have been furnished as ~~required~~ provided in ~~Subparagraph 7.5.1,~~ Paragraph 7.5, the written consent of the surety to the payment of the balance due for that portion of the Work fully completed and accepted shall be submitted by the Contractor to the Architect prior to certification of such payment. Such payment shall be made under the terms and conditions governing final payment, except that it shall not constitute a waiver of claims.

9.7.5 The making of final payment shall constitute a waiver of all claims by the Owner except those arising from:

.1 unsettled liens,

.2 faulty or defective Work appearing after Substantial Completion,

.3 failure of the Work to comply with the requirements of the Contract Documents, or

.4 terms of any special guarantees required by the Contract Documents.

9.9.4 The making of final payment shall constitute a waiver of all claims by the Owner except those arising from:

.1 unsettled liens,

.2 faulty or defective Work appearing after Substantial Completion,

.3 failure of the Work to comply with the requirements of the Contract Documents, or

.4 terms of any special ~~guarantees~~ warranties required by the Contract Documents.

9.7.6 The acceptance of final payment shall constitute a waiver of all claims by the Contractor except those previously made in writing and still unsettled.

9.9.5 The acceptance of final payment shall constitute a waiver of all claims by the Contractor except those previously made in writing and ~~still~~ identified by the Contractor as unsettled at the time of the final Application for Payment.

ARTICLE 10

PROTECTION OF PERSONS AND PROPERTY

10.1 SAFETY PRECAUTIONS AND PROGRAMS

10.1.1 The Contractor shall be responsible for initiating, maintaining and supervising all safety precautions and programs in connection with the Work.

ARTICLE 10

PROTECTION OF PERSONS AND PROPERTY

10.1 SAFETY PRECAUTIONS AND PROGRAMS

10.1.1 The Contractor shall be responsible for initiating, maintaining and supervising all safety precautions and programs in connection with the Work.

10.2 SAFETY OF PERSONS AND PROPERTY

10.2.1 The Contractor shall take all reasonable precautions for the safety of, and shall provide all reasonable protection to prevent damage, injury or loss to:

.1 all employees on the Work and all other persons who may be affected thereby;

.2 all the Work and all materials and equipment to be incorporated therein, whether in storage on or off the site, under the care, custody or control of the Contractor or any of his Subcontractors or Sub-subcontractors; and

.3 other property at the site or adjacent thereto, including trees, shrubs, lawns, walks, pavements, roadways, structures and utilities not designated for removal, relocation or replacement in the course of construction.

10.2 SAFETY OF PERSONS AND PROPERTY

10.2.1 The Contractor shall take all reasonable precautions for the safety of, and shall provide all reasonable protection to prevent damage, injury or loss to:

.1 all employees on the Work and all other persons who may be affected thereby;

.2 all the Work and all materials and equipment to be incorporated therein, whether in storage on or off the site, under the care, custody or control of the Contractor or any of his Subcontractors or Sub-subcontractors; and

.3 other property at the site or adjacent thereto, including trees, shrubs, lawns, walks, pavements, roadways, structures and utilities not designated for removal, relocation or replacement in the course of construction.

10.2.2 The Contractor shall comply with all applicable laws, ordinances, rules, regulations and lawful orders of any public authority having jurisdiction for the safety of persons or property or to protect them from damage, injury or loss. He shall erect and maintain, as required by existing conditions and progress of the Work, all reasonable safeguards for safety and protection, including posting danger signs and other warnings against hazards, promulgating safety regulations and notifying owners and users of adjacent utilities.

New 10.2.3 replaces last sentence of former 10.2.2 ⟶

10.2.2 The Contractor shall give all notices and comply with all applicable laws, ordinances, rules, regulations and lawful orders of any public authority ~~having jurisdiction for~~ bearing on the safety of persons or property or ~~to protect them~~ their protection from damage, injury or loss.

10.2.3 ~~He~~ The Contractor shall erect and maintain, as required by existing conditions and progress of the Work, all reasonable safeguards for safety and protection, including posting danger signs and other warnings against hazards, promulgating safety regulations and notifying owners and users of adjacent utilities.

10.2.3 When the use or storage of explosives or other hazardous materials or equipment is necessary for the execution of the Work, the Contractor shall exercise the utmost care and shall carry on such activities under the supervision of properly qualified personnel.

10.2.4 When the use or storage of explosives or other hazardous materials or equipment is necessary for the execution of the Work, the Contractor shall exercise the utmost care and shall carry on such activities under the supervision of properly qualified personnel.

10.2.4 All damage or loss to any property referred to in Clauses 10.2.1.2 and 10.2.1.3 caused in whole or in part by the Contractor, any Subcontractor, any Sub-subcontractor, or anyone directly or indirectly employed by any of them, or by anyone for whose acts any of them may be liable, shall be remedied by the Contractor, except damage or loss attributable to faulty Drawings or Specifications or to the acts or omissions of the Owner or Architect or anyone employed by either of them or for whose acts either of them may be liable, and not attributable to the fault or negligence of the Contractor.

10.2.5 The Contractor shall promptly remedy all damage or loss (other than damage or loss insured under Paragraph 11.3) to any property referred to in Clauses 10.2.1.2 and 10.2.1.3 caused in whole or in part by the Contractor, any Subcontractor, any Sub-subcontractor, or anyone directly or indirectly employed by any of them, or by anyone for whose acts any of them may be liable ~~shall be remedied by the Contractor~~ and for which the Contractor is responsible under Clauses 10.2.1.2 and 10.2.1.3, except damage or loss attributable to ~~faulty Drawings or Specifications or~~ the acts or omissions of the Owner or Architect or anyone directly or indirectly employed by either of them, or by anyone for whose acts either of them may be liable, and not attributable to the fault or negligence of the Contractor. The foregoing obligations of the Contractor are in addition to his obligations under Paragraph 4.18.

10.2.5 The Contractor shall designate a responsible member of his organization at the site whose duty shall be the prevention of accidents. This person shall be the Contractor's superintendent unless otherwise designated in writing by the Contractor to the Owner and the Architect.

10.2.6 The Contractor shall designate a responsible member of his organization at the site whose duty shall be the prevention of accidents. This person shall be the Contractor's superintendent unless otherwise designated ~~in writing~~ by the Contractor in writing to the Owner and the Architect.

10.2.6 The Contractor shall not load or permit any part of the Work to be loaded so as to endanger its safety.

10.2.7 The Contractor shall not load or permit any part of the Work to be loaded so as to endanger its safety.

10.3 EMERGENCIES

10.3.1 In any emergency affecting the safety of persons or property, the Contractor shall act, at his discretion, to prevent threatened damage, injury or loss. Any additional compensation or extension of time claimed by the Contractor on account of emergency work shall be determined as provided in Article 12 for Changes in the Work.

10.3 EMERGENCIES

10.3.1 In any emergency affecting the safety of persons or property, the Contractor shall act, at his discretion, to prevent threatened damage, injury or loss. Any additional compensation or extension of time claimed by the Contractor on account of emergency work shall be determined as provided in Article 12 for Changes in the Work.

ARTICLE 11

INSURANCE
11.1 CONTRACTOR'S LIABILITY INSURANCE

11.1.1 The Contractor shall purchase and maintain such insurance as will protect him from claims set forth below which may arise out of or result from the Contractor's operations under the Contract, whether such operations be by himself or by any Subcontractor or by anyone directly or indirectly employed by any of them, or by anyone for whose acts any of them may be liable:

ARTICLE 11

INSURANCE

11.1   CONTRACTOR'S LIABILITY INSURANCE

11.1.1 The Contractor shall purchase and maintain such insurance as will protect him from claims set forth below which may arise out of or result from the Contractor's operations under the Contract, whether such operations be by himself or by any Subcontractor or by anyone directly or indirectly employed by any of them, or by anyone for whose acts any of them may be liable:

.1 claims under workmen's compensation, disability benefit and other similar employee benefit acts;

.2 claims for damages because of bodily injury, occupational sickness or disease, or death of his employees;

.3 claims for damages because of bodily injury, sickness or disease, or death of any person other than his employees;

.4 claims for damages insured by usual personal injury liability coverage which are sustained (1) by any person as a result of an offense directly or indirectly related to the employment of such person by the Contractor, or (2) by any other person; and

.5 claims for damages because of injury to or destruction of tangible property, including loss of use resulting therefrom.

New Clause ⟶

11.1.2 The insurance required by Subparagraph 11.1.1 shall be written for not less than any limits of liability specified in the Contract Documents, or required by law, whichever is greater, and shall include contractual liability insurance as applicable to the Contractor's obligations under Paragraph 4.18.

New 11.1.3 replaces last portion of former 11.1.2 ⟶

11.1.3 Certificates of Insurance acceptable to the Owner shall be filed with the Owner prior to commencement of the Work. These Certificates shall contain a provision that coverages afforded under the policies will not be cancelled until at least fifteen days' prior written notice has been given to the Owner.

11.2 OWNER'S LIABILITY INSURANCE

11.2.1 The Owner shall be responsible for purchasing and maintaining his own liability insurance and, at his option, may purchase and maintain such insurance as will protect him against claims which may arise from operations under the Contract.

11.3 PROPERTY INSURANCE

11.3.1 Unless otherwise provided, the Owner shall purchase and maintain property insurance upon the entire Work at the site to the full insurable value thereof. This insurance shall include the interests of the Owner, the Contractor, Subcontractors and Sub-subcontractors in the Work and shall insure against the perils of Fire, Extended Coverage, Vandalism and Malicious Mischief.

.1 claims under workers' or workmen's compensation, disability benefit and other similar employee benefit acts;

.2 claims for damages because of bodily injury, occupational sickness or disease, or death of his employees;

.3 claims for damages because of bodily injury, sickness or disease, or death of any person other than his employees;

.4 claims for damages insured by usual personal injury liability coverage which are sustained (1) by any person as a result of an offense directly or indirectly related to the employment of such person by the Contractor, or (2) by any other person;

.5 claims for damages, other than to the Work itself, because of injury to or destruction of tangible property, including loss of use resulting therefrom; and

.6 claims for damages because of bodily injury or death of any person or property damage arising out of the ownership, maintenance or use of any motor vehicle.

11.1.2 The insurance required by Subparagraph 11.1.1 shall be written for not less than any limits of liability specified in the Contract Documents, or required by law, whichever is greater . ~, and~

11.1.3 The insurance required by Subparagraph 11.1.1 shall include contractual liability insurance ~as~ applicable to the Contractor's obligations under Paragraph 4.18.

11.1.4 Certificates of Insurance acceptable to the Owner shall be filed with the Owner prior to commencement of the Work. These Certificates shall contain a provision that coverages afforded under the policies will not be cancelled until at least ~fifteen~ thirty days' prior written notice has been given to the Owner.

11.2 OWNER'S LIABILITY INSURANCE

11.2.1 The Owner shall be responsible for purchasing and maintaining his own liability insurance and, at his option, may purchase and maintain such insurance as will protect him against claims which may arise from operations under the Contract.

11.3 PROPERTY INSURANCE

11.3.1 Unless otherwise provided, the Owner shall purchase and maintain property insurance upon the entire Work at the site to the full insurable value thereof. This insurance shall include the interests of the Owner, the Contractor, Subcontractors and Sub-subcontractors in the Work and shall insure against the perils of fire and extended coverage and shall include "all risk" insurance for physical loss or damage including, without duplication of coverage, theft, vandalism, and malicious mischief. If the

New 11.3.1 replaces last portion
of former 11.3.4 ———————→

Owner does not intend to purchase such insurance for the full insurable value of the entire Work, he shall inform the Contractor in writing prior to commencement of the Work. The Contractor may then effect insurance which will protect the interests of himself, his Subcontractors and the Sub-subcontractors in the Work, and by appropriate Change Order the cost thereof shall be charged to the Owner. If the Contractor is damaged by failure of the Owner to purchase or maintain such insurance and to so notify the Contractor, then the Owner shall bear all reasonable costs properly attributable thereto. If not covered under the all risk insurance or otherwise provided in the Contract Documents, the Contractor shall effect and maintain similar property insurance on portions of the Work stored off the site or in transit when such portions of the Work are to be included in an Application for Payment under Subparagraph 9.3.2.

11.3.2 The Owner shall purchase and maintain such steam boiler and machinery insurance as may be required by the Contract Documents or by law. This insurance shall include the interests of the Owner, the Contractor, Subcontractors and Sub-subcontractors in the Work.

11.3.2 The Owner shall purchase and maintain such ~~steam~~ boiler and machinery insurance as may be required by the Contract Documents or by law. This insurance shall include the interests of the Owner, the Contractor, Subcontractors and Sub-subcontractors in the Work.

11.3.3 Any insured loss is to be adjusted with the Owner and made payable to the Owner as trustee for the insureds, as their interests may appear, subject to the requirements of any applicable mortgagee clause and of Subparagraph 11.3.8.

11.3.3 Any ~~insured~~ loss insured under Subparagraph 11.3.1 is to be adjusted with the Owner and made payable to the Owner as trustee for the insureds, as their interests may appear, subject to the requirements of any applicable mortgagee clause and of Subparagraph 11.3.8. The Contractor shall pay each Subcontractor a just share of any insurance moneys received by the Contractor, ~~under Article 11~~ and by appropriate agreement, written where legally required for validity, ~~he~~ shall require each Subcontractor to make ~~similar~~ payments to his Sub-subcontractors in similar manner.

New 11.3.3 includes former 5.4.3 ———————→

11.3.4 The Owner shall file a copy of all policies with the Contractor before an exposure to loss may occur. If the Owner does not intend to purchase such insurance, he shall inform the Contractor in writing prior to commencement of the Work. The Contractor may then effect insurance which will protect the interests of himself, his Subcontractors and the Sub-subcontractors in the Work, and by appropriate Change Order the cost thereof shall be charged to the Owner. If the Contractor is damaged by failure of the Owner to purchase or maintain such insurance and so to notify the Contractor, then the Owner shall bear all reasonable costs properly attributable thereto.

11.3.4 The Owner shall file a copy of all policies with the Contractor before an exposure to loss may occur.

←——————— Last portion of former 11.3.4
included in new 11.3.1

11.3.5 If the Contractor requests in writing that insurance for special hazards be included in the property insurance policy, the Owner shall, if possible, include such insurance, and the cost thereof shall be charged to the Contractor by appropriate Change Order.

11.3.5 If the Contractor requests in writing that insurance for risks other than those described in Subparagraphs 11.3.1 and 11.3.2 or other special hazards be included in the property insurance policy, the Owner shall, if possible, include such insurance, and the cost thereof shall be charged to the Contractor by appropriate Change Order.

11.3.6 The Owner and Contractor waive all rights against each other for damages caused by fire or other perils to the extent covered by insurance provided under this Paragraph 11.3, except such rights as they may have to the proceeds of such insurance held by the Owner as trustee. The Contractor shall require similar waivers by Subcontractors and Sub-subcontractors in accordance with Clause 5.3.1.5.

11.3.6 The Owner and Contractor waive all rights against 1) each other and the subcontractors, sub-subcontractors, agents and employees each of the other, and 2) the Architect and Separate Contractors, if any, and their subcontractors, sub-subcontractors, agents and employees, for damages caused by fire or other perils to the extent covered by insurance obtained pursuant to this Paragraph 11.3 or any other property insurance applicable to the Work, except such rights as they may have to the proceeds of such insurance held by the Owner as trustee. The foregoing waiver afforded the Architect, his agents and employees shall not extend to the liability imposed by Subparagraph 4.18.3. The Owner or the Contractor, as appropriate, shall require of the Architect, Separate Contractors, subcontractors and sub-subcontractors by appropriate agreements, written where legally required for validity, similar waivers each in favor of all other parties enumerated in this Subparagraph 11.3.6.

11.3.7 If required in writing by any party in interest, the Owner as trustee shall, upon the occurrence of an Insured loss, give bond for the proper performance of his duties. He shall deposit in a separate account any money so received, and he shall distribute it in accordance with such agreement as the parties in interest may reach, or in accordance with an award by arbitration in which case the procedure shall be as provided in Paragraph 7.10. If after such loss no other special agreement is made, replacement of damaged work shall be covered by an appropriate Change Order.

11.3.7 If required in writing by any party in interest, the Owner as trustee shall, upon the occurrence of an insured loss, give bond for the proper performance of his duties. He shall deposit in a separate account any money so received, and he shall distribute it in accordance with such agreement as the parties in interest may reach, or in accordance with an award by arbitration in which case the procedure shall be as provided in Paragraph ~~7.10~~ 7.9. If after such loss no other special agreement is made, replacement of damaged work shall be covered by an appropriate Change Order.

11.3.8 The Owner as trustee shall have power to adjust and settle any loss with the insurers unless one of the parties in interest shall object in writing within five days after the occurrence of loss to the Owner's exercise of this power, and if such objection be made, arbitrators shall be chosen as provided in Paragraph 7.10. The Owner as trustee shall, in that case, make settlement with the insurers in accordance with the directions of such arbitrators. If distribution of the insurance proceeds by arbitration is required, the arbitrators will direct such distribution.

11.3.8 The Owner as trustee shall have power to adjust and settle any loss with the insurers unless one of the parties in interest shall object in writing within five days after the occurrence of loss to the Owner's exercise of this power, and if such objection be made, arbitrators shall be chosen as provided in Paragraph ~~7.10~~ 7.9. The Owner as trustee shall, in that case, make settlement with the insurers in accordance with the directions of such arbitrators. If distribution of the insurance proceeds by arbitration is required, the arbitrators will direct such distribution.

New Subparagraph ⟶

11.3.9 If the Owner finds it necessary to occupy or use a portion or portions of the Work prior to Substantial Completion thereof, such occupancy shall not commence prior to a time mutually agreed to by the Owner and Contractor and to which the insurance company or companies providing the property insurance have consented by endorsement to the policy or policies. This insurance shall not be cancelled or lapsed on account of such partial occupancy. Consent of the Contractor and of the insurance company or companies to such occupancy or use shall not be unreasonably withheld.

11.4 LOSS OF USE INSURANCE

11.4.1 The Owner, at his option, may purchase and maintain such insurance as will insure him against loss of use of his property due to fire or other hazards, however caused.

11.4 LOSS OF USE INSURANCE

11.4.1 The Owner, at his option, may purchase and maintain such insurance as will insure him against loss of use of his property due to fire or other hazards, however caused. The Owner waives all rights of action against the Contractor for loss of use of his property, including consequential losses due to fire or other hazards however caused, to the extent covered by insurance under this Paragraph 11.4.

ARTICLE 12

CHANGES IN THE WORK

12.1 CHANGE ORDERS

ARTICLE 12

CHANGES IN THE WORK

12.1 CHANGE ORDERS

New 12.1.1 replaces former 12.1.2 →

12.1.1 A Change Order is a written order to the Contractor signed by the Owner and the Architect, issued after execution of the Contract, authorizing a change in the Work or an adjustment in the Contract Sum or the Contract Time. ~~Alternatively, the Change Order may be signed by the Architect alone, provided he has written authority from the Owner for such procedure and that a copy of such written authority is furnished to the Contractor upon request.~~ The Contract Sum and the Contract Time may be changed only by Change Order. A Change Order ~~may also be~~ signed by the Contractor ~~if he agrees to~~ indicates his agreement therewith, including the adjustment in the Contract Sum or the Contract Time. ~~The Contract Sum and the Contract Time may be changed only by Change Order.~~

12.1.1 The Owner, without invalidating the Contract, may order Changes in the Work within the general scope of the Contract consisting of additions, deletions or other revisions, the Contract Sum and the Contract Time being adjusted accordingly. All such Changes in the Work shall be authorized by Change Order, and shall be executed under the applicable conditions of the Contract Documents.

12.1.2 The Owner, without invalidating the Contract, may order changes in the Work within the general scope of the Contract consisting of additions, deletions or other revisions, the Contract Sum and the Contract Time being adjusted accordingly. All such changes in the Work shall be authorized by Change Order, and shall be ~~executed~~ performed under the applicable conditions of the Contract Documents.

12.1.2 A Change Order is a written order to the Contractor signed by the Owner and the Architect, issued after the execution of the Contract, authorizing a Change in the Work or an adjustment in the Contract Sum or the Contract Time. Alternatively, the Change Order may be signed by the Architect alone, provided he has written authority from the Owner for such procedure and that a copy of such written authority is furnished to the Contractor upon request. A Change Order may also be signed by the Contractor if he agrees to the adjustment in the Contract Sum or the Contract Time. The Contract Sum and the Contract Time may be changed only by Change Order.

← Former 12.1.2 becomes new 12.1.1

12.1.3 The cost or credit to the Owner resulting from a Change in the Work shall be determined in one or more of the following ways:

   .1 by mutual acceptance of a lump sum properly itemized;

   .2 by unit prices stated in the Contract Documents or subsequently agreed upon; or

   .3 by cost and a mutually acceptable fixed or percentage fee.

New Clause   ⟶

12.1.4 If none of the methods set forth in Subparagraph 12.1.3 is agreed upon, the Contractor, provided he receives a Change Order, shall promptly proceed with the Work involved. The cost of such Work shall then be determined by the Architect on the basis of the Contractor's reasonable expenditures and savings, including, in the case of an increase in the Contract Sum, a reasonable allowance for overhead and profit. In such case, and also under Clause 12.1.3.3 above, the Contractor shall keep and present, in such form as the Architect may prescribe, an itemized accounting together with appropriate supporting data. Pending final determination of cost to the Owner, payments on account shall be made on the Architect's Certificate for Payment. The amount of credit to be allowed by the Contractor to the Owner for any deletion or change which results in a net decrease in cost will be the amount of the actual net decrease as confirmed by the Architect. When both additions and credits are involved in any one change, the allowance for overhead and profit shall be figured on the basis of net increase, if any.

12.1.5 If unit prices are stated in the Contract Documents or subsequently agreed upon, and if the quantities originally contemplated are so changed in a proposed Change Order that application of the agreed unit prices to the quantities of Work proposed will create a hardship on the Owner or the Contractor, the applicable unit prices shall be equitably adjusted to prevent such hardship.

12.1.3 The cost or credit to the Owner resulting from a change in the Work shall be determined in one or more of the following ways:

   .1 by mutual acceptance of a lump sum properly itemized <u>and supported by sufficient substantiating data to permit evaluation</u>;

   .2 by unit prices stated in the Contract Documents or subsequently agreed upon;

   .3 by cost <u>to be determined in a manner agreed upon by the parties</u> and a mutually acceptable fixed or percentage fee <u>; or</u>

   <u>.4</u> <u>by the method provided in Subparagraph 12.1.4.</u>

12.1.4 If none of the methods set forth in ~~Subparagraph 12.1.3~~ <u>Clauses 12.1.3.1, 12.1.3.2 or 12.1.3.3</u> is agreed upon, the Contractor, provided he receives a ~~Change Order~~ <u>written order signed by the</u> Owner, shall promptly proceed with the Work involved. The cost of such Work shall then be determined by the Architect on the basis of the ~~Contractor's~~ reasonable expenditures and savings <u>of those performing the Work attributable to the change</u>, including, in the case of an increase in the Contract Sum, a reasonable allowance for overhead and profit. In such case, and also under ~~Clause~~ <u>Clauses</u> 12.1.3.3 <u>and 12.1.3.4</u> above, the Contractor shall keep and present, in such form as the Architect may prescribe, an itemized accounting together with appropriate supporting data <u>for inclusion</u> in a Change Order. <u>Unless otherwise provided in the Contract Documents, cost shall be limited to the following: cost of materials, including sales tax and cost of delivery; cost of labor, including social security, old age and unemployment insurance, and fringe benefits required by agreement or custom; workers' or workmen's compensation insurance; bond premiums; rental value of equipment and machinery; and the additional costs of supervision and field office personnel directly attributable</u> to the change. Pending final determination of cost to the Owner, payments on account shall be made on the Architect's Certificate for Payment. The amount of credit to be allowed by the Contractor to the Owner for any deletion or change which results in a net decrease in ~~cost~~ <u>the Contract Sum</u> will be the amount of the actual net ~~decrease~~ <u>cost</u> as confirmed by the Architect. When both additions and credits <u>covering related Work or substitutions</u> are involved in any one change <u>,</u> the allowance for overhead and profit shall be figured on the basis of the net increase, if any <u>, with respect to that change.</u>

12.1.5 If unit prices are stated in the Contract Documents or subsequently agreed upon, and if the quantities originally contemplated are so changed in a proposed Change Order that application of the agreed unit prices to the quantities of Work proposed will ~~create a hardship on~~ <u>cause substantial inequity to</u> the Owner or the Contractor, the applicable unit prices shall be equitably adjusted. ~~to prevent such hardship.~~

12.2 CONCEALED CONDITIONS

12.1.6 Should concealed conditions encountered in the performance of the Work below the surface of the ground be at variance with the conditions indicated by the Contract Documents or should unknown physical conditions below the surface of the ground of an unusual nature, differing materially from those ordinarily encountered and generally recognized as inherent in work of the character provided for in this Contract, be encountered, the Contract Sum shall be equitably adjusted by Change Order upon claim by either party made within twenty days after the first observance of the conditions.

12.2.1 Should concealed conditions encountered in the performance of the Work below the surface of the ground or should concealed or unknown conditions in an existing structure be at variance with the conditions indicated by the Contract Documents, or should unknown physical conditions below the surface of the ground or should concealed or unknown conditions in an existing structure of an unusual nature, differing materially from those ordinarily encountered and generally recognized as inherent in work of the character provided for in this Contract, be encountered, the Contract Sum shall be equitably adjusted by Change Order upon claim by either party made within twenty days after the first observance of the conditions.

12.1.7 If the Contractor claims that additional cost is involved because of (1) any written interpretation issued pursuant to Subparagraph 1.2.5, (2) any order by the Owner to stop the Work pursuant to Paragraph 3.3 where the Contractor was not at fault, or (3) any written order for a minor change in the Work issued pursuant to Paragraph 12.3, the Contractor shall make such claim as provided in Paragraph 12.2.

← Former 12.1.7 becomes new 12.3.2

12.2 CLAIMS FOR ADDITIONAL COST

12.2.1 If the Contractor wishes to make a claim for an increase in the Contract Sum, he shall give the Architect written notice thereof within twenty days after the occurrence of the event giving rise to such claim. This notice shall be given by the Contractor before proceeding to execute the Work, except in an emergency endangering life or property in which case the Contractor shall proceed in accordance with Subparagraph 10.3.1. No such claim shall be valid unless so made. If the Owner and the Contractor cannot agree on the amount of the adjustment in the Contract Sum, it shall be determined by the Architect. Any change in the Contract Sum resulting from such claim shall be authorized by Change Order.

12.3 CLAIMS FOR ADDITIONAL COST

12.3.1 If the Contractor wishes to make a claim for an increase in the Contract Sum, he shall give the Architect written notice thereof within twenty days after the occurrence of the event giving rise to such claim. This notice shall be given by the Contractor before proceeding to execute the Work, except in an emergency endangering life or property in which case the Contractor shall proceed in accordance with ~~Subparagraph 10.3.1~~ Paragraph 10.3. No such claim shall be valid unless so made. If the Owner and the Contractor cannot agree on the amount of the adjustment in the Contract Sum, it shall be determined by the Architect. Any change in the Contract Sum resulting from such claim shall be authorized by Change Order.

New 12.3.2 replaces former 12.1.7 →

12.3.2 If the Contractor claims that additional cost is involved because of , but not limited to, (1) any written interpretation pursuant to Subparagraph ~~1.2.5~~ 2.2.8, (2) any order by the Owner to stop the Work pursuant to Paragraph 3.3 where the Contractor was not at fault, ~~or~~ (3) any written order for a minor change in the Work issued pursuant to Paragraph ~~12.3~~ 12.4, or (4) failure of payment by the Owner pursuant to Paragraph 9.7, the Contractor shall make such claim as provided in ~~Paragraph 12.2~~ Subparagraph 12.3.1.

12.3 MINOR CHANGES IN THE WORK

12.3.1 The Architect shall have authority to order minor changes in the Work not involving an adjustment in the Contract Sum or an extension of the Contract Time and not inconsistent with the intent of the Contract Documents. Such changes may be effected by Field Order or by other written order. Such changes shall be binding on the Owner and the Contractor.

12.4   MINOR CHANGES IN THE WORK

12.4.1 The Architect ~~shall~~ will have authority to order minor changes in the Work not involving an adjustment in the Contract Sum or an extension of the Contract Time and not inconsistent with the intent of the Contract Documents. Such changes ~~may~~ shall be effected by ~~Field Order or by other~~ written order , and ~~such changes~~ shall be binding on the Owner and the Contractor. The Contractor shall carry out such written orders promptly.

12.4 FIELD ORDERS

12.4.1 The Architect may issue written Field Orders which interpret the Contract Documents in accordance with Subparagraph 1.2.5 or which order minor changes in the Work in accordance with Paragraph 12.3 without change in Contract Sum or Contract Time. The Contractor shall carry out such Field Orders promptly.

← ——————

Former 12.4.1 Deleted

ARTICLE 13

UNCOVERING AND CORRECTION OF WORK

13.1 UNCOVERING OF WORK

13.1.1 If any Work should be covered contrary to the request of the Architect, it must, if required by the Architect, be uncovered for his observation and replaced, at the Contractor's expense.

13.1.2 If any other Work has been covered which the Architect has not specifically requested to observe prior to being covered, the Architect may request to see such Work and it shall be uncovered by the Contractor. If such Work be found in accordance with the Contract Documents, the cost of uncovering and replacement shall, by appropriate Change Order, be charged to the Owner. If such Work be found not in accordance with the Contract Documents, the Contractor shall pay such costs unless it be found that this condition was caused by a separate contractor employed as provided in Article 6, and in that event the Owner shall be responsible for the payment of such costs.

13.2 CORRECTION OF WORK

13.2.1 The Contractor shall promptly correct all Work rejected by the Architect as defective or as failing to conform to the Contract Documents whether observed before or after Substantial Completion and whether or not fabricated, installed or completed. The Contractor shall bear all cost of correcting such rejected Work, including the cost of the Architect's additional services thereby made necessary.

ARTICLE 13

UNCOVERING AND CORRECTION OF WORK

13.1   UNCOVERING OF WORK

13.1.1 If any portion of the Work should be covered contrary to the request of the Architect or to requirements specifically expressed in the Contract Documents, it must, if required in writing by the Architect, be uncovered for his observation and shall be replaced at the Contractor's expense.

13.1.2 If any other portion of the Work has been covered which the Architect has not specifically requested to observe prior to being covered, the Architect may request to see such Work and it shall be uncovered by the Contractor. If such Work be found in accordance with the Contract Documents, the cost of uncovering and replacement shall, by appropriate Change Order, be charged to the Owner. If such Work be found not in accordance with the Contract Documents, the Contractor shall pay such costs unless it be found that this condition was caused by the Owner or a separate contractor ~~employed~~ as provided in Article 6, ~~and~~ in ~~that~~ which event the Owner shall be responsible for the payment of such costs.

13.2   CORRECTION OF WORK

13.2.1 The Contractor shall promptly correct all Work rejected by the Architect as defective or as failing to conform to the Contract Documents whether observed before or after Substantial Completion and whether or not fabricated, installed or completed. The Contractor shall bear all ~~cost~~ costs of correcting such rejected Work, including ~~the cost of~~ compensation for the Architect's additional services ~~thereby~~ made necessary thereby.

13.2.2 If, within one year after the Date of Substantial Completion or within such longer period of time as may be prescribed by law or by the terms of any applicable special guarantee required by the Contract Documents, any of the Work is found to be defective or not in accordance with the Contract Documents, the Contractor shall correct it promptly after receipt of a written notice from the Owner to do so unless the Owner has previously given the Contractor a written acceptance of such condition. The Owner shall give such notice promptly after discovery of the condition.

13.2.3 All such defective or non-conforming Work under Subparagraphs 13.2.1 and 13.2.2 shall be removed from the site if necessary, and the Work shall be corrected to comply with the Contract Documents without cost to the Owner.

13.2.4 The Contractor shall bear the cost of making good all work of separate contractors destroyed or damaged by such removal or correction.

New 13.2.4 replaces former 13.2.6

13.2.5 If the Contractor does not remove such defective or non-conforming Work within a reasonable time fixed by written notice from the Architect, the Owner may remove it and may store the materials or equipment at the expense of the Contractor. If the Contractor does not pay the cost of such removal and storage within ten days thereafter, the Owner may upon ten additional days' written notice sell such Work at auction or at private sale and shall account for the net proceeds thereof, after deducting all the costs that should have been borne by the Contractor including compensation for additional architectural services. If such proceeds of sale do not cover all costs which the Contractor should have borne, the difference shall be charged to the Contractor and an appropriate Change Order shall be issued. If the payments then or thereafter due the Contractor are not sufficient to cover such amount, the Contractor shall pay the difference to the Owner.

13.2.6 If the Contractor fails to correct such defective or non-conforming Work, the Owner may correct it in accordance with Paragraph 3.4.

13.2.2 If, within one year after the Date of Substantial Completion of the Work or designated portion thereof or within one year after acceptance by the Owner of designated equipment or within such longer period of time as may be prescribed by law or by the terms of any applicable special guarantee warranty required by the Contract Documents, any of the Work is found to be defective or not in accordance with the Contract Documents, the Contractor shall correct it promptly after receipt of a written notice from the Owner to do so unless the Owner has previously given the Contractor a written acceptance of such condition. This obligation shall survive termination of the Contract. The Owner shall give such notice promptly after discovery of the condition.

13.2.3 The Contractor shall remove from the site all portions of the such defective or non-conforming Work which are defective or non-conforming and which have not been corrected under Subparagraphs 4.5.1, 13.2.1 and 13.2.2, unless removal is waived by the Owner. shall be removed from the site if necessary, and the Work shall be corrected to comply with the Contract Documents without cost to the Owner.

Former 13.2.4 becomes new 13.2.6

13.2.4 If the Contractor fails to correct such defective or non-conforming Work as provided in Subparagraphs 4.5.1, 13.2.1 and 13.2.2, the Owner may correct it in accordance with Paragraph 3.4.

13.2.5 If the Contractor does not remove proceed with the correction of such defective or non-conforming Work within a reasonable time fixed by written notice from the Architect, the Owner may remove it and may store the materials or equipment at the expense of the Contractor. If the Contractor does not pay the cost of such removal and storage within ten days thereafter, the Owner may upon ten additional days' written notice sell such Work at auction or at private sale and shall account for the net proceeds thereof, after deducting all the costs that should have been borne by the Contractor, including compensation for the Architect's additional architectural services made necessary thereby. If such proceeds of sale do not cover all costs which the Contractor should have borne, the difference shall be charged to the Contractor and an appropriate Change Order shall be issued. If the payments then or thereafter due the Contractor are not sufficient to cover such amount, the Contractor shall pay the difference to the Owner.

Former 13.2.6 becomes new 13.2.4

New 13.2.6 replaces former 13.214 ➡

13.2.6 The Contractor shall bear the cost of making good all work of ~~the Owner or~~ separate contractors destroyed or damaged by such ~~removal or~~ correction or removal.

New Subparagraph ➡

13.2.7 Nothing contained in this Paragraph 13.2 shall be construed to establish a period of limitation with respect to any other obligation which the Contractor might have under the Contract Documents, including Paragraph 4.5 hereof. The establishment of the time period of one year after the Date of Substantial Completion or such longer period of time as may be prescribed by law or by the terms of any warranty required by the Contract Documents relates only to the specific obligation of the Contractor to correct the Work, and has no relationship to the time within which his obligation to comply with the Contract Documents may be sought to be enforced, nor to the time within which proceedings may be commenced to establish the Contractor's liability with respect to his obligations other than specifically to correct the Work.

13.3 ACCEPTANCE OF DEFECTIVE OR NON-CONFORMING WORK

13.3.1 If the Owner prefers to accept defective or non-conforming Work, he may do so instead of requiring its removal and correction, in which case a Change Order will be issued to reflect an appropriate reduction in the Contract Sum, or, if the amount is determined after final payment, it shall be paid by the Contractor.

13.3 ACCEPTANCE OF DEFECTIVE OR NON-CONFORMING WORK

13.3.1 If the Owner prefers to accept defective or non-conforming Work, he may do so instead of requiring its removal and correction, in which case a Change Order will be issued to reflect ~~an appropriate~~ a reduction in the Contract Sum ~~where appropriate and equitable. or if the amount is determined after final payment, it shall be paid by the Contractor.~~ Such adjustment shall be effected whether or not final payment has been made.

ARTICLE 14

TERMINATION OF THE CONTRACT

14.1 TERMINATION BY THE CONTRACTOR

14.1.1 If the Work is stopped for a period of thirty days under an order of any court or other public authority having jurisdiction, or as a result of an act of government, such as a declaration of a national emergency making materials unavailable, through no act or fault of the Contractor or a Subcontractor or their agents or employees or any other persons performing any of the Work under a contract with the Contractor, or if the Work should be stopped for a period of thirty days by the Contractor for the Architect's failure to issue a Certificate for Payment as provided in Paragraph 9.6 or for the Owner's failure to make payment thereon as provided in Paragraph 9.6, then the Contractor may, upon seven days' written notice to the Owner and the Architect, terminate the Contract and recover from the Owner payment for all Work executed and for any proven loss sustained upon any materials, equipment, tools, construction equipment and machinery, including reasonable profit and damages.

ARTICLE 14

TERMINATION OF THE CONTRACT

14.1   TERMINATION BY THE CONTRACTOR

14.1.1 If the Work is stopped for a period of thirty days under an order of any court or other public authority having jurisdiction, or as a result of an act of government, such as a declaration of a national emergency making materials unavailable, through no act or fault of the Contractor or a Subcontractor or their agents or employees or any other persons performing any of the Work under a contract with the Contractor, or if the Work should be stopped for a period of thirty days by the Contractor ~~for~~ because the ~~Architect's failure to issue~~ Architect has not issued a Certificate for Payment as provided in Paragraph ~~9.6~~ 9.7 or ~~for~~ because the ~~Owner's failure to make~~ Owner has not made payment thereon as provided in Paragraph ~~9.6,~~ 9.7, then the Contractor may, upon seven additional days' written notice to the Owner and the Architect, terminate the Contract and recover from the Owner payment for all Work executed and for any proven loss sustained upon any materials, equipment, tools, construction equipment and machinery, including reasonable profit and damages.

14.2   TERMINATION BY THE OWNER

14.2.1 If the Contractor is adjudged a bankrupt, or if he makes a general assignment for the benefit of his creditors, or if a receiver is appointed on account of his insolvency, or if he persistently or repeatedly refuses or fails, except in cases for which extension of time is provided, to supply enough properly skilled workmen or proper materials, or if he fails to make prompt payment to Subcontractors or for materials or labor, or persistently disregards laws, ordinances, rules, regulations or orders of any public authority having jurisdiction, or otherwise is guilty of a substantial violation of a provision of the Contract Documents, then the Owner, upon certification by the Architect that sufficient cause exists to justify such action, may, without prejudice to any right or remedy and after giving the Contractor and his surety, if any, seven days' written notice, terminate the employment of the Contractor and take possession of the site and of all materials, equipment, tools, construction equipment and machinery thereon owned by the Contractor and may finish the Work by whatever method he may deem expedient. In such case the Contractor shall not be entitled to receive any further payment until the Work is finished.

14.2.2 If the unpaid balance of the Contract Sum exceeds the costs of finishing the Work, including compensation for the Architect's additional services, such excess shall be paid to the Contractor. If such costs exceed such unpaid balance, the Contractor shall pay the difference to the Owner. The costs incurred by the Owner as herein provided shall be certified by the Architect.

14.2   TERMINATION BY THE OWNER

14.2.1   If the Contractor is adjudged a bankrupt, or if he makes a general assignment for the benefit of his creditors, or if a receiver is appointed on account of his insolvency, or if he persistently or repeatedly refuses or fails, except in cases for which extension of time is provided, to supply enough properly skilled workmen or proper materials, or if he fails to make prompt payment to Subcontractors or for materials or labor, or persistently disregards laws, ordinances, rules, regulations or orders of any public authority having jurisdiction, or otherwise is guilty of a substantial violation of a provision of the Contract Documents, then the Owner, upon certification by the Architect that sufficient cause exists to justify such action, may, without prejudice to any right or remedy and after giving the Contractor and his surety, if any, seven days' written notice, terminate the employment of the Contractor and take possession of the site and of all materials, equipment, tools, construction equipment and machinery thereon owned by the Contractor and may finish the Work by whatever method he may deem expedient. In such case the Contractor shall not be entitled to receive any further payment until the Work is finished.

14.2.2   If the unpaid balance of the Contract Sum exceeds the costs of finishing the Work, including compensation for the Architect's additional services made necessary thereby, such excess shall be paid to the Contractor. If such costs exceed the unpaid balance, the Contractor shall pay the difference to the Owner. ~~The costs incurred by~~ The amount to be paid to the Contractor or to the Owner , as ~~herein provided~~ the case may be, shall be certified by the Architect, upon application, in the manner provided in Paragraph 9.4, and this obligation for payment shall survive the termination of the Contract.

APPENDIX B

THE AMERICAN INSTITUTE OF ARCHITECTS

AIA Document C141

Standard Form of Agreement Between Architect and Engineer

THIS DOCUMENT HAS IMPORTANT LEGAL CONSEQUENCES; CONSULTATION WITH AN ATTORNEY IS ENCOURAGED WITH RESPECT TO ITS COMPLETION OR MODIFICATION

AGREEMENT

made this   day of   in the year of Nineteen
Hundred and

BETWEEN the Architect:

and the Engineer:

The Architect has made an agreement dated   with
the Owner:

which is hereinafter referred to as the Prime Agreement and which provides for furnishing professional services in connection with the Project described therein. A copy of the Terms and Conditions and all other portions of the Prime Agreement pertinent to the Engineer's responsibilities, compensation and timing of services hereunder is attached, made a part hereof and marked Exhibit A.

The Engineer has been furnished the Owner's latest program for the Project to date.

The Architect and Engineer agree as set forth below.

AIA DOCUMENT C141 • ARCHITECT-ENGINEER AGREEMENT • JANUARY 1974 EDITION • AIA® • ©1974
THE AMERICAN INSTITUTE OF ARCHITECTS, 1735 NEW YORK AVE., N.W., WASHINGTON, D. C. 20006

1

I. THE ENGINEER shall provide the following professional services for the Architect, in accordance with the Terms and Conditions of this Agreement, which the Architect is required to provide for the Owner under the Prime Agreement:
(Describe services)

The part of the Project for which the Engineer is to provide such services is hereinafter called This Part of the Project.

II. THE ARCHITECT shall compensate the Engineer, in accordance with the Terms and Conditions of this Agreement.

A. *FOR BASIC SERVICES* described in Paragraph 1.2, compensation shall be computed as follows:

1. On the basis of a MULTIPLE of DIRECT PERSONNEL EXPENSE

Principals' time at the fixed rate of   dollars ($) per hour.
For the purposes of this Agreement, the Principals are:

Employees' time (other than Principals) at a multiple of   (   )
times the employees' Direct Personnel Expense as defined in Article 4.

Services of professional consultants engaged by the Engineer at a multiple of
(   ) times the amount billed to the Engineer for such services. The terms and conditions of employment of such consultants shall be subject to prior written approval of the Architect.

OR 2. On the basis of a PROFESSIONAL FEE PLUS EXPENSES

(1) a Professional Fee of

dollars ($)

plus (2) expenses computed as follows:

Principals' time at the fixed rate of   ($) per hour.
For the purposes of this Agreement, the Principals are:

Employees' time (other than Principals) at a multiple of   (   )
times the employees' Direct Personnel Expense as defined in Article 4.

Services of professional consultants engaged by the Engineer at a multiple of
(   ) times the amount billed to the Engineer for such services. The terms and conditions of employment of such consultants shall be subject to prior written approval of the Architect.

OR   3. On the basis of one of the following PERCENTAGES OF CONSTRUCTION COST.

The Construction Cost of the Project as defined in Article 3, under

a) A Single Stipulated Sum Contract   percent (%)

b) Separate Stipulated Sum Contracts   percent (%)

c) A Single Cost Plus Fee Contract   percent (%)

d) Separate Cost Plus Fee Contracts   percent (%)

OR

The Construction Cost of This Part of the Project, as defined in Article 3, under

e) A Single Stipulated Sum Contract   percent (%)

f) Separate Stipulated Sum Contracts   percent (%)

g) A Single Cost Plus Fee Contract   percent (%)

h) Separate Cost Plus Fee Contracts   percent (%)

OR   4.   On the basis of a FIXED FEE

                                      dollars ($)

B. *FOR ADDITIONAL SERVICES* as described in Paragraph 1.4, compensation computed as follows:

Principals' time at the fixed rate of   dollars ($) per hour.
For the purposes of this Agreement, the Principals are:

Employees' time (other than Principals) at a multiple of   (   )
times the employees' Direct Personnel Expense as defined in Article 4.

Services of professional consultants engaged by the Engineer at a multiple of
            (   ) times the amount billed to the Engineer for such services.
The terms and conditions of employment of such consultants shall be subject to prior written
approval of the Architect.

C. *FOR THE ENGINEER'S REIMBURSABLE EXPENSES,* amounts expended as defined in Article 5
when authorized in advance by the Architect.

III. THE ARCHITECT AND ENGINEER agree in accordance with the Terms and Conditions of this Agreement that:

A. PAYMENTS FOR BASIC SERVICES based on Subparagraphs IIA2(1), IIA3 or IIA4, shall equal the following percentages of the total compensation at the completion of each Phase of This Part of the Project:

Schematic Design	(%)
Design Development	(%)
Construction Documents	(%)
Bidding or Negotiation	(%)
Construction	(100%)

B. IF THE SCOPE of the Project is changed materially, compensation for Basic Services shall be subject to renegotiation.

C. IF THE SERVICES covered by this Agreement have not been completed within (   ) months of the date hereof, the rates, multiples and amounts of compensation set forth in Paragraph II shall be subject to renegotiation.

IV. THE ARCHITECT shall be the Project administrator. The Engineer is the Architect's independent professional consultant for This Part of the Project, responsible for methods and means used in performing his services under this Agreement, and is not a joint venturer with the Architect.

ARTICLE 1

ENGINEER'S SERVICES

1.1 GENERAL

1.1.1 The Engineer shall collaborate with the Architect for This Part of the Project and shall be bound to perform the services undertaken hereunder for the Architect in the same manner and to the same extent that the Architect is bound by the Prime Agreement to perform such services for the Owner. Except as set forth herein, the Engineer shall not have any duties or responsibilities for any other part of the Project.

1.1.2 The Engineer shall perform his work in character, sequence and timing so that it will be coordinated with that of the Architect and all other consultants for the Project and in accordance with a schedule to be provided by the Architect and accepted by the Engineer. The Engineer agrees to a mutual exchange of Drawings and Specifications for the Project with the Architect.

1.1.3 The Engineer shall cooperate with the Architect in determining the proper share of the construction budget which shall be allocated to This Part of the Project.

1.1.4 The Engineer shall not be responsible for the acts or omissions of the Architect, the Architect's other consultants, the Contractor, or any Subcontractors or any of their agents or employees, or any other persons performing any of the Work.

1.2 BASIC SERVICES

The Engineer's Basic Services consist of the five phases described below and any other services included in Article 15 as Basic Services.

SCHEMATIC DESIGN PHASE

1.2.1 The Engineer shall consult with the Architect to ascertain the requirements for This Part of the Project and shall confirm such requirements to the Architect.

1.2.2 The Engineer shall make recommendations regarding basic systems, attend necessary conferences, prepare necessary analyses and be available for general consultation. When necessary, the Engineer shall consult with public agencies and other organizations concerning utility services and requirements.

1.2.3 The Engineer shall prepare and submit to the Architect a Statement of Probable Construction Cost of This Part of the Project based on current area, volume or other unit costs.

1.2.4 The Engineer shall recommend to the Architect the obtaining of such investigations, surveys, tests and analyses as may be necessary for the proper execution of the Engineer's work.

DESIGN DEVELOPMENT PHASE

1.2.5 When authorized by the Architect, the Engineer shall prepare from the Schematic Design Studies approved by the Owner and confirmed by the Architect, the Design Development Documents. These shall consist of drawings and other documents to fix and describe This Part of the Project, including materials, equipment, component systems and types of construction as may be appropriate, all of which are to be approved by the Owner and the Architect.

1.2.6 The Engineer shall submit to the Architect a further Statement of Probable Construction Cost of This Part of the Project.

CONSTRUCTION DOCUMENTS PHASE

1.2.7 When authorized by the Architect, the Engineer shall prepare from the Design Development Documents approved by the Owner and confirmed by the Architect, Drawings and Specifications setting forth in detail the requirements for the construction of This Part of the Project, all of which are to be approved by the Owner and the Architect. The Engineer shall make available to the Architect the Drawings and Specifications in such form as the Architect may reasonably require.

1.2.8 The Engineer shall advise the Architect of any adjustments to previous Statements of Probable Construction Cost of This Part of the Project indicated by changes in requirements or general market conditions.

1.2.9 The Engineer shall assist the Architect in filing the required documents with respect to This Part of the Project for the approval of governmental authorities having jurisdiction over the Project.

BIDDING OR NEGOTIATION PHASE

1.2.10 If required by the Architect, the Engineer shall assist the Architect and the Owner in obtaining bids or negotiated proposals, and in awarding and preparing construction contracts.

CONSTRUCTION PHASE — ADMINISTRATION OF THE CONSTRUCTION CONTRACT

1.2.11 The Construction Phase will commence with the award of the Construction Contract for This Part of the Project and will terminate when the final Certificate for Payment is issued to the Owner by the Architect.

1.2.12 The Engineer shall assist the Architect in the Administration of the Construction Contract with respect to This Part of the Project.

1.2.13 The Engineer shall at all times have access to the Work of This Part of the Project wherever it is in preparation or progress.

1.2.14 The Engineer shall make periodic visits to the site to familiarize himself generally with the progress and quality of the Work for This Part of the Project to determine in general if such Work is proceeding in accordance with the Contract Documents. On the basis of his on-site observations as an engineer, he shall endeavor to guard the Owner and the Architect against defects and deficiencies in the Work of the Contractor. The Engineer shall not be required to make exhaustive or continuous on-site inspections to check the quality or quantity of the Work. The Engineer shall not be responsible for construction means, methods, techniques, sequences or procedures, or for safety precautions and programs in connection with the Work, and he shall not be responsible for the Contractor's failure to carry out the Work for This Part of the Project in accordance with the Contract Documents.

1.2.15 Based on such observations at the site and on the Contractor's Applications for Payment, the Engineer shall assist the Architect in determining the amount owing to the Contractor for This Part of the Project. If requested, he shall certify such amounts to the Architect. A certification by the Engineer to the Architect of an amount owing to the Contractor shall constitute a representation by the Engineer to the Architect that, based on the Engineer's observations at the site as provided in Subparagraph 1.2.14 and the data comprising the Application for Payment, the Work for This Part of the Project has progressed to the point indicated; that to the best of his knowledge, information and belief, the quality of such Work is in accordance with the Contract Documents (subject to an evaluation of such Work for conformance with the Contract Documents upon Substantial Completion, to the results of any subsequent tests called for in the Contract Documents, to minor deviations from the Contract Documents correctable prior to completion, and to any specific qualifications stated by the Engineer); and that the Contractor is entitled to payment in the amount certified.

1.2.16 The Engineer shall assist the Architect in making decisions on all claims of the Owner or Contractor relating to the execution and progress of the Work on This Part of the Project and on all other matters or questions related thereto. The Engineer shall not be liable for the results of any interpretation or decision rendered in good faith.

1.2.17 The Engineer shall assist the Architect in determining whether the Architect shall reject Work for This Part of the Project which does not conform to the Contract Documents or whether special inspection or testing is required.

1.2.18 The Engineer shall review and approve shop drawings, samples and other submissions of the Contractor with respect to This Part of the Project only for conformance with the design concept and for compliance with the information given in the Contract Documents. All comments and approvals shall be submitted to the Architect.

1.2.19 The Engineer shall assist the Architect in preparing Change Orders for This Part of the Project.

1.2.20 The Engineer shall assist the Architect in conducting inspections with respect to This Part of the Project to determine the Dates of Substantial Completion and final completion and in receiving and reviewing written guarantees and related documents assembled by the Contractor with respect to This Part of the Project.

1.3 PROJECT REPRESENTATION BEYOND BASIC SERVICES

1.3.1 If more extensive representation at the site than is described under Subparagraphs 1.2.11 through 1.2.20 inclusive is required for This Part of the Project and if the Architect and Engineer agree, the Engineer shall provide one or more Full-Time Project Representatives to assist the Engineer.

1.3.2 Such Full-Time Project Representatives shall be selected, employed and directed by the Engineer, and the Engineer shall be compensated therefor as mutually agreed between the Architect and the Engineer.

1.3.3 The duties, responsibilities and limitations of authority of such Full-Time Project Representatives shall be as set forth in an exhibit appended to this Agreement.

1.3.4 Through the on-site observations by Full-Time Project Representatives of the Work in progress, the Engineer shall endeavor to provide further protection for the Owner and the Architect against defects in the Work for This Part of the Project, but the furnishing of such project representation shall not make the Engineer responsible for construction means, methods, techniques, sequences or procedures, or for safety precautions and programs, or for the Contractor's failure to perform the Work in accordance with the Contract Documents.

1.4 ADDITIONAL SERVICES

The following services shall be provided when authorized in writing by the Architect, and they shall be paid for by the Architect as hereinbefore provided.

1.4.1 Providing special analyses of the Owner's needs and programming the requirements for This Part of the Project.

1.4.2 Providing financial feasibility or other special studies.

1.4.3 Providing planning surveys, site evaluations, environmental studies, or comparative studies of prospective sites.

1.4.4 Providing design services relative to future facilities, systems and equipment which are not intended to be constructed as part of the Project.

1.4.5 Providing services to investigate existing conditions or facilities or to make measured drawings thereof, or to verify the accuracy of drawings or other information furnished by the Owner.

1.4.6 Preparing documents for alternate bids or out-of-sequence services requested by the Owner.

1.4.7 Providing Detailed Estimates of Construction Cost or detailed quantity surveys or inventories of material, equipment and labor.

AIA DOCUMENT C141 • ARCHITECT-ENGINEER AGREEMENT • JANUARY 1974 EDITION • AIA® • ©1974
THE AMERICAN INSTITUTE OF ARCHITECTS, 1735 NEW YORK AVE., N.W., WASHINGTON, D. C. 20006

1.4.8 Providing interior design and other services required for or in connection with the selection of furniture and furnishings.

1.4.9 Providing services for planning tenant or rental spaces.

1.4.10 Making revisions in Drawings, Specifications or other documents when such revisions are inconsistent with written approvals or instructions previously given and are due to causes beyond the control of the Engineer.

1.4.11 Preparing supporting data and other services in connection with Change Orders if the change in the Basic Compensation resulting from the adjusted Contract sum is not commensurate with the services required of the Engineer.

1.4.12 Making investigations involving detailed appraisals and valuations of existing facilities, and surveys or inventories required in connection with construction performed by the Owner.

1.4.13 Providing consultation concerning replacement of any Work damaged by fire or other cause during construction, and furnishing professional services of the type set forth in Paragraph 1.2 as may be required in connection with the replacement of such Work.

1.4.14 Providing professional services made necessary by the default of the Contractor or by major defects in the Work of the Contractor in the performance of the Construction Contract.

1.4.15 Preparing a set of reproducible record prints of drawings showing significant changes in This Part of the Work made during the construction process, based on marked-up prints, drawings and other data furnished by the Contractor.

1.4.16 Providing extensive assistance in the utilization of any equipment or system such as initial start-up or testing, adjusting and balancing, preparation of operating and maintenance manuals, training personnel for operation and maintenance, and consultation during operation.

1.4.17 Providing Contract Administration and observation of construction after the time for completion of This Part of the Project has exceeded or extended by more than 30 days through no fault of the Engineer.

1.4.18 Providing services after issuance to the Owner of the final Certificate for Payment.

1.4.19 Preparing to serve or serving as an expert witness in connection with any public hearing, arbitration proceeding or legal proceeding.

1.4.20 Providing services of professional consultants for other than the normal engineering services for This Part of the Project.

1.4.21 Providing any other services not otherwise included in this Agreement or not customarily furnished in accordance with generally accepted engineering practice.

ARTICLE 2

THE ARCHITECT'S RESPONSIBILITIES

2.1 The Architect shall provide all available information regarding the requirements for This Part of the Project.

2.2 The Architect shall designate, when necessary, a representative authorized to act in his behalf with respect to This Part of the Project. The Architect or his representative shall examine documents submitted by the Engineer and shall render decisions pertaining thereto promptly, to avoid unreasonable delay in the progress of the Engineer's services.

2.3 The Architect shall furnish to the Engineer a copy of the Owner's certified land survey of the site giving, as applicable, grades and lines of streets, alleys, pavements and adjoining property; rights-of-way, restrictions, easements, encroachments, zoning, deed restrictions, boundaries and contours of the site; locations, dimensions and complete data pertaining to existing buildings, other improvements and trees; and full information concerning available service and utility lines both public and private, above and below grade, including inverts and depths.

2.4 The Architect shall furnish to the Engineer (1) detailed layouts showing the location of connections and (2) tabulations giving sizes and loads of all equipment to be incorporated in other parts of the Project.

2.5 The Architect shall request the Owner to furnish the services of a soils engineer or other consultant when such services are deemed necessary for This Part of the Project by the Engineer, including reports, test borings, test pits, soil bearing values, percolation tests, air and water pollution tests, ground corrosion and resistivity tests and other necessary operations for determining subsoil, air and water conditions, with appropriate professional recommendations.

2.6 The services, information, surveys and reports required by Paragraphs 2.3, 2.4 and 2.5 shall be furnished at no expense to the Engineer, who shall be entitled to rely upon the accuracy and completeness thereof.

2.7 If the Architect becomes aware of any fault or defect with respect to This Part of the Project or nonconformance with the Contract Documents, he shall give prompt written notice thereof to the Engineer.

2.8 The Architect shall consult with the Engineer before issuing interpretations or clarifications of the Engineer's Drawings and Specifications and shall obtain the prompt written consent of the Engineer before acting upon Shop Drawings, Samples or other submissions of the Contractor or Change Orders affecting This Part of the Project.

2.9 The Architect shall furnish to the Engineer a copy of Estimates of Construction Cost as submitted to the Owner, bidding documents, bid tabulations, and Contract Documents (including Change Orders as issued) to the extent that they pertain to This Part of the Project.

2.10 The Architect shall advise the Engineer of the identity of other consultants participating in the Project and the scope of their services.

ARTICLE 3

CONSTRUCTION COST

3.1. If the Construction Cost of the Project is to be used as the basis for determining the Engineer's Compensation for Basic Services, it shall be the total cost or estimated cost to the Owner of all Work. If the Construction Cost of This Part of the Project is to be used as the basis for determining the Engineer's Basic Compensation, it shall be the total cost or estimated cost to the Owner of all Work designed or specified by the Engineer. The Construction Cost of the Project or the Construction Cost of This Part of the Project shall be determined as follows, with precedence in the order listed:

3.1.1 For completed construction, the cost of all such Work;

3.1.2 For Work not constructed, (1) the lowest bona fide bid received from a qualified bidder for any or all of such Work, or (2) if the Work is not bid, the bona fide negotiated proposal submitted for any or all of such Work; or

3.1.3 For Work for which no such bid or proposal is received, (1) the latest Detailed Estimate of Construction Cost if one is available, or (2) the latest Statement of Probable Construction Cost.

3.2 Construction Cost does not include the compensation of the Architect, the Engineer and other consultants, the cost of the land, rights-of-way, or cost of other items which are to be furnished as provided in Paragraphs 2.3 through 2.5 inclusive.

3.3 The cost of labor, materials and equipment furnished by the Owner for the Project shall be included in the Construction Cost at current market rates including a reasonable allowance for overhead and profit.

3.4 Statements of Probable Construction Cost and Detailed Cost Estimates prepared by the Engineer represent his best judgment as a design professional familiar with the construction industry. It is recognized, however, that neither the Engineer nor the Architect has any control over the cost of labor, materials or equipment, over the contractor's methods of determining bid prices, or over competitive bidding or market conditions. Accordingly, the Engineer cannot and does not guarantee that bids will not vary from any Statement of Probable Construction Cost or other cost estimate prepared by him.

3.5 When a fixed limit of Construction Cost is established as a condition of the Prime Agreement, the Architect and the Engineer shall establish, if practicable, a fixed limit of Construction Cost for This Part of the Project which shall include a bidding contingency of ten percent unless another amount is agreed upon in writing. When such a fixed limit is established, the Engineer, after consultation with the Architect, shall be permitted to determine what materials, equipment, component systems and types of construction are to be included in the Contract Documents with respect to This Part of the Project, and to make reasonable adjustments in the scope of This Part of the Project to bring it within the fixed limit. If

required, the Engineer shall assist the Architect in including in the Contract Documents alternate bids to adjust the Construction Cost to the fixed limit.

3.5.1 If the Bidding or Negotiating Phase for This Part of the Project has not commenced within six months after submission by the Architect of the Construction Documents to the Owner, any fixed limit of Construction Cost established as a condition of this Agreement shall be adjusted. The adjustment shall reflect any change in the general level of prices which may have occurred in the construction industry for the area in which the Project is located between the date of submission of the Construction Documents to the Owner and the date on which proposals are sought.

3.5.2 When a fixed limit of Construction Cost, including the bidding contingency (adjusted as provided in Subparagraph 3.5.1, if applicable) is established as a condition of this Agreement and is exceeded by the lowest bona fide bid or negotiated proposal, the Detailed Estimate of Construction Cost or the Statement of Probable Construction Cost, the Architect may require the Engineer, without additional charge, to modify the Drawings and Specifications for This Part of the Project as necessary to bring the Construction Cost thereof within such fixed limit for This Part of the Project. If it was not practicable to establish a fixed limit of Construction Cost for This Part of the Project, and if the lowest bona fide bid or negotiated proposal, the Detailed Estimate of Construction Cost or the Statement of Probable Construction Cost established for the entire Project (including the bidding contingency) exceeds the fixed limit of Construction Cost for the entire Project, the Architect may require the Engineer without additional compensation to modify his Drawings and Specifications as necessary to make them bear a reasonable portion of reducing the Construction Cost to the fixed limit of Construction Cost for the entire Project. The providing of such service shall be the limit of the Engineer's responsibility in this regard, and having done so, he shall be entitled to compensation.

ARTICLE 4

DIRECT PERSONNEL EXPENSE

Direct Personnel Expense is defined as the salaries of professional, technical and clerical employees engaged on the Project by the Engineer, and the cost of their mandatory and customary benefits such as statutory employee benefits, insurance, sick leave, holidays, vacations, pensions and similar benefits.

ARTICLE 5

REIMBURSABLE EXPENSES

5.1 Reimbursable Expenses are in addition to Compensation for Basic and Additional Services and include actual expenditures made by the Engineer, his employees, or his professional consultants in the interest of This Part of the Project for the expenses listed in the following Subparagraphs:

5.1.1 Expense of transportation and living when traveling in connection with This Part of the Project, long distance calls and telegrams, and fees paid for securing approval of authorities having jurisdiction over the Project.

5.1.2 Expense of reproductions, postage and handling of Drawings and Specifications excluding duplicate sets at the completion of each Phase for the Owner's review and approval.

5.1.3 Expense of overtime work requiring higher than regular rates.

5.1.4 Expense of models for the Owner's use.

5.1.5 Expense of computer time for professional services when compensation is computed on the basis of Subparagraph IIA1 or IIA2.

5.1.6 Expense of computer time when used in connection with Additional Services.

ARTICLE 6

PAYMENT TO THE ENGINEER

6.1 Payments for Basic Services shall be made monthly upon presentation of the Engineer's statement of services rendered in proportion to services performed so that the compensation at the completion of each Phase of This Part of the Project shall equal the percentages of the total Basic Compensation set forth in Paragraph IIIA.

6.2 Payments for Additional Services of the Engineer as defined in Paragraph 1.4, and for Reimbursable Expenses as defined in Article 5, shall be made monthly upon presentation of the Engineer's statement.

6.3 If the Architect objects to any statement submitted by the Engineer, he shall so advise the Engineer in writing, giving his reasons within 14 days of receipt of such statement.

6.4 No deductions shall be made from the Engineer's compensation on account of penalty, liquidated damages, or other sums withheld from payments to contractors.

6.5 If the Architect does not receive full payment when due from the Owner for any cause which is not the fault of the Architect or the Engineer, the Architect shall pay the Engineer for his accepted statements of services in the same proportion that payments received from the Owner bear to the total payments due to the Architect.

6.6 The Architect shall exert all reasonable and diligent efforts to collect payment from the Owner until the Engineer has been paid in full.

6.7 If the Architect fails to pay the Engineer within 60 days after receipt and acceptance of the Engineer's statement, the Engineer may, after giving seven days' written notice to the Architect, suspend services under this Agreement until his outstanding statements which have been accepted by the Architect have been paid in full.

6.8 If the Project is suspended for more than three months or abandoned in whole or in part, the Engineer shall be paid his compensation for services performed

prior to receipt of written notice from the Architect of such suspension or abandonment, together with Reimbursable Expenses then due and all termination expenses resulting from such suspension or abandonment. If the Project is resumed after being suspended for more than three months, the Engineer's Compensation shall be subject to renegotiation.

ARTICLE 7

ENGINEER'S RECORDS

7.1 Records of Reimbursable Expenses and expenses pertaining to Additional Services on the Project and for services performed on the basis of a Multiple of Direct Personnel Expense shall be kept on a generally recognized accounting basis and shall be available to the Architect or his authorized representative at mutually convenient times.

7.2 The Engineer shall maintain on file and available to the Architect his design calculations in legible form.

ARTICLE 8

TERMINATION OF AGREEMENT

8.1 This Agreement is terminated if and when the Prime Agreement is terminated. The Architect shall promptly notify the Engineer of such termination.

8.2 This Agreement may be terminated by either party upon seven days' written notice should the other party fail substantially to perform in accordance with its terms through no fault of the party initiating the termination.

8.3 In the event of termination due to the fault of parties other than the Engineer, the Engineer shall be paid his compensation plus Reimbursable Expenses for services performed to termination date and all termination expenses, contingent upon comparable adjustment by the Owner of the Architect's compensation.

8.4 Termination Expenses are defined as Reimbursable Expenses directly attributable to termination, plus an amount computed as a percentage of the total compensation earned at the time of termination, as follows:

20 percent if termination occurs during the Schematic Design Phase; or
10 percent if termination occurs during the Design Development Phase; or
5 percent if termination occurs during any subsequent phase.

ARTICLE 9

OWNERSHIP OF DOCUMENTS

Drawings and Specifications prepared by the Engineer as instruments of service are and shall remain the property of the Engineer whether the Project for which they are made is executed or not. They are not to be used on other projects or extensions to this Project except by agreement in writing and with appropriate compensation to the Engineer.

ARTICLE 10
INSURANCE

The Architect and the Engineer shall each effect and maintain insurance to protect himself from claims under workmen's compensation acts; claims for damages because of bodily injury including personal injury, sickness or disease, or death of any of his employees or of any person other than his employees; and from claims for damages because of injury to or destruction of tangible property including loss of use resulting therefrom; and from claims arising out of the performance of professional services caused by any errors, omissions or negligent acts for which he is legally liable.

ARTICLE 11
SUCCESSORS AND ASSIGNS

The Architect and the Engineer each binds himself, his partners, successors, assigns and legal representatives to the other party to this Agreement and to the partners, successors, assigns and legal representatives of such other party with respect to all covenants of this Agreement. Neither the Architect nor the Engineer shall assign, sublet or transfer his interest in this Agreement without the written consent of the other.

ARTICLE 12
ARBITRATION

12.1 All claims, disputes and other matters in question arising out of, or relating to, this Agreement, or the breach thereof, shall be decided by arbitration in accordance with the Construction Industry Arbitration Rules of the American Arbitration Association then obtaining unless the parties mutually agree otherwise. This agreement to arbitrate shall be specifically enforceable under the prevailing arbitration law.

12.2 Notice of the demand for arbitration shall be filed in writing with the other party to this Agreement and with the American Arbitration Association. The demand shall be made within a reasonable time after the claim, dispute or other matter in question has arisen. In no event shall the demand for arbitration be made after the date when institution of legal or equitable proceedings based on such claim, dispute or other matter in question would be barred by the applicable statute of limitations.

12.3 The award rendered by the arbitrators shall be final, and judgment may be entered upon it in accordance with applicable law in any court having jurisdiction thereof.

ARTICLE 13
EXTENT OF AGREEMENT

This Agreement represents the entire and integrated agreement between the Architect and the Engineer and supersedes all prior negotiations, representations or agreements, either written or oral. This Agreement may be amended only by written instrument signed by both Architect and Engineer.

ARTICLE 14
GOVERNING LAW

Unless otherwise specified, this Agreement shall be governed by the law of the principal place of business of the Architect.

ARTICLE 15
OTHER CONDITIONS OR SERVICES

AIA DOCUMENT C141 • ARCHITECT-ENGINEER AGREEMENT • JANUARY 1974 EDITION • AIA® • ©1974
THE AMERICAN INSTITUTE OF ARCHITECTS, 1735 NEW YORK AVE., N.W., WASHINGTON, D. C. 20006

This Agreement executed the day and year first written above.

ARCHITECT   ENGINEER

_____   _____

STANDARD
GENERAL CONDITIONS
OF THE
CONSTRUCTION CONTRACT

PROFESSIONAL ENGINEERS IN PRIVATE PRACTICE

a practice division of the

NATIONAL SOCIETY OF PROFESSIONAL ENGINEERS

These General Conditions have been prepared for use with the NSPE Owner-Contractor Agreement (Document 1910-8-A-1 or 1910-8-A-2, 1974 edition) and with the NSPE Instructions to Bidders (Document 1910-12, 1974 edition). Their provisions are interrelated and a change in one may necessitate a change in the others.

This document has been approved by the American Consulting Engineers Council, and recommended for use by its members.

© 1974 NATIONAL SOCIETY OF PROFESSIONAL ENGINEERS
2029 K STREET, N.W., WASHINGTON, D.C. 20006

NSPE 1910-8 (May 1974 10M)
Reprinted 4-77 8M

TABLE OF CONTENTS OF GENERAL CONDITIONS

INDEX TO GENERAL CONDITIONS

4

5

6

GENERAL CONDITIONS

ARTICLE 1—DEFINITIONS

Wherever used in these General Conditions or in the other Contract Documents, the following terms have the meanings indicated which are applicable to both the singular and plural thereof:

Agreement—The written agreement between OWNER and CONTRACTOR covering the Work to be performed; other Contract Documents are attached to the Agreement.

Application for Payment—The form furnished by ENGINEER which is to be used by CONTRACTOR in requesting progress payments and which is to include the schedule of values required by paragraph 14.1 and an affidavit of CONTRACTOR that progress payments theretofore received on account of the Work have been applied by CONTRACTOR to discharge in full all of CONTRACTOR's obligations reflected in prior Applications for Payment.

Bid—The offer or proposal of the Bidder submitted on the prescribed form setting forth the prices for the Work to be performed.

Bidder—Any person, firm or corporation submitting a Bid for the Work.

Bonds—Bid, performance and payment bonds and other instruments of security, furnished by CONTRACTOR and his surety in accordance with the Contract Documents.

Change Order—A written order to CONTRACTOR signed by OWNER authorizing an addition, deletion or revision in the Work, or an adjustment in the Contract Price or the Contract Time issued after execution of the Agreement.

Contract Documents—The Agreement, Addenda (whether issued prior to the opening of Bids or the execution of the Agreement), Instructions to Bidders, CONTRACTOR's Bid, the Bonds, the Notice of Award, these General Conditions, the Supplementary Conditions, the Specifications, Drawings and Modifications.

Contract Price—The total moneys payable to CONTRACTOR under the Contract Documents.

Contract Time—The number of days stated in the Agreement for the completion of the Work, computed as provided in paragraph 17.2.

CONTRACTOR—The person, firm or corporation with whom OWNER has executed the Agreement.

Day—A calendar day of twenty-four hours measured from midnight to the next midnight.

Drawings—The drawings which show the character and scope of the Work to be performed and which have been prepared or approved by ENGINEER and are referred to in the Contract Documents.

ENGINEER—The person, firm or corporation named as such in the Agreement.

Field Order—A written order issued by ENGINEER which clarifies or interprets the Contract Documents in accordance with paragraph 9.3 or orders minor changes in the Work in accordance with paragraph 10.2.

Modification—(a) A written amendment of the Contract Documents signed by both parties, (b) a Change Order, (c) a written clarification or interpretation issued by ENGINEER in accordance with paragraph 9.3 or (d) a written order for a minor change or alteration in the Work issued by ENGINEER pursuant to paragraph 10.2. A Modification may only be issued after execution of the Agreement.

Notice of Award—The written notice by OWNER to the apparent successful Bidder stating that upon compliance with the conditions precedent to be fulfilled by him within the time specified, OWNER will execute and deliver the Agreement to him.

Notice to Proceed—A written notice given by OWNER to CONTRACTOR (with a copy to ENGINEER) fixing the date on which the Contract Time will commence to run and on which CONTRACTOR shall start to perform his obligations under the Contract Documents.

OWNER—A public body or authority, corporation, association, partnership, or individual for whom the Work is to be performed.

Project—The entire construction to be performed as provided in the Contract Documents.

Resident Project Representative—The authorized representative of ENGINEER who is assigned to the Project site or any part thereof.

Shop Drawings—All drawings, diagrams, illustrations, brochures, schedules and other data which are prepared by CONTRACTOR, a Subcontractor, manufacturer, supplier or distributor and which illustrate the equipment, material or some portion of the Work.

Specifications—Those portions of the Contract Documents consisting of written technical descriptions of materials, equipment, construction systems, standards and workmanship as applied to the Work. The specifications are customarily organized in 16 divisions in accordance with the Uniform System for Construction Specifications endorsed by the Construction Specifications Institute. [*Note: the term "Technical Provisions" formerly described what is now referred to as the Specifications. For uniformity with the usage of other professional societies the term "Project Manual" is used to describe the volume formerly referred to as "The Specifications." The Project Manual contains documents concerning bidding requirements which in general govern relationships prior to the execution of the Agreement (such as the Invitation to Bid, Instructions to Bidders, Bid Bonds and Notice of Award) and the other portions of the Contract Documents.*]

Subcontractor—An individual, firm or corporation having a direct contract with CONTRACTOR or with any other Subcontractor for the performance of a part of the Work at the site.

8

Substantial Completion—The date as certified by ENGI-NEER when the construction of the Project or a specified part thereof is sufficiently completed, in accordance with the Contract Documents, so that the Project or specified part can be utilized for the purposes for which it was intended; or if there be no such certification, the date when final payment is due in accordance with paragraph 14.13.

Work—Any and all obligations, duties and responsibilities necessary to the successful completion of the Project assigned to or undertaken by CONTRACTOR under the Contract Documents, including all labor, materials, equipment and other incidentals, and the furnishing thereof.

ARTICLE 2—PRELIMINARY MATTERS

Execution of Agreement:

2.1. At least three counterparts of the Agreement and such other Contract Documents as practicable will be executed and delivered by CONTRACTOR to OWNER within fifteen days of the Notice of Award; and OWNER will execute and deliver one counterpart to CONTRACTOR within ten days of receipt of the executed Agreement from CONTRACTOR. ENGINEER will identify those portions of the Contract Documents not so signed and such identification will be binding on all parties. OWNER, CONTRACTOR and ENGINEER shall each receive an executed counterpart of the Contract Documents and additional conformed copies as required.

Delivery of Bonds:

2.2. When he delivers the executed Agreements to OWNER, CONTRACTOR shall also deliver to OWNER such Bonds as he may be required to furnish in accordance with paragraph 5.1.

Copies of Documents:

2.3. OWNER shall furnish to CONTRACTOR up to ten copies (unless otherwise provided in the Supplementary Conditions) of the Contract Documents as are reasonably necessary for the execution of the Work. Additional copies will be furnished, upon request, at the cost of reproduction.

Contractor's Pre-Start Representations:

2.4. CONTRACTOR represents that he has familiarized himself with, and assumes full responsibility for having familiarized himself with, the nature and extent of the Contract Documents, Work, locality, and with all local conditions and federal, state and local laws, ordinances, rules and regulations that may in any manner affect performance of the Work, and represents that he has correlated his study and observations with the requirements of the Contract Documents. CONTRACTOR also represents that he has studied all surveys and investigation reports of subsurface and latent physical conditions referred to in the General Requirements (Division 1) of the Specifications and made such additional surveys and

investigations as he deems necessary for the performance of the Work at the Contract Price in accordance with the requirements of the Contract Documents and that he has correlated the results of all such data with the requirements of the Contract Documents.

Commencement of Contract Time; Notice to Proceed:

2.5. The Contract Time will commence to run on the thirtieth day after the day on which the executed Agreement is delivered by OWNER to CONTRACTOR; or, if a Notice to Proceed is given, on the day indicated in the Notice to Proceed; but in no event shall the Contract Time commence to run later than the ninetieth day after the day of Bid opening or the thirtieth day after the day on which OWNER delivers the executed Agreement to CONTRACTOR. A Notice to Proceed may be given at any time within thirty days after the day on which OWNER delivers the executed Agreement to CONTRACTOR.

Starting the Project:

2.6. CONTRACTOR shall start to perform his obligations under the Contract Documents on the date when the Contract Time commences to run. No Work shall be done at the site prior to the date on which the Contract Time commences to run.

Before Starting Construction:

2.7. Before undertaking each part of the Work, CONTRACTOR shall carefully study and compare the Contract Documents and check and verify pertinent figures shown thereon and all applicable field measurements. He shall at once report in writing to ENGINEER any conflict, error or discrepancy which he may discover; however, he shall not be liable to OWNER or ENGINEER for his failure to discover any conflict, error or discrepancy in the Drawings or Specifications.

2.8. Within ten days after delivery of the executed Agreement by OWNER to CONTRACTOR, CONTRACTOR shall submit to ENGINEER for approval, an estimated progress schedule indicating the starting and completion dates of the various stages of the Work, and a preliminary schedule of Shop Drawing submissions.

2.9. Before starting the Work at the site, CONTRACTOR shall furnish OWNER and ENGINEER certificates of insurance as required by Article 5. Within twenty days after delivery of the executed Agreement by OWNER to CONTRACTOR, but before starting the Work at the site, a conference will be held to review the above schedules, to establish procedures for handling Shop Drawings and other submissions and for processing Applications for Payment, and to establish a working understanding between the parties as to the Project. Present at the conference will be OWNER or his representative, ENGINEER, Resident Project Representatives, CONTRACTOR and his Superintendent.

9

ARTICLE 3—CORRELATION, INTERPRETATION AND INTENT OF CONTRACT DOCUMENTS

3.1. It is the intent of the Specifications and Drawings to describe a complete Project to be constructed in accordance with the Contract Documents. The Contract Documents comprise the entire Agreement between OWNER and CONTRACTOR. They may be altered only by a Modification.

3.2. The Contract Documents are complementary; what is called for by one is as binding as if called for by all. If CONTRACTOR finds a conflict, error or discrepancy in the Contract Documents, he shall call it to ENGI-NEER's attention in writing at once and before proceeding with the Work affected thereby; however, he shall not be liable to OWNER or ENGINEER for his failure to discover any conflict, error or discrepancy in the Specifications or Drawings. In resolving such conflicts, errors and discrepancies, the documents shall be given precedence in the following order: Agreement, Modifications, Addenda, Supplementary Conditions, Instructions to Bidders, General Conditions, Specifications and Drawings. Figure dimensions on Drawings shall govern over scale dimensions, and detailed Drawings shall govern over general Drawings. Any Work that may reasonably be inferred from the Specifications or Drawings as being required to produce the intended result shall be supplied whether or not it is specifically called for. Work, materials or equipment described in words which so applied have a well-known technical or trade meaning shall be deemed to refer to such recognized standards.

ARTICLE 4—AVAILABILITY OF LANDS; PHYSICAL CONDITIONS; REFERENCE POINTS

Availability of Lands:

4.1. OWNER shall furnish, as indicated in the Contract Documents and not later than the date when needed by CONTRACTOR, the lands upon which the Work is to be done, rights-of-way for access thereto, and such other lands which are designated for the use of CONTRACTOR. Easements for permanent structures or permanent changes in existing facilities will be obtained and paid for by OWNER, unless otherwise specified in the Contract Documents. If CONTRACTOR believes that any delay in OWNER's furnishing these lands or easements entitles him to an extension of the Contract Time, he may make a claim therefor as provided in Article 12. CONTRACTOR shall provide for all additional lands and access thereto that may be required for temporary construction facilities or storage of materials and equipment.

Physical Conditions—Surveys and Reports:

4.2. Reference is made to the General Requirements (Division 1) of the Specifications for identification of those surveys and investigation reports of subsurface and latent physical conditions at the Project site or otherwise affecting performance of the Work which have been relied upon by ENGINEER in preparation of the Drawings and Specifications.

Unforeseen Physical Conditions:

4.3. CONTRACTOR shall promptly notify OWNER and ENGINEER in writing of any subsurface or latent physical conditions at the site differing materially from those indicated in the Contract Documents. ENGINEER will promptly investigate those conditions and advise OWNER in writing if further surveys or subsurface tests are necessary. Promptly thereafter, OWNER shall obtain the necessary additional surveys and tests and furnish copies to ENGINEER and CONTRACTOR. If ENGINEER finds that the results of such surveys or tests indicate that there are subsurface or latent physical conditions which differ materially from those intended in the Contract Documents, and which could not reasonably have been anticipated by CONTRACTOR, a Change Order shall be issued incorporating the necessary revisions.

Reference Points:

4.4. OWNER shall provide engineering surveys for construction to establish reference points which in his judgment are necessary to enable CONTRACTOR to proceed with the Work. CONTRACTOR shall be responsible for surveying and laying out the Work (unless otherwise provided in the Supplementary Conditions), and shall protect and preserve the established reference points and shall make no changes or relocations without the prior written approval of OWNER. He shall report to ENGINEER whenever any reference point is lost or destroyed or requires relocation because of necessary changes in grades or locations. CONTRACTOR shall replace and accurately relocate all reference points so lost, destroyed or moved.

ARTICLE 5—BONDS AND INSURANCE

Performance, Payment and Other Bonds:

5.1. CONTRACTOR shall furnish performance and payment Bonds as security for the faithful performance and payment of all his obligations under the Contract Documents. These Bonds shall be in amounts at least equal to the Contract Price, and (except as otherwise provided in the Supplementary Conditions) in such form and with such sureties as are licensed to conduct business in the state where the Project is located and are named in the current list of "Surety Companies Acceptable on Federal Bonds" as published in the Federal Register by the Audit Staff Bureau of Accounts, U.S. Treasury Department.

5.2. If the surety on any Bond furnished by CONTRACTOR is declared a bankrupt or becomes insolvent or its right to do business is terminated in any state where any part of the Project is located is revoked, CONTRACTOR shall within five days thereafter substitute another Bond and surety, both of which shall be acceptable to OWNER.

10

Contractor's Liability Insurance:

5.3. CONTRACTOR shall purchase and maintain such insurance as will protect him from claims under workmen's compensation laws, disability benefit laws or other similar employee benefit laws; from claims for damages because of bodily injury, occupational sickness or disease, or death of his employees, and claims insured by usual personal injury liability coverage; from claims for damages because of bodily injury, sickness or disease, or death of any person other than his employees including claims insured by usual personal injury liability coverage; and from claims for injury to or destruction of tangible property, including loss of use resulting therefrom—any or all of which may arise out of or result from CONTRACTOR's operations under the Contract Documents, whether such operations be by himself or by any Subcontractor or anyone directly or indirectly employed by any of them or for whose acts any of them may be legally liable. This insurance shall include the specific coverages and be written for not less than any limits of liability and maximum deductibles specified in the Supplementary Conditions or General Requirements (Division 1) or required by law, whichever is greater, shall include contractual liability insurance and shall include OWNER and ENGINEER as additional insured parties. Before starting the Work, CONTRACTOR shall file with OWNER and ENGINEER certificates of such insurance, acceptable to OWNER; these certificates shall contain a provision that the coverage afforded under the policies will not be cancelled or materially changed until at least fifteen days' prior written notice has been given to OWNER and ENGINEER.

Owner's Liability Insurance:

5.4. OWNER shall be responsible for purchasing and maintaining his own liability insurance and, at his option, may purchase and maintain such insurance as will protect him against claims which may arise from operations under the Contract Documents.

Property Insurance:

5.5. Unless otherwise provided, OWNER shall purchase and maintain property insurance upon the Project to the full insurable value thereof. This insurance shall include the interests of OWNER, CONTRACTOR and Subcontractors in the Work, shall insure against the perils of Fire, Extended Coverage, Vandalism and Malicious Mischief and such other perils as may be specified in the Supplementary Conditions or General Requirements (Division 1), and shall include damages, losses and expenses arising out of or resulting from any insured loss or incurred in the repair or replacement of any insured property (including fees and charges of engineers, architects, attorneys and other professionals).

5.6. OWNER shall purchase and maintain such steam boiler and machinery insurance as may be required by the Supplementary Conditions or by law. This insurance shall include the interests of OWNER, CONTRACTOR and Subcontractors in the Work.

5.7. Any insured loss under the policies of insurance required by paragraphs 5.5 and 5.6 is to be adjusted with OWNER and made payable to OWNER as trustee for the insureds, as their interests may appear, subject to the requirements of any applicable mortgage clause and of paragraph 5.11.

5.8. OWNER shall file a copy of all policies with CONTRACTOR before an exposure to loss may occur. If OWNER does not intend to purchase such insurance, he shall inform CONTRACTOR in writing prior to commencement of the Work. CONTRACTOR may then effect insurance which will protect the interests of himself and his Subcontractors in the Work, and by appropriate Change Order the cost thereof shall be charged to OWNER. If CONTRACTOR is damaged by failure of OWNER to purchase or maintain such insurance and so to notify CONTRACTOR, then OWNER shall bear reasonable costs properly attributable thereto.

5.9. If CONTRACTOR requests in writing that other special insurance be included in the property insurance policy, OWNER shall, if possible, include such insurance, and the cost thereof shall be charged to CONTRACTOR by appropriate Change Order.

5.10. OWNER and CONTRACTOR waive all rights against each other for damages caused by fire or other perils to the extent covered by insurance provided under paragraphs 5.5 through 5.11, inclusive, except such rights as they may have to the proceeds of such insurance held by OWNER as trustee. CONTRACTOR shall require similar waivers by Subcontractors in accordance with paragraph 6.12.

5.11. OWNER as trustee shall have power to adjust and settle any loss with the insurers unless one of the parties in interest shall object in writing within five days after the occurrence of loss to OWNER's exercise of this power, and if such objection be made, arbitrators shall be chosen as provided in Article 16. OWNER as trustee shall, in that case, make settlement with the insurers in accordance with the directions of such arbitrators. If distribution of the insurance proceeds by arbitration is required, the arbitrators will direct such distribution.

Additional Bonds and Insurance:

5.12. Prior to delivery of the executed Agreement by OWNER to CONTRACTOR, OWNER may require CONTRACTOR to furnish such other Bonds and such additional insurance, in such form and with such sureties or insurers as OWNER may require. If such other Bonds or such other insurance are specified by written instructions given prior to opening of Bids, the premiums shall be paid by CONTRACTOR; if subsequent thereto, they shall be paid by OWNER (except as otherwise provided in paragraph 6.7).

11

321

Supervision and Superintendence:

6.1. CONTRACTOR shall supervise and direct the Work efficiently and with his best skill and attention. He shall be solely responsible for the means, methods, techniques, sequences and procedures of construction, but he shall not be solely responsible for the negligence of others in the design or selection of a specific means, method, technique, sequence or procedure of construction which is indicated in and required by the Contract Documents. CONTRACTOR shall be responsible to see that the finished Work complies accurately with the Contract Documents.

6.2. CONTRACTOR shall keep on the Work at all times during its progress a competent resident superintendent, who shall not be replaced without written notice to OWNER and ENGINEER except under extraordinary circumstances. The superintendent will be CONTRACTOR's representative at the site and shall have authority to act on behalf of CONTRACTOR. All communications given to the superintendent shall be as binding as if given to CONTRACTOR.

Labor, Materials and Equipment:

6.3. CONTRACTOR shall provide competent, suitably qualified personnel to survey and lay out the Work and perform construction as required by the Contract Documents. He shall at all times maintain good discipline and order at the site.

6.4. CONTRACTOR shall furnish all materials, equipment, labor, transportation, construction equipment and machinery, tools, appliances, fuel, power, light, heat, telephone, water and sanitary facilities and all other facilities and incidentals necessary for the execution, testing, initial operation and completion of the Work.

6.5. All materials and equipment shall be new, except as otherwise provided in the Contract Documents. If required by ENGINEER, CONTRACTOR shall furnish satisfactory evidence as to the kind and quality of materials and equipment.

6.6. All materials and equipment shall be applied, installed, connected, erected, used, cleaned and conditioned in accordance with the instructions of the applicable manufacturer, fabricator or processors, except as otherwise provided in the Contract Documents.

Substitute Materials or Equipment:

6.7. If the General Requirements (Division 1) of the Specifications, law, ordinance or applicable rules or regulations permit CONTRACTOR to furnish or use a substitute that is equal to any material or equipment specified, and if CONTRACTOR wishes to furnish or use a proposed substitute, he shall, prior to the conference called for by paragraph 2.9 (unless another time is provided in the General Requirements), make written application to ENGINEER for approval of such a substitute certifying in writing that the proposed substitute will perform adequately the functions called for by the general design, be similar and of equal substance to that specified and be suited to the same use and capable of performing the same function as that specified; stating whether or not its incorporation in or use in connection with the Project is subject to the payment of any license fee or royalty; and identifying all variations of the proposed substitute from that specified and indicating available maintenance service. No substitute shall be ordered or installed without the written approval of ENGINEER who will be the judge of equality and may require CONTRACTOR to furnish such other data about the proposed substitute as he considers pertinent. No substitute shall be ordered or installed without such performance guarantee and bonds as OWNER may require which shall be furnished at CONTRACTOR's expense.

Concerning Subcontractors:

6.8. CONTRACTOR shall not employ any Subcontractor or other person or organization (including those who are to furnish the principal items of materials or equipment), whether initially or as a substitute, against whom OWNER or ENGINEER may have reasonable objection. A Subcontractor or other person or organization identified in writing to OWNER and ENGINEER by CONTRACTOR prior to the Notice of Award and not objected to in writing by OWNER or ENGINEER prior to the Notice of Award will be deemed acceptable to OWNER and ENGINEER. Acceptance of any Subcontractor, other person or organization by OWNER or ENGINEER shall not constitute a waiver of any right of OWNER or ENGINEER to reject defective Work or Work not in conformance with the Contract Documents. If OWNER or ENGINEER after due investigation has reasonable objection to any Subcontractor, other person or organization proposed by CONTRACTOR after the Notice of Award, CONTRACTOR shall submit an acceptable substitute and the Contract Price shall be increased or decreased by the difference in cost occasioned by such substitution, and an appropriate Change Order shall be issued. CONTRACTOR shall not be required to employ any Subcontractor, other person or organization against whom he has reasonable objection. CONTRACTOR shall not without the consent of OWNER and ENGINEER make any substitution for any Contractor, other person or organization who has been accepted by OWNER and ENGINEER unless ENGINEER determines that there is good cause for doing so.

6.9. CONTRACTOR shall be fully responsible for all acts and omissions of his Subcontractors and of persons and organizations directly or indirectly employed by them and of persons and organizations for whose acts any of them may be liable to the same extent that he is responsible for the acts and omissions of persons directly employed by him. Nothing in the Contract Documents shall create any contractual relationship between OWNER or ENGINEER and any Subcontractor or other person or organization having a direct contract with CONTRACTOR, nor shall it create any obligation on the part of

12

OWNER or ENGINEER to pay or to see to the payment of any moneys due any Subcontractor or other person or organization, except as may otherwise be required by law. OWNER or ENGINEER may furnish to any Subcontractor or other person or organization, to the extent practicable, evidence of amounts paid to CONTRACTOR on account of specific Work done in accordance with the schedule of values.

6.10. The divisions and sections of the Specifications and the identifications of any Drawings shall not control CONTRACTOR in dividing the Work among Subcontractors or delineating the Work to be performed by any specific trade.

6.11. CONTRACTOR agrees to bind specifically every Subcontractor to the applicable terms and conditions of the Contract Documents for the benefit of OWNER.

6.12. All Work performed for CONTRACTOR by a Subcontractor shall be pursuant to an appropriate agreement between CONTRACTOR and the Subcontractor which shall contain provisions that waive all rights the contracting parties may have against one another for damages caused by fire or other perils covered by insurance provided in accordance with paragraphs 5.5 through 5.11, inclusive, except such rights as they may have to the proceeds of such insurance held by OWNER as trustee under paragraph 5.9. CONTRACTOR shall pay each Subcontractor a just share of any insurance moneys received by CONTRACTOR under paragraphs 5.5 through 5.11, inclusive.

Patent Fees and Royalties:

6.13. CONTRACTOR shall pay all license fees and royalties and assume all costs incident to the use in the performance of the Work of any invention, design, process, product or device which is the subject of patent rights or copyrights held by others. If a particular invention, design, process, product or device is specified in the Contract Documents for use in the performance of the Work and if to the actual knowledge of OWNER or ENGINEER its use is subject to patent rights or copyrights calling for the payment of any license fee or royalty to others, the existence of such rights shall be disclosed by OWNER in the Contract Documents. CONTRACTOR shall indemnify and hold harmless OWNER and ENGINEER and anyone directly or indirectly employed by either of them from and against all claims, damages, losses and expenses (including attorneys' fees) arising out of any infringement of patent rights or copyrights incident to the use in the performance of the Work or resulting from the incorporation in the Work of any invention, design, process, product or device not specified in the Contract Documents, and shall defend all such claims in connection with any alleged infringement of such rights.

Permits:

6.14. CONTRACTOR shall obtain and pay for all construction permits and licenses and shall pay all govern-

mental charges and inspection fees necessary for the prosecution of the Work, which are applicable at the time of his Bid. OWNER shall assist CONTRACTOR, when necessary, in obtaining such permits and licenses. CONTRACTOR shall also pay all public utility charges.

Laws and Regulations:

6.15. CONTRACTOR shall give all notices and comply with all laws, ordinances, rules and regulations applicable to the Work. If CONTRACTOR observes that the Specifications or Drawings are at variance therewith, he shall give ENGINEER prompt written notice thereof, and any necessary changes shall be adjusted by an appropriate Modification. If CONTRACTOR performs any Work knowing it to be contrary to such laws, ordinances, rules and regulations, and without such notice to ENGINEER, he shall bear all costs arising therefrom; however, it shall not be his primary responsibility to make certain that the Specifications and Drawings are in accordance with such laws, ordinances, rules and regulations.

Taxes:

6.16. CONTRACTOR shall pay all sales, consumer, use and other similar taxes required to be paid by him in accordance with the law of the place where the Work is to be performed.

Use of Premises:

6.17. CONTRACTOR shall confine his equipment, the storage of materials and equipment and the operations of his workmen to areas permitted by law, ordinances, permits, or the requirements of the Contract Documents, and shall not unreasonably encumber the premises with materials or equipment.

6.18. CONTRACTOR shall not load nor permit any part of any structure to be loaded with weights that will endanger the structure, nor shall he subject any part of the Work to stresses or pressures that will endanger it.

Record Drawings:

6.19. CONTRACTOR shall keep one record copy of all Specifications, Drawings, Addenda, Modifications, and Shop Drawings at the site in good order and annotated to show all changes made during the construction process. These shall be available to ENGINEER and shall be delivered to him for OWNER upon completion of the Project. [*Note: Further provisions in respect of such record drawings may be included in the General Requirements (Division 1).*]

Safety and Protection:

6.20. CONTRACTOR shall be responsible for initiating, maintaining and supervising all safety precautions and programs in connection with the Work. He shall take all necessary precautions for the safety of, and shall provide the necessary protection to prevent damage, injury or loss to:

6.20.1. all employees on the Work and other persons who may be affected thereby,

6.20.2. all the Work and all materials or equipment to be incorporated therein, whether in storage on or off the site, and

6.20.3. other property at the site or adjacent thereto, including trees, shrubs, lawns, walks, pavements, roadways, structures and utilities not designated for removal, relocation or replacement in the course of construction.

CONTRACTOR shall comply with all applicable laws, ordinances, rules, regulations and orders of any public body having jurisdiction for the safety of persons or property or to protect them from damage, injury or loss. He shall erect and maintain, as required by the conditions and progress of the Work, all necessary safeguards for its safety and protection. He shall notify owners of adjacent utilities when prosecution of the Work may affect them. All damage, injury or loss to any property referred to in paragraph 6.20.2 or 6.20.3 caused, directly or indirectly, in whole or in part, by CONTRACTOR, any Subcontractor or anyone directly or indirectly employed by any of them or anyone for whose acts any of them may be liable, shall be remedied by CONTRACTOR: except damage or loss attributable to the fault of Drawings or Specifications or to the acts or omissions of OWNER or ENGINEER or anyone employed by either of them or anyone for whose acts either of them may be liable, and not attributable, directly or indirectly, in whole or in part, to the fault or negligence of CONTRACTOR. CONTRACTOR's duties and responsibilities for the safety and protection of the Work shall continue until such time as all the Work is completed and ENGINEER has issued a notice to OWNER and CONTRACTOR in accordance with paragraph 14.13 that Work is acceptable.

6.21. CONTRACTOR shall designate a responsible member of his organization at the site whose duty shall be the prevention of accidents. This person shall be CONTRACTOR's superintendent unless otherwise designated in writing by CONTRACTOR to OWNER.

Emergencies:

6.22. In emergencies affecting the safety of persons or the Work or property at the site or adjacent thereto, CONTRACTOR, without special instruction or authorization from ENGINEER or OWNER, is obligated to act, at his discretion, to prevent threatened damage, injury or loss. He shall give ENGINEER prompt written notice of any significant changes in the Work or deviations from the Contract Documents caused thereby, and a Change Order shall thereupon be issued covering the changes and deviations involved. If CONTRACTOR believes that additional work done by him in an emergency which arose from causes beyond his control entitles him to an increase in the Contract Price or an extension of the Contract Time, he may make a claim therefor as provided in Articles 11 and 12.

Shop Drawings and Samples:

6.23. After checking and verifying all field measurements, CONTRACTOR shall submit to ENGINEER for approval, in accordance with the accepted schedule of Shop Drawing submissions (see paragraph 2.8) five copies (or at ENGINEER's option, one reproducible copy) of all Shop Drawings, which shall have been checked by and stamped with the approval of CONTRACTOR and identified as ENGINEER may require. The data shown on the Shop Drawings will be complete with respect to dimensions, design criteria, materials of construction and the like to enable ENGINEER to review the information as required.

6.24. CONTRACTOR shall also submit to ENGINEER for approval with such promptness as to cause no delay in Work, all samples required by the Contract Documents. All samples will have been checked by and stamped with the approval of CONTRACTOR, identified clearly as to material, manufacturer, any pertinent catalog numbers and the use for which intended.

6.25. At the time of each submission, CONTRACTOR shall in writing call ENGINEER's attention to any deviations that the Shop Drawing or sample may have from the requirements of the Contract Documents.

6.26. ENGINEER will review and approve with reasonable promptness Shop Drawings and samples, but his review and approval shall be only for conformance with the design concept of the Project and for compliance with the information given in the Contract Documents. The approval of a separate item as such will not indicate approval of the assembly in which the item functions. CONTRACTOR shall make any corrections required by ENGINEER and shall return the required number of corrected copies of Shop Drawings and resubmit new samples until approved. CONTRACTOR shall direct specific attention in writing or on resubmitted Shop Drawings to revisions other than the corrections called for by ENGINEER on previous submissions. CONTRACTOR's stamp of approval on any Shop Drawing or sample shall constitute a representation to OWNER and ENGINEER that CONTRACTOR has either determined and verified all quantities, dimensions, field construction criteria, materials, catalog numbers, and similar data or he assumes full responsibility for doing so, and that he has reviewed or coordinated each Shop Drawing or sample with the requirements of the Work and the Contract Documents.

6.27. Where a Shop Drawing or sample submission is required by the Specifications, no related Work shall be commenced until the submission has been approved by ENGINEER. A copy of each approved Shop Drawing and each approved sample shall be kept in good order by CONTRACTOR at the site and shall be available to ENGINEER.

6.28. ENGINEER's approval of Shop Drawings or samples shall not relieve CONTRACTOR from his responsibility for any deviations from the requirements of the Contract Documents unless CONTRACTOR has in writing called ENGINEER's attention to such deviation at the time of submission and ENGINEER has given written approval to the specific deviation, nor shall any approval by ENGINEER relieve CONTRACTOR from responsibility for errors or omissions in the Shop Drawings.

[*Note: Further provisions in respect to Shop Drawings and samples may be included in the General Requirements (Division 1).*]

Cleaning:

6.29. CONTRACTOR shall keep the premises free from accumulations of waste materials, rubbish and other debris resulting from the Work, and at the completion of the Work he shall remove all waste materials, rubbish and debris from and about the premises as well as all tools, construction equipment and machinery, and surplus materials, and shall leave the site clean and ready for occupancy by OWNER. CONTRACTOR shall restore to their original condition those portions of the site not designated for alteration by the Contract Documents. [*Note: Further provisions in respect of cleaning may be included in the General Requirements (Division 1).*]

Indemnification:

6.30. CONTRACTOR shall indemnify and hold harmless OWNER and ENGINEER and their agents and employees from and against all claims, damages, losses and expenses including attorneys' fees arising out of or resulting from the performance of the Work, provided that any such claim, damage, loss or expense (a) is attributable to bodily injury, sickness, disease or death, or to injury to or destruction of tangible property (other than the Work itself) including the loss of use resulting therefrom and (b) is caused in whole or in part by any negligent act or omission of CONTRACTOR, any Subcontractor, anyone directly or indirectly employed by any of them or anyone for whose acts any of them may be liable, regardless of whether or not it is caused in part by a party indemnified hereunder.

6.31. In any and all claims against OWNER or ENGINEER or any of their agents or employees by any employee of CONTRACTOR, any Subcontractor, anyone directly or indirectly employed by any of them or anyone for whose acts any of them may be liable, the indemnification obligation under paragraph 6.30 shall not be limited in any way by any limitation on the amount or type of damages, compensation or benefits payable by or for CONTRACTOR or any Subcontractor under workmen's compensation acts, disability benefit acts or other employee benefit acts.

6.32. The obligations of CONTRACTOR under paragraph 6.30 shall not extend to the liability of ENGINEER, his agents or employees arising out of (a) the preparation or approval of maps, drawings, opinions, reports, surveys, Change Orders, designs or specifications or (b) the giving of or the failure to give directions or instructions by ENGINEER, his agents or employees provided such giving or failure to give is the primary cause of injury or damage.

ARTICLE 7—WORK BY OTHERS

7.1. OWNER may perform additional work related to the Project by himself, or he may let other direct contracts therefor which shall contain General Conditions similar to these. CONTRACTOR shall afford the other contractors who are parties to such direct contracts (or OWNER, if he is performing the additional work himself), reasonable opportunity for the introduction and storage of materials and equipment and the execution of work, and shall properly connect and coordinate his Work with theirs.

7.2. If any part of CONTRACTOR's Work depends for proper execution or results upon the work of any such other contractor (or OWNER), CONTRACTOR shall inspect and promptly report to ENGINEER in writing any defects or deficiencies in such work that render it unsuitable for such proper execution and results. His failure so to report shall constitute an acceptance of the other work as fit and proper for the relationship of his Work except as to defects and deficiencies which may appear in the other work after the execution of his Work.

7.3. CONTRACTOR shall do all cutting, fitting and patching of his Work that may be required to make its several parts come together properly and fit it to receive or be received by such other work. CONTRACTOR shall not endanger any work of others by cutting, excavating or otherwise altering their work and will only cut or alter their work with the written consent of ENGINEER and of the other contractors whose work will be affected.

7.4. If the performance of additional work by other contractors or OWNER is not noted in the Contract Documents prior to the execution of the contract, written notice thereof shall be given to CONTRACTOR prior to starting any such additional work. If CONTRACTOR believes that the performance of such additional work by OWNER or others involves him in additional expense or entitles him to an extension of the Contract Time, he may make a claim therefor as provided in Articles 11 and 12.

ARTICLE 8—OWNER'S RESPONSIBILITIES

8.1. OWNER shall issue all communications to CONTRACTOR through ENGINEER.

8.2. In case of termination of the employment of ENGINEER, OWNER shall appoint an engineer against whom CONTRACTOR makes no reasonable objection, whose status under the Contract Documents shall be that of the former ENGINEER. Any dispute in connection with such appointment shall be subject to arbitration.

8.3. OWNER shall furnish the data required of him under the Contract Documents promptly and shall make payments to CONTRACTOR promptly after they are due as provided in paragraphs 14.4 and 14.13.

8.4. OWNER's duties in respect of providing lands and easements and providing engineering surveys to establish

15

reference points are set forth in paragraphs 4.1 and 4.4. Paragraph 4.2 refers to OWNER's identifying and making available to CONTRACTOR copies of surveys and investigation reports of subsurface and latent physical conditions at the site or otherwise affecting performance of the Work which have been relied upon by ENGINEER in preparing the Drawings and Specifications.

8.5. OWNER's responsibilities in respect of liability and property insurance are set forth in paragraph 5.4 and 5.5.

8.6. In addition to his rights to request changes in the Work in accordance with Article 10, OWNER (especially in certain instances as provided in paragraph 10.4) shall be obligated to execute Change Orders.

8.7. OWNER's responsibility in respect of certain inspections, tests and approvals is set forth in paragraph 13.2.

8.8. In connection with OWNER's right to stop Work or suspend Work, see paragraphs 13.8 and 15.1. Paragraph 15.2 deals with OWNER's right to terminate services of CONTRACTOR under certain circumstances.

ARTICLE 9—ENGINEER'S STATUS DURING CONSTRUCTION

Owner's Representative:

9.1. ENGINEER will be OWNER's representative during the construction period. The duties and responsibilities and the limitations of authority of ENGINEER as OWNER's representative during construction are set forth in Articles 1 through 17 of these General Conditions and shall not be extended without written consent of OWNER and ENGINEER.

Visits to Site:

9.2. ENGINEER will make periodic visits to the site to observe the progress and quality of the executed Work and to determine, in general, if the Work is proceeding in accordance with the Contract Documents. He will not be required to make exhaustive or continuous on-site inspections to check the quality or quantity of the Work. His efforts will be directed toward providing assurance for OWNER that the completed Project will conform to the requirements of the Contract Documents. On the basis of his on-site observations as an experienced and qualified design professional, he will keep OWNER informed of the progress of the Work and will endeavor to guard OWNER against defects and deficiencies in the Work of contractors.

Clarifications and Interpretations:

9.3. ENGINEER will issue with reasonable promptness such written clarifications or interpretations of the Contract Documents (in the form of Drawings or otherwise) as he may determine necessary, which shall be consistent with or reasonably inferable from the overall intent of the Contract Documents. If CONTRACTOR believes that a written clarification and interpretation entitles him to an increase in the Contract Price, he may make a claim therefor as provided in Article 11.

Rejecting Defective Work:

9.4. ENGINEER will have authority to disapprove or reject Work which is "defective" (which term is hereinafter used to describe Work that is unsatisfactory, faulty or defective, or does not conform to the requirements of the Contract Documents or does not meet the requirements of any inspection, test or approval referred to in paragraph 13.2 or has been damaged prior to approval of final payment). He will also have authority to require special inspection or testing of the Work as provided in paragraph 13.7, whether or not the Work is fabricated, installed or completed.

Shop Drawings, Change Orders and Payments:

9.5. In connection with ENGINEER's responsibility for Shop Drawings and samples, see paragraphs 6.23 through 6.28 inclusive.

9.6. In connection with ENGINEER's responsibility for Change Orders, see Articles 10, 11 and 12.

9.7. In connection with ENGINEER's responsibilities in respect of Applications for Payment, etc., see Article 14.

Resident Project Representatives:

9.8. If OWNER and ENGINEER agree, ENGINEER will furnish a Resident Project Representative and assistants to assist ENGINEER in carrying out his responsibilities at the site. The duties, responsibilities and limitations of authority of any such Resident Project Representative and assistants shall be as set forth in an exhibit to be incorporated in the Contract Documents.

Decisions on Disagreements:

9.9. ENGINEER will be the interpreter of the requirements of the Contract Documents and the judge of the performance thereunder. In his capacity as interpreter and judge he will exercise his best efforts to insure faithful performance by both OWNER and CONTRACTOR. He will not show partiality to either and will not be liable for the result of any interpretation or decision rendered in good faith. Claims, disputes and other matters relating to the execution and progress of the Work or the interpretation of or performance under the Contract Documents shall be referred to ENGINEER for decision; which he will render in writing within a reasonable time.

9.10. Either OWNER or CONTRACTOR may demand arbitration with respect to any such claim, dispute or other matter that has been referred to ENGINEER, except any which have been waived by the making and acceptance of final payment as provided in paragraph 14.16, such arbitration to be in accordance with Article 16. However, no demand for arbitration of any such claim, dispute or other matter shall be made until the earlier of (a) the date on which ENGINEER has rendered his decision or (b) the tenth day after the parties have presented their evidence to ENGINEER if he has not rendered his written decision before that date. No demand for arbitration shall be made later than thirty days after the date on which ENGINEER rendered his written decision in respect of the claim, dispute or other matter as to which arbitration is sought; and the failure to demand arbitration within said thirty days' period shall result in ENGINEER's decision being final and binding upon OWNER and CONTRACTOR. If ENGINEER renders a decision after arbitration proceedings have been initiated, such decision may be entered as evidence but shall not supersede the arbitration proceedings, except where the decision is acceptable to the parties concerned.

Limitations on ENGINEER's Responsibilities:

9.11. Neither ENGINEER's authority to act under this Article 9 or elsewhere in the Contract Documents nor any decision made by him in good faith either to exercise or not exercise such authority shall give rise to any duty or responsibility of ENGINEER to CONTRACTOR, any Subcontractor, any materialman, fabricator, supplier or any of their agents or employees or any other person performing any of the Work.

9.12. ENGINEER will not be responsible for CONTRACTOR's means, methods, techniques, sequences or procedures of construction, or the safety precautions and programs incident thereto, and he will not be responsible for CONTRACTOR's failure to perform the Work in accordance with the Contract Documents.

9.13. ENGINEER will not be responsible for the acts or omissions of CONTRACTOR, or any Subcontractors, or any of his or their agents or employees, or any other persons at the site or otherwise performing any of the Work.

ARTICLE 10—CHANGES IN THE WORK

10.1. Without invalidating the Agreement, OWNER may, at any time or from time to time, order additions, deletions or revisions in the Work; these will be authorized by Change Orders. Upon receipt of a Change Order, CONTRACTOR shall proceed with the Work involved. All such Work shall be executed under the applicable conditions of the Contract Documents. If any Change Order causes an increase or decrease in the Contract Price or an extension or shortening of the Contract Time, an equitable adjustment will be made as provided in Article 11 or Article 12 on the basis of a claim made by either party.

10.2. ENGINEER may authorize minor changes or alterations in the Work not involving extra cost and not inconsistent with the overall intent of the Contract Documents. These may be accomplished by a Field Order. If CONTRACTOR believes that any minor change or alteration authorized by ENGINEER entitles him to an increase in the Contract Price, he may make a claim therefor as provided in Article 11.

10.3. Additional Work performed by CONTRACTOR without authorization of a Change Order will not entitle him to an increase in the Contract Price or an extension of the Contract Time, except in the case of an emergency as provided in paragraph 6.22 and except as provided in paragraphs 10.2 and 13.7.

10.4. OWNER shall execute appropriate Change Orders prepared by ENGINEER covering changes in the Work to be performed as provided in paragraph 4.3, and Work performed in an emergency as provided in paragraph 6.22 and any other claim of CONTRACTOR for a change in the Contract Time or the Contract Price which is approved by ENGINEER.

10.5. It is CONTRACTOR's responsibility to notify his Surety of any changes affecting the general scope of the Work or change in the Contract Price and the amount of the applicable Bonds shall be adjusted accordingly. CONTRACTOR shall furnish proof of such adjustment to OWNER.

ARTICLE 11—CHANGE OF CONTRACT PRICE

11.1. The Contract Price constitutes the total compensation payable to CONTRACTOR for performing the Work. All duties, responsibilities and obligations assigned to or undertaken by CONTRACTOR shall be at his expense without change in the Contract Price.

11.2. The Contract Price may only be changed by a Change Order. Any claim for an increase in the Contract Price shall be based on written notice delivered to OWNER and ENGINEER within fifteen days of the occurrence of the event giving rise to the claim. Notice of the amount of the claim with supporting data shall be delivered within forty-five days of such occurrence unless ENGINEER allows an additional period of time to ascertain accurate cost data. All claims for adjustments in the Contract Price shall be determined by ENGINEER if OWNER and CONTRACTOR cannot otherwise agree on the amount involved. Any change in the Contract Price resulting from any such claim shall be incorporated in a Change Order.

11.3. The value of any Work covered by a Change Order or of any claim for an increase or decrease in the Contract Price shall be determined in one of the following ways:

11.3.1. Where the Work involved is covered by unit prices contained in the Contract Documents, by application of unit prices to the quantities of the items involved.

11.3.2. By mutual acceptance of a lump sum.

11.3.3. On the basis of the Cost of the Work (determined as provided in paragraphs 11.4 and 11.5) plus a Contractor's Fee for overhead and profit (determined as provided in paragraph 11.6).

Cost of the Work:

11.4. The term Cost of the Work means the sum of all costs necessarily incurred and paid by the CONTRACTOR in the proper performance of the Work. Except as otherwise may be agreed to in writing by OWNER, such costs shall be in amounts no higher than those prevailing in the locality of the Project, shall include only the following items and shall not include any of the costs itemized in paragraph 11.5:

11.4.1. Payroll costs for employees in the direct employ of CONTRACTOR in the performance of the Work under schedules of job classifications agreed upon by OWNER and CONTRACTOR. Payroll costs for employees not employed full time on the Work shall be apportioned on the basis of their time spent on the Work. Payroll costs shall include, but not be limited to, salaries and wages plus the cost of fringe benefits which shall include social security contributions, unemployment, excise and payroll taxes, workmen's compensation, health and retirement benefits, bonuses, sick leave, vacation and holiday pay applicable thereto. Such employees shall include superintendents and foremen at the site. The expenses of performing work after regular working hours, on Sunday or legal holidays shall be included in the above to the extent authorized by OWNER.

11.4.2. Cost of all materials and equipment furnished and incorporated in the Work, including costs of transportation and storage thereof, and manufacturers' field services required in connection therewith. All cash discounts shall accrue to CONTRACTOR unless OWNER deposits funds with CONTRACTOR with which to make payments, in which case the cash discounts shall accrue to OWNER. All trade discounts, rebates and refunds, and all returns from sale of surplus materials and equipment shall accrue to OWNER, and CONTRACTOR shall make provisions so that they may be obtained.

11.4.3. Payments made by CONTRACTOR to the Subcontractors for Work performed by Subcontractors. If required by OWNER, CONTRACTOR shall obtain competitive bids from Subcontractors acceptable to him and shall deliver such bids to OWNER who will then determine with the advice of ENGINEER, which bids will be accepted. If a subcontract provides that the Subcontractor is to be paid on the basis of Cost of the Work Plus a Fee, the Cost of the Work shall be determined in accordance with paragraphs 11.4 and 11.5. All subcontracts shall be subject to the other provisions of the Contract Documents insofar as applicable.

11.4.4. Costs of special consultants (including, but not limited to, engineers, architects, testing laboratories, surveyors, lawyers and accountants) employed for services specifically related to the Work.

11.4.5. Supplemental costs including the following:

11.4.5.1. The proportion of necessary transportation, traveling and subsistence expenses of CONTRACTOR's employees incurred in discharge of duties connected with the Work.

11.4.5.2. Cost, including transportation and maintenance, of all materials, supplies, equipment, machinery, appliances, office and temporary facilities at the site and hand tools not owned by the workmen, which are consumed in the performance of the Work, and cost less market value of such items used but not consumed which remain the property of CONTRACTOR.

11.4.5.3. Rentals of all construction equipment and machinery and the parts thereof whether rented from CONTRACTOR or others in accordance with rental agreements approved by OWNER with the advice of ENGINEER, and the costs of transportation, loading, unloading, installation, dismantling and removal thereof—all in accordance with terms of said rental agreements. The rental of any such equipment, machinery or parts shall cease when the use thereof is no longer necessary for the Work.

11.4.5.4. Sales, use or similar taxes related to the Work, and for which CONTRACTOR is liable, imposed by any governmental authority.

11.4.5.5. Deposits lost for causes other than CONTRACTOR's negligence, royalty payments and fees for permits and licenses.

11.4.5.6. Losses, damages and expenses, not compensated by insurance or otherwise, sustained by CONTRACTOR in connection with the execution of, and to, the Work, provided they have resulted from causes other than the negligence of CONTRACTOR, any Subcontractor, or anyone directly or indirectly employed by any of them or for whose acts any of them may be liable. Such losses shall include settlements made with the written consent and approval of OWNER. No such losses, damages and expenses shall be included in the Cost of the Work for the purpose of determining Contractor's Fee. If, however, any such loss or damage requires reconstruction and CONTRACTOR is placed in charge thereof, he shall be paid for his services a fee proportionate to that stated in paragraph 11.6.2.

11.4.5.7. The cost of utilities, fuel and sanitary facilities at the site.

11.4.5.8. Minor expenses such as telegrams, long distance telephone calls, telephone service at the site, expressage and similar petty cash items in connection with the Work.

11.4.5.9. Cost of premiums for bonds and insurance which OWNER is required to pay in accordance with paragraph 5.12.

18

11.5. The term Cost of the Work shall not include any of the following:

11.5.1. Payroll costs and other compensation of CONTRACTOR's officers, executives, principals (of partnership and sole proprietorships), general managers, engineers, architects, estimators, lawyers, auditors, accountants, purchasing and contracting agents, expeditors, timekeepers, clerks and other personnel employed by CONTRACTOR whether at the site or in his principal or a branch office for general administration of the Work and not specifically included in the schedule referred to in subparagraph 11.4.1—all of which are to be considered administrative costs covered by the Contractor's Fee.

11.5.2. Expenses of CONTRACTOR's principal and branch offices other than his office at the site.

11.5.3. Any part of CONTRACTOR's capital expenses, including interest on CONTRACTOR's capital employed for the Work and charges against CONTRACTOR for delinquent payments.

11.5.4. Cost of premiums for all bonds and for all insurance policies whether or not CONTRACTOR is required by the Contract Documents to purchase and maintain the same (except as otherwise provided in subparagraph 11.4.5.9).

11.5.5. Costs due to the negligence of CONTRACTOR, any Subcontractor, or anyone directly or indirectly employed by any of them or for whose acts any of them may be liable, including but not limited to, the correction of defective work, disposal of materials or equipment wrongly supplied and making good any damage to property.

11.5.6. Other overhead or general expense costs of any kind and the costs of any item not specifically and expressly included in paragraph 11.4.

Contractor's Fee:

11.6. The Contractor's Fee which shall be allowed to CONTRACTOR for his overhead and profit shall be determined as follows:

11.6.1. a mutually acceptable fixed fee; or if none can be agreed upon,

11.6.2. a fee based on the following percentages of the various portions of the Cost of the Work:

11.6.2.1. for costs incurred under paragraphs 11.4.1 and 11.4.2, the Contractor's Fee shall be ten percent,

11.6.2.2. for costs incurred under paragraph 11.4.3, the Contractor's Fee shall be five percent; and if a subcontract is on the basis of Cost of the Work Plus a Fee, the maximum allowable to the Subcontractor as a fee for overhead and profit shall be ten percent, and

11.6.2.3. no fee shall be payable on the basis of costs itemized under paragraphs 11.4.4, 11.4.5 and 11.5.

11.7. The amount of credit to be allowed by CONTRACTOR to OWNER for any such change which results in a net decrease in cost, will be the amount of the actual net decrease. When both additions and credits are involved in any one change, the combined overhead and profit shall be figured on the basis of the net increase, if any.

11.8. Whenever the cost of any Work is to be determined pursuant to paragraphs 11.4. and 11.5, CONTRACTOR will submit in form prescribed by ENGINEER an itemized cost breakdown together with supporting data.

Cash Allowances:

11.9. It is understood that CONTRACTOR has included in the Contract Price all allowances so named in the Contract Documents and shall cause the Work so covered to be done by such materialmen, suppliers or Subcontractors and for such sums within the limit of the allowances as ENGINEER may approve. Upon final payment, the Contract Price shall be adjusted as required and an appropriate Change Order issued. CONTRACTOR agrees that the original Contract Price includes such sums as he deems proper for costs and profit on account of cash allowances. No demand for additional cost or profit in connection therewith will be allowed.

ARTICLE 12—CHANGE OF THE CONTRACT TIME

12.1. The Contract Time may only be changed by a Change Order. Any claim for an extension in the Contract Time shall be based on written notice delivered to OWNER and ENGINEER within fifteen days of the occurrence of the event giving rise to the claim. Notice of the extent of the claim with supporting data shall be delivered within forty-five days of such occurrence unless ENGINEER allows an additional period of time to ascertain more accurate data. All claims for adjustment in the Contract Time shall be determined by ENGINEER if OWNER and CONTRACTOR cannot otherwise agree. Any change in the Contract Time resulting from any such claim shall be incorporated in a Change Order.

12.2. The Contract Time will be extended in an amount equal to time lost due to delays beyond the control of CONTRACTOR if he makes a claim therefor as provided in paragraph 12.1. Such delays shall include, but not be restricted to, acts or neglect by any separate contractor employed by OWNER, fires, floods, labor disputes, epidemics, abnormal weather conditions, or acts of God.

12.3. All time limits stated in the Contract Documents are of the essence of the Agreement. The provisions of this Article 12 shall not exclude recovery for damages (including compensation for additional professional services) for delay by either party.

19

ARTICLE 13—WARRANTY AND GUARANTEE; TESTS AND INSPECTIONS; CORRECTION, REMOVAL OR ACCEPTANCE OF DEFECTIVE WORK

Warranty and Guarantee:

13.1. CONTRACTOR warrants and guarantees to OWNER and ENGINEER that all materials and equipment will be new unless otherwise specified and that all Work will be of good quality and free from faults or defects and in accordance with the requirements of the Contract Documents and of any inspections, tests or approvals referred to in paragraph 13.2. All unsatisfactory Work, all faulty or defective Work, and all Work not conforming to the requirements of the Contract Documents at the time of acceptance thereof or of such inspections, tests or approvals, shall be considered defective. Prompt notice of all defects shall be given to CONTRACTOR. All defective Work, whether or not in place, may be rejected, corrected or accepted as provided in this Article 13.

Tests and Inspections:

13.2. If the Contract Documents, laws, ordinances, rules, regulations or orders of any public authority having jurisdiction require any Work to specifically be inspected, tested, or approved by some public body, CONTRACTOR shall assume full responsibility therefor, pay all costs in connection therewith and furnish ENGINEER the required certificates of inspection, testing or approval. All other inspections, tests and approvals required by the Contract Documents shall be performed by organizations acceptable to OWNER and CONTRACTOR and the costs thereof shall be borne by OWNER unless otherwise specified.

13.3. CONTRACTOR shall give ENGINEER timely notice of readiness of the Work for all inspections, tests or approvals. If any such Work required so to be inspected, tested or approved is covered without written approval of ENGINEER, it must, if requested by ENGINEER, be uncovered for observation, and such uncovering shall be at CONTRACTOR's expense unless CONTRACTOR has given ENGINEER timely notice of his intention to cover such Work and ENGINEER has not acted with reasonable promptness in response to such notice.

13.4. Neither observations by ENGINEER nor inspections, tests or approvals by persons other than CONTRACTOR shall relieve CONTRACTOR from his obligations to perform the Work in accordance with the requirements of the Contract Documents.

Access to Work:

13.5. ENGINEER and his representatives and other representatives of OWNER will at reasonable times have access to the Work. CONTRACTOR shall provide proper and safe facilities for such access and observation of the Work and also for any inspection or testing thereof by others.

Uncovering Work:

13.6. If any Work is covered contrary to the written request of ENGINEER, it must, if requested by ENGINEER, be uncovered for his observation and replaced at CONTRACTOR's expense.

13.7. If any Work has been covered which ENGINEER has not specifically requested to observe prior to its being covered, or if ENGINEER considers it necessary or advisable that covered Work be inspected or tested by others, CONTRACTOR, at ENGINEER's request, shall uncover, expose or otherwise make available for observation, inspection or testing as ENGINEER may require, that portion of the Work in question, furnishing all necessary labor, material and equipment. If it is found that such Work is defective, CONTRACTOR shall bear all the expenses of such uncovering, exposure, observation, inspection and testing and of satisfactory reconstruction, including compensation for additional professional services, and an appropriate deductive Change Order shall be issued. If, however, such Work is not found to be defective, CONTRACTOR shall be allowed an increase in the Contract Price or an extension of the Contract Time, or both, directly attributable to such uncovering, exposure, observation, inspection, testing and reconstruction if he makes a claim therefor as provided in Articles 11 and 12.

Owner May Stop the Work:

13.8. If the Work is defective, or CONTRACTOR fails to supply sufficient skilled workmen or suitable materials or equipment, or if CONTRACTOR fails to make prompt payments to Subcontractors or for labor, materials or equipment, OWNER may order CONTRACTOR to stop the Work, or any portion thereof, until the cause for such order has been eliminated; however, this right of OWNER to stop the Work shall not give rise to any duty on the part of OWNER to exercise this right for the benefit of CONTRACTOR or any other party.

Correction or Removal of Defective Work:

13.9. If required by ENGINEER prior to approval of final payment, CONTRACTOR shall promptly, without cost to OWNER and as specified by ENGINEER, either correct any defective Work, whether or not fabricated, installed or completed, or, if the Work has been rejected by ENGINEER, remove it from the site and replace it with nondefective Work. If CONTRACTOR does not correct such defective Work or remove and replace such rejected Work within a reasonable time, all as specified in a written notice from ENGINEER, OWNER may have the deficiency corrected or the rejected Work removed and replaced. All direct or indirect costs of such correction or removal and replacement, including compensation for additional professional services, shall be paid by CONTRACTOR, and an appropriate deductive Change Order shall be issued. CONTRACTOR shall also bear the expenses of making good all Work of others destroyed or damaged by his correction, removal or replacement of his defective Work.

One Year Correction Period:

13.10. If, after the approval of final payment and prior to the expiration of one year after the date of Substantial Completion or such longer period of time as may be prescribed by law or by the terms of any applicable special guarantee required by the Contract Documents, any Work is found to be defective, CONTRACTOR shall promptly, without cost to OWNER and in accordance with OWNER's written instructions, either correct such defective Work, or, if it has been rejected by OWNER, remove it from the site and replace it with nondefective Work. If CONTRACTOR does not promptly comply with the terms of such instructions, OWNER may have the defective Work corrected or the rejected Work removed and replaced, and all direct and indirect costs of such removal and replacement, including compensation for additional professional services, shall be paid by CONTRACTOR.

Acceptance of Defective Work:

13.11. If, instead of requiring correction or removal and replacement of defective Work, OWNER (and, prior to approval of final payment, also ENGINEER) prefers to accept it, he may do so. In such case, if acceptance occurs prior to approval of final payment, a Change Order shall be issued incorporating the necessary revisions in the Contract Documents, including appropriate reduction in the Contract Price; or, if the acceptance occurs after approval of final payment, an appropriate amount shall be paid by CONTRACTOR to OWNER.

Neglected Work by Contractor:

13.12. If CONTRACTOR should fail to prosecute the Work in accordance with the Contract Documents, including any requirements of the progress schedule, OWNER, after seven days' written notice to CONTRACTOR may, without prejudice to any other remedy he may have, make good such deficiences and the cost thereof (including compensation for additional professional services) shall be charged against CONTRACTOR if ENGINEER approves such action, in which case a Change Order shall be issued incorporating the necessary revisions in the Contract Documents including an appropriate reduction in the Contract Price. If the payments then or thereafter due CONTRACTOR are not sufficient to cover such amount, CONTRACTOR shall pay the difference to OWNER.

ARTICLE 14—PAYMENTS AND COMPLETION

Schedules:

14.1. At least ten days prior to submitting the first Application for a progress payment, CONTRACTOR shall submit a progress schedule, a final schedule of Shop Drawing submission and a schedule of values of the Work. These schedules shall be satisfactory in form and substance to ENGINEER. The schedule of values shall include quantities and unit prices aggregating the Contract Price, and shall subdivide the Work into component parts in sufficient detail to serve as the basis for progress payments during construction. Upon approval of the schedules of values by ENGINEER, it shall be incorporated into the form of Application for Payment furnished by ENGINEER.

Application for Progress Payment:

14.2. At least ten days before each progress payment falls due (but not more often than once a month), CONTRACTOR shall submit to ENGINEER for review an Application for Payment filled out and signed by CONTRACTOR covering the Work completed as of the date of the Application and accompanied by such data and schedules as ENGINEER may reasonably require. If payment is requested on the basis of materials and equipment not incorporated in the Work but delivered and suitably stored at the site or at another location agreed to in writing, the Application for Payment shall also be accompanied by such data, satisfactory to OWNER, as will establish OWNER's title to the material and equipment and protect his interest therein, including applicable insurance. Each subsequent Application for Payment shall include an affidavit of CONTRACTOR stating that all previous progress payments received on account of the Work have been applied to discharge in full all of CONTRACTOR's obligations reflected in prior Applications for Payment.

Contractor's Warranty of Title:

14.3. CONTRACTOR warrants and guarantees that title to all Work, materials and equipment covered by any Application for Payment, whether incorporated in the Project or not, will pass to OWNER at the time of payment free and clear of all liens, claims, security interests and encumbrances (hereafter in these General Conditions referred to as "Liens").

Approval of Payments:

14.4. ENGINEER will, within ten days after receipt of each Application for Payment, either indicate in writing his approval of payment and present the Application to OWNER, or return the Application to CONTRACTOR indicating in writing his reasons for refusing to approve payment. In the latter case, CONTRACTOR may make the necessary corrections and resubmit the Application. OWNER shall, within ten days of presentation to him of an approved Application for Payment, pay CONTRACTOR the amount approved by ENGINEER.

14.5. ENGINEER's approval of any payment requested in an Application for Payment will constitute a representation by him to OWNER, based on ENGINEER's on-site observations of the Work in progress as an experienced and qualified design professional and on his review of the Application for Payment and the accompanying data and schedules that the Work has progressed to the point indicated; that, to the best of his knowledge, information and belief, the quality of the Work is in accordance with the Contract Documents (subject to an evaluation of the Work as a functioning Project upon Substantial Completion, to the results of any subsequent tests called for in the Contract Documents and any qualifications stated in his approval); and that CONTRACTOR is entitled to

21

331

payment of the amount approved. However, by approving any such payment ENGINEER will not thereby be deemed to have represented that he made exhaustive or continuous on-site inspections to check the quality or the quantity of the Work, or that he has reviewed the means, methods, techniques, sequences, and procedures of construction, or that he has made any examination to ascertain how or for what purpose CONTRACTOR has used the moneys paid or to be paid to him on account of the Contract Price, or that title to any Work, materials or equipment has passed to OWNER free and clear of any Liens.

14.6. ENGINEER's approval of final payment will constitute an additional representation by him to OWNER that the conditions precedent to CONTRACTOR's being entitled to final payment as set forth in paragraph 14.13 have been fulfilled.

14.7. ENGINEER may refuse to approve the whole or any part of any payment if, in his opinion, it would be incorrect to make such representations to OWNER. He may also refuse to approve any such payment, or, because of subsequently discovered evidence or the results of subsequent inspections or tests, nullify any such payment previously approved, to such extent as may be necessary in his opinion to protect OWNER from loss because:

14.7.1. the Work is defective, or completed Work has been damaged requiring correction or replacement,

14.7.2. claims or Liens have been filed or there is reasonable cause to believe such may be filed,

14.7.3. the Contract Price has been reduced because of Modifications,

14.7.4. OWNER has been required to correct defective Work or complete the Work in accordance with paragraph 13.11, or

14.7.5. of unsatisfactory prosecution of the Work, including failure to furnish acceptable submittals or to clean up.

Substantial Completion:

14.8. Prior to final payment, CONTRACTOR may, in writing to OWNER and ENGINEER, certify that the entire Project is substantially complete and request that ENGINEER issue a certificate of Substantial Completion. Within a reasonable time thereafter, OWNER, CONTRACTOR and ENGINEER shall make an inspection of the Project to determine the status of completion. If ENGINEER does not consider the Project substantially complete, he will notify CONTRACTOR in writing giving his reasons therefor. If ENGINEER considers the Project substantially complete, he will prepare and deliver to OWNER a tentative certificate of Substantial Completion which shall fix the date of Substantial Completion and the responsibilities between OWNER and CONTRACTOR for maintenance, heat and utilities. There shall be attached to the certificate a tentative list of items to be completed or corrected before final payment, and the certificate shall fix the

time within which such items shall be completed or corrected, said time to be within the Contract Time. OWNER shall have seven days after receipt of the tentative certificate during which he may make written objection to ENGINEER as to any provisions of the certificate or attached list. If, after considering such objections, ENGINEER concludes that the Project is not substantially complete, he will within fourteen days after submission of the tentative certificate to OWNER notify CONTRACTOR in writing, stating his reasons therefor. If, after consideration of OWNER's objections, ENGINEER considers the PROJECT substantially complete, he will within said fourteen days execute and deliver to OWNER and CONTRACTOR a definitive certificate of Substantial Completion (with a revised tentative list of items to be completed or corrected) reflecting such changes from the tentative certificate as he believes justified after consideration of any objections from OWNER.

14.9. OWNER shall have the right to exclude CONTRACTOR from the Project after the date of Substantial Completion, but OWNER shall allow CONTRACTOR reasonable access to complete or correct items on the tentative list.

Partial Utilization:

14.10. Prior to final payment, OWNER may request CONTRACTOR in writing to permit him to use a specified part of the Project which he believes he may use without significant interference with construction of the other parts of the Project. If CONTRACTOR agrees, he will certify to OWNER and ENGINEER that said part of the Project is substantially complete and request ENGINEER to issue a certificate of Substantial Completion for that part of the Project. Within a reasonable time thereafter OWNER, CONTRACTOR and ENGINEER shall make an inspection of that part of the Project to determine its status of completion. If ENGINEER does not consider that it is substantially complete, he will notify OWNER and CONTRACTOR in writing giving his reasons therefor. If ENGINEER considers that part of the Project to be substantially complete, he will execute and deliver to OWNER and CONTRACTOR a certificate to that effect, fixing the date of Substantial Completion as to that part of the Project, attaching thereto a tentative list of items to be completed or corrected before final payment and fixing the responsibility between OWNER and CONTRACTOR for maintenance, heat and utilities as to that part of the Project. OWNER shall have the right to exclude CONTRACTOR from any part of the Project which ENGINEER has so certified to be substantially complete, but OWNER shall allow CONTRACTOR reasonable access to complete or correct items on the tentative list.

Final Inspection:

14.11. Upon written notice from CONTRACTOR that the Project is complete, ENGINEER will make a final inspection with OWNER and CONTRACTOR and will notify CONTRACTOR in writing of all particulars in which this inspection reveals that the Work is incomplete or defective. CONTRACTOR shall immediately take such measures as are necessary to remedy such deficiences.

Final Application for Payment:

14.12. After CONTRACTOR has completed all such corrections to the satisfaction of ENGINEER and delivered all maintenance and operating instructions, schedules, guarantees, Bonds, certificates of inspection and other documents—all as required by the Contract Documents, he may make application for final payment following the procedure for progress payments. The final Application for Payment shall be accompanied by such data and schedules as ENGINEER may reasonably require, together with complete and legally effective releases or waivers (satisfactory to OWNER) of all Liens arising out of the Contract Documents and the labor and services performed and the material and equipment furnished thereunder. In lieu thereof and as approved by OWNER, CONTRACTOR may furnish receipts or releases in full; an affidavit of CONTRACTOR that the releases and receipts include all labor, services, material and equipment for which a Lien could be filed, and that all payrolls, material and equipment bills, and other indebtedness connected with the Work for which OWNER or his property might in any way be responsible, have been paid or otherwise satisfied; and consent of the Surety, if any, to final payment. If any Subcontractor materialman, fabricator or supplier fails to furnish a release or receipt in full, CONTRACTOR may furnish a Bond or other collateral satisfactory to OWNER to indemnify him against any Lien.

Approval of Final Payment:

14.13. If, on the basis of his observation and review of the Work during construction, his final inspection and his review of the final Application for Payment—all as required by the Contract Documents, ENGINEER is satisfied that the Work has been completed and CONTRACTOR has fulfilled all of his obligations under the Contract Documents, he will, within ten days after receipt of the final Application for Payment, indicate in writing his approval of payment and present the Application to OWNER for payment. Thereupon ENGINEER will give written notice to OWNER and CONTRACTOR that the Work is acceptable subject to the provisions of paragraph 14.16. Otherwise, he will return the Application to CONTRACTOR, indicating in writing his reasons for refusing to approve final payment, in which case CONTRACTOR shall make the necessary corrections and resubmit the Application. OWNER shall, within ten days of presentation to him of an approved final Application for Payment, pay CONTRACTOR the amount approved by ENGINEER.

14.14. If after Substantial Completion of the Work final completion thereof is materially delayed through no fault of CONTRACTOR, and ENGINEER so confirms, OWNER shall, upon certification by ENGINEER, and without terminating the Agreement, make payment of the balance due for that portion of the Work fully completed and accepted. If the remaining balance for Work not fully completed or corrected is less than the retainage stipulated in the Agreement, and if Bonds have been furnished as required in paragraph 5.1, the written consent of the Surety to the payment of the balance due for that portion of the Work fully completed and accepted shall be submitted by the CONTRACTOR to the ENGINEER prior to certification of such payment. Such payment shall be made under the terms and conditions governing final payment, except that it shall not constitute a waiver of claims.

Contractor's Continuing Obligation:

14.15. CONTRACTOR's obligation to perform the Work and complete the Project in accordance with the Contract Documents shall be absolute. Neither approval of any progress or final payment by ENGINEER, nor the issuance of a certificate of Substantial Completion, nor any payment by OWNER to CONTRACTOR under the Contract Documents, nor any use or occupancy of the Project or any part thereof by OWNER, nor any act of acceptance by OWNER nor any failure to do so, nor any correction of defective work by OWNER shall constitute an acceptance of Work not in accordance with the Contract Documents.

Waiver of Claims:

14.16. The making and acceptance of final payment shall constitute:

14.16.1. a waiver of all claims by OWNER against CONTRACTOR other than those arising from unsettled Liens, from defective work appearing after final inspection pursuant to paragraph 14.11 or from failure to comply with the requirements of the Contract Documents or the terms of any special guarantees specified therein, and

14.16.2. a waiver of all claims by CONTRACTOR against OWNER other than those previously made in writing and still unsettled.

ARTICLE 15—SUSPENSION OF WORK AND TERMINATION

Owner May Suspend Work:

15.1. OWNER may, at any time and without cause, suspend the Work or any portion thereof for a period of not more than ninety days by notice in writing to CONTRACTOR and ENGINEER which shall fix the date on which Work shall be resumed. CONTRACTOR shall resume the Work on the date so fixed. CONTRACTOR will be allowed an increase in the Contract Price or an extension of the Contract Time, or both, directly attributable to any suspension if he makes a claim therefor as provided in Articles 11 and 12.

Owner May Terminate:

15.2. If CONTRACTOR is adjudged a bankrupt or insolvent, or if he makes a general assignment for the benefit of his creditors, or if a trustee or receiver is appointed for CONTRACTOR or for any of his property, or if he files a petition to take advantage of any debtor's act, or to reorganize under the bankruptcy or similar laws, or if he repeatedly fails to supply sufficient skilled workmen or suitable materials or equipment, or if he repeatedly fails to make prompt payments to Subcontractors or for labor, materials or equipment or if he disregards laws, ordinances, rules, regulations or orders of any public body having jurisdiction, or if he disregards the authority of ENGINEER, or if he otherwise violates any provision of the Contract Documents, then OWNER may, without prejudice to any other right or remedy and after giving CONTRACTOR and his Surety seven days' written notice, terminate the services of CONTRACTOR and take possession of the Project and of all materials, equipment, tools, construction equipment and machinery thereon owned by CONTRACTOR, and finish the Work by whatever method he may deem expedient. In such case CONTRACTOR shall not be entitled to receive any further payment until the Work is finished. If the unpaid balance of the Contract Price exceeds the direct and indirect costs of completing the Project, including compensation for additional professional services, such excess shall be paid to CONTRACTOR. If such costs exceed such unpaid balance, CONTRACTOR shall pay the difference to OWNER. Such costs incurred by OWNER shall be determined by ENGINEER and incorporated in a Change Order.

15.3. Where CONTRACTOR's services have been so terminated by OWNER, said terminations shall not affect any rights of OWNER against CONTRACTOR then existing or which may thereafter accrue. Any retention or payment of moneys by OWNER due CONTRACTOR will not release CONTRACTOR from liability.

15.4. Upon seven days' written notice to CONTRACTOR and ENGINEER, OWNER may, without cause and without prejudice to any other right or remedy, elect to abandon the Project and terminate the Agreement. In such case, CONTRACTOR shall be paid for all Work executed and any expense sustained plus a reasonable profit.

Contractor May Stop Work or Terminate:

15.5. If, through no act or fault of CONTRACTOR, the Work is suspended for a period of more than ninety days by OWNER or under an order of court or other public authority, or ENGINEER fails to act on any Application for Payment within thirty days after it is submitted, or OWNER fails to pay CONTRACTOR any sum approved by ENGINEER or awarded by arbitrators within thirty days of its approval and presentation, then CONTRACTOR may, upon seven days' written notice to OWNER and ENGINEER, terminate the Agreement and recover from OWNER payment for all Work executed and any expense sustained plus a reasonable profit. In addition and in lieu of terminating the Agreement, if ENGINEER has failed to act on an Application for Payment or OWNER has failed to make any payment as afore-

said, CONTRACTOR may upon seven days' notice to OWNER and ENGINEER stop the Work until he has been paid all amounts then due.

ARTICLE 16—ARBITRATION

16.1. All claims, disputes and other matters in question arising out of, or relating to, this Agreement or the breach thereof except for claims which have been waived by the making or acceptance of final payment as provided by paragraph 14.16, shall be decided by arbitration in accordance with the Construction Industry Arbitration Rules of the American Arbitration Association then obtaining. This agreement so to arbitrate shall be specifically enforceable under the prevailing arbitration law. The award rendered by the arbitrators shall be final, and judgment may be entered upon it in any court having jurisdiction thereof.

16.2. Notice of the demand for arbitration shall be filed in writing with the other party to the Agreement and with the American Arbitration Association, and a copy shall be filed with ENGINEER. The demand for arbitration shall be made within the thirty-day period specified in paragraph 9.10 where applicable, and in all other cases within a reasonable time after the claim, dispute or other matter in question has arisen, and in no event shall it be made after institution of legal or equitable proceedings based on such claim, dispute or other matter in question would be barred by the applicable statute of limitations.

16.3. CONTRACTOR will carry on the Work and maintain the progress schedule during any arbitration proceedings, unless otherwise agreed by him and OWNER in writing.

ARTICLE 17—MISCELLANEOUS

Giving Notice:

17.1. Whenever any provision of the Contract Documents requires the giving of written notice it shall be deemed to have been validly given if delivered in person to the individual or to a member of the firm or to an officer of the corporation for whom it is intended, or if delivered at or sent by registered or certified mail, postage prepaid, to the last business address known to him who gives the notice.

Computation of Time:

17.2. When any period of time is referred to in the Contract Documents by days, it shall be computed to exclude the first and include the last day of such period. If the last day of any such period falls on a Saturday or Sunday or on a day made a legal holiday by the law of the applicable jurisdiction, such day shall be omitted from the computation.

24

General:

17.3. All moneys not paid when due hereunder shall bear interest at the maximum rate allowed by law at the place of the Project.

17.4. All Specifications, Drawings and copies thereof furnished by ENGINEER shall remain his property. They shall not be used on another Project, and, with the exception of those sets which have been signed in connection with the execution of the Agreement, shall be returned to him on request upon completion of the Project.

17.5. The duties and obligations imposed by these General Conditions and the rights and remedies available hereunder, and, in particular but without limitation, the warranties, guarantees and obligations imposed upon CONTRACTOR by paragraphs 6.30, 13.1, 13.10 and 14.3 and the rights and remedies available to OWNER and ENGINEER thereunder, shall be in addition to, and shall not be construed in any way as a limitation of, any rights and remedies available to them which are otherwise imposed or available by law, by special guarantee or by other provisions of the Contract Documents.

17.6. Should OWNER or CONTRACTOR suffer injury or damage to his person or property because of any error, omission or act of the other or of any of his employees or agents or others for whose acts he is legally liable, claim shall be made in writing to the other party within a reasonable time of the first observance of such injury or damage.

17.7. The Contract Documents shall be governed by the law of the place of the Project.

STANDARD
FORM OF AGREEMENT
BETWEEN OWNER AND ENGINEER
FOR
PROFESSIONAL SERVICES

Issued by

PROFESSIONAL ENGINEERS IN PRIVATE PRACTICE

A practice division of the

NATIONAL SOCIETY OF PROFESSIONAL ENGINEERS

NSPE 1910-1 (1974 Edition)
Reprinted 10-13 4M

Guide Sheet for Completing Standard Form of Agreement Between Owner and Engineer for Professional Services

1. Page 1—Insert proper name of partnership, corporation or governmental body that is the Owner on first page and execution page. Make certain person signing for Owner has authority to do so and that there is an indication of the capacity in which he signs. (See Commentary, paragraph 7.)

2. Page 1—Insert as complete a description of the Project as possible. Include, to the extent known, a description of the land where the Project is to be located, any special requirements as to performance, capacity or function, budgetary limitations and any special source of funds for which the Project must qualify. Identify studies, reports or analyses previously prepared which are being furnished by Owner to Engineer for his guidance, such as reports and studies referred to in paragraph 3.3. Identify other special aspects or peculiarities of the Project. (See Commentary, paragraph 8.)

3. Section 4—Period of Service. This Section has been prepared in recognition of thought expressed in paragraph 4.1 that there will be a continuous period of service through completion of the Construction Phase with timely responses from the Owner to the Engineer's submission and with prompt authority to proceed with each Phase of services after the preceding Phase has been completed. The blank spaces provided in paragraphs 4.2, 4.3 and 4.4 should be filled in recognizing this understanding. The blank space in paragraph 4.10 should be filled in with whatever is considered a reasonable time in the particular Project for the taking of bids or receiving proposals, awarding the contract and starting construction. The three-month and one-year periods provided in paragraph 4.11 may not be appropriate for all Projects and should be modified as circumstances indicate.

4. Paragraph 5.1.1—Methods of Payment for Services (See Commentary, paragraph 12). One of the applicable Methods of Compensation provided in paragraphs 5.1.1.1 through 5.1.1.5 should pertain; otherwise use paragraph 5.1.1.6. Cross out the inapplicable paragraphs and insert initials of both parties opposite the crossing out. When payment is on the basis of a lump sum or a percentage fee, it is customary to provide for higher compensation for the Engineer in a Project involving several prime contracts or in the case of Construction Contracts which contain cost-plus or incentive-savings provisions. Accordingly, blank spaces have been provided in paragraphs 5.1.1.1 through 5.1.1.5. The blank spaces for the number of prime contractors is intended to be filled in with the number normally anticipated for a Project of the type involved, such as a governmentally financed one where the law requires separate prime contracts. Paragraph 2.1.8 is intended to cover an additional number, if any, not anticipated at the time the Engineer makes his fee commitment with the Owner. In filling in the blank spaces with respect to the anticipated extra services due to cost-plus or incentive-compensation arrangements with Contractors, bear in mind the provision of paragraph 3.8 which requires the Owner to furnish auditing services if he wishes to ascertain how or for what purposes the Contractor has used the moneys paid to him under the Construction Contract. Before inserting the factor in paragraphs 5.1.1.5, 5.1.2.1 and 5.1.2.4 review the definition of payroll costs contained in paragraph 5.3.2.

If none of the suggested Methods of Compensation are applicable, paragraph 5.1.1.6 may be used and an appropriately identified exhibit attached to the Agreement. Typical use of paragraph 5.1.1.6 would be in the situation where a percentage fee arrangement is to be converted to a fixed fee at the conclusion of the Preliminary Design Phase or where there is a provision for a guaranteed maximum fee. When paragraph 5.1.1.6 is used, paragraph 5.2.3 may also be used to cover special arrangements with respect to the times of payment.

5. Paragraph 5.1.2—Payment for Additional Services (See Commentary, paragraph 12). Space has been provided in paragraph 5.1.2.1 through 5.1.2.4 for different rates of compensation for different types of Additional Services. Any inapplicable paragraph should be crossed out and initials inserted in the margin by both parties. If one method of compensation is to apply to all Additional Services, be certain that the cross references are correctly adjusted.

6. Paragraph 5.3.2 contemplates the identification of the key personnel in the Engineer's organization who are to be assigned to the Project. (See Commentary, paragraph 12 as to the proper meaning of the term "principals.") In developing the factor for the payroll cost method of compensation, bear in mind that paragraph 5.3.2 is set up so that the factor is applied to the salary, wages and fringe benefits of all personnel, and payroll costs of principals (whether they be a corporate officer at a fixed salary or a partner with a drawing account who shares in the profits) are to be itemized. In lieu of the detailed accounting required to substantiate the amount paid for customary and statutory benefits of personnel, a percentage of salaries and wages may be agreed to in advance and inserted in the blank space at the end of paragraph 5.3.2; otherwise the sentence should be crossed out and initialed by both parties.

7. The blank space in paragraph 5.3.5 should be filled in after consultation with the Owner.

338

TABLE OF CONTENTS

STANDARD FORM OF AGREEMENT
BETWEEN
OWNER AND ENGINEER
FOR
PROFESSIONAL SERVICES

THIS IS AN AGREEMENT made as of .... day of ....

in the year Nineteen Hundred and Seventy .... by and between

....

.... (hereinafter called OWNER) and

....

.... (hereinafter called ENGINEER).

OWNER intends to ....

....

....

....

....

.... (hereinafter called the Project).

OWNER and ENGINEER in consideration of their mutual covenants herein agree in respect of the performance of professional engineering services by ENGINEER and the payment for those services by OWNER, as set forth below.

ENGINEER shall serve as OWNER's professional engineering representative in those phases of the Project to which this Agreement applies, and will give consultation and advice to OWNER during the performance of his services.

SECTION 1—BASIC SERVICES OF ENGINEER

1.1. General.

1.1.1. ENGINEER shall perform professional services as hereinafter stated which include normal civil, structural, mechanical and electrical engineering services and normal architectural services incidental thereto.

1.2. Study and Report Phase.

After written authorization to proceed, ENGINEER shall:

1.2.1. Consult with OWNER to determine his requirements for the Project and review available data.

1.2.2. Advise OWNER as to the necessity of his providing or obtaining from others data or services of the types described in paragraph 3.3, and act as OWNER's representative in connection with any such services.

1.2.3. Provide special analyses of OWNER's needs, planning surveys, site evaluations and comparative studies of prospective sites and solutions.

1.2.4. Provide general economic analysis of OWNER's requirements applicable to various alternatives.

1.2.5. Prepare a Report with appropriate exhibits indicating clearly the considerations involved and the alternative solutions available to OWNER, and setting forth ENGINEER's findings and recommendations with opinions of probable costs.

Page 1 of .... pages

1.2.6. Furnish five copies of the Report and present and review it in person with OWNER.

1.3. Preliminary Design Phase.

After written authorization to proceed with the Preliminary Design Phase, ENGINEER shall:

1.3.1. In consultation with OWNER and on the basis of the accepted Report, determine the scope of the Project.

1.3.2. Prepare preliminary design documents consisting of final design criteria, preliminary drawings and outline specifications.

1.3.3. Based on the information contained in the preliminary design documents, submit a revised opinion of probable cost for the Project including Construction Cost, contingencies, compensation for all professionals and consultants, costs of land, rights-of-way and compensation for or damages to properties and interest and financing charges (all of which are hereinafter called "Project Costs").

1.3.4. Furnish five copies of the above preliminary design documents and present and review them in person with OWNER.

1.4. Final Design Phase.

After written authorization to proceed with the Final Design Phase, ENGINEER shall:

1.4.1. On the basis of the accepted preliminary design documents prepare for incorporation in the Contract Documents, final drawings to show the character and scope of the work to be performed by contractors on the Project (hereinafter called "Drawings"), and Specifications.

1.4.2. Furnish to OWNER such documents and design data as may be required for, and assist in the preparation of, the required documents so that OWNER may obtain approvals of such governmental authorities as have jurisdiction over design criteria applicable to the Project, and assist in obtaining such approvals by participating in submissions to and negotiations with appropriate authorities.

1.4.3. Advise OWNER of any adjustments to his latest opinion of probable Project Cost caused by changes in scope, design requirements or Construction Costs and furnish a revised opinion of probable Project Cost based on the Drawings and Specifications.

1.4.4. Prepare bid forms, notice to bidders, instructions to bidders, general conditions and supplementary conditions, and assist in the preparation of other related documents.

1.4.5. Furnish five copies of the above documents and present and review them in person with OWNER.

1.5. Bidding or Negotiating Phase.

After written authorization to proceed with the Bidding or Negotiating Phase, ENGINEER shall:

1.5.1. Assist OWNER in obtaining bids or negotiating proposals for each separate prime contract for construction or equipment.

1.5.2. Consult with and advise OWNER as to the acceptability of subcontractors and other persons and organizations proposed by the prime contractor(s) (hereinafter called "Contractor(s)") for those portions of the work as to which such acceptability is required by the Contract Documents.

1.5.3. Consult with and advise OWNER as to the acceptability of substitute materials and equipment proposed by

Contractor(s) when substitution is permitted by the Contract Documents.

1.5.4. Assist OWNER in evaluating bids or proposals and in assembling and awarding contracts.

1.6. Construction Phase.

During the Construction Phase ENGINEER shall:

1.6.1. Consult with and advise OWNER and act as his representative as provided in Articles 1 through 17, inclusive, of the Standard General Conditions of the Construction Contract, National Society of Professional Engineers document 1910-8, 1974 edition; the extent and limitations of the duties, responsibilities and authority of ENGINEER as assigned in said Standard General Conditions shall not be modified without ENGINEER's written consent; all of OWNER's instructions to Contractor(s) will be issued through ENGINEER who will have authority to act on behalf of OWNER to the extent provided in said Standard General Conditions except as otherwise provided in writing.

1.6.2. Make periodic visits to the site to observe as an experienced and qualified design professional the progress and quality of the executed work and to determine in general if the work is proceeding in accordance with the Contract Documents; he shall not be required to make exhaustive or continuous on-site inspections to check the quality or quantity of work; he shall not be responsible for the means, methods, techniques, sequences or procedures of construction selected by Contractor(s) or the safety precautions and programs incident to the work of Contractor(s). His efforts will be directed toward providing assurance for OWNER that the completed Project will conform to the Contract Documents, but he shall not be responsible for the failure of Contractor(s) to perform the construction work in accordance with the Contract Documents. During such visits and on the basis of his on-site observations he shall keep OWNER informed of the progress of the work, shall endeavor to guard OWNER against defects and deficiencies in the work of Contractor(s) and may disapprove or reject work as failing to conform to the Contract Documents.

1.6.3. Review and approve Shop Drawings (as that term is defined in the aforesaid Standard General Conditions) and samples, the results of tests and inspections and other data which any Contractor is required to submit, but only for conformance with the design concept of the Project and compliance with the information given in the Contract Documents; determine the acceptability of substitute materials and equipment proposed by Contractor(s); and receive and review (for general content as required by the Specifications) maintenance and operating instructions, schedules, guarantees, bonds and certificates of inspection which are to be assembled by Contractor(s) in accordance with the Contract Documents.

1.6.4. Issue all instructions of OWNER to Contractor(s); prepare routine change orders as required; he may, as OWNER's representative, require special inspection or testing of the work; he shall act as interpreter of the requirements of the Contract Documents and judge of the performance thereunder by the parties thereto and shall make decisions on all claims of OWNER and Contractor(s) relating to the execution and progress of the work and all other matters and questions related thereto; but ENGINEER shall not be liable for the results of any such interpretations or decisions rendered by him in good faith.

1.6.5. Based on his on-site observations as an experienced and qualified design professional and on his review of Contractor(s)' applications for payment and the accompanying data and schedules, determine the amounts owing to Contractor(s) and approve in writing payments to Contractor(s) in such amounts; such approvals of payment will constitute a representation to OWNER, based on such observations and review, that the work has progressed to the point indicated and that, to the best of his knowledge, information and belief, the quality of the work is in accordance with the Contract Documents (subject to an evaluation of the work as a functioning Project upon Substantial Completion, to the results of any subsequent tests called for in the Contract Documents, and to any qualifications stated in his approval), but by approving an application for payment ENGINEER will not be deemed to have represented that he has made any examination to determine how or for what purposes any Contractor has used the moneys paid on account of the Contract Price, or that title to any of the Contractor(s)' work, materials or equipment has passed to OWNER free and clear of any lien, claims, security interests or encumbrances.

1.6.6. Conduct an inspection to determine if the Project is substantially complete and a final inspection to determine if the Project has been completed in accordance with the Contract Documents and if each Contractor has fulfilled all of his obligations thereunder so that ENGINEER may approve, in writing, final payment to each Contractor.

1.6.7. ENGINEER shall not be responsible for the acts or omissions of any Contractor, any subcontractor or any of the Contractor(s)' or subcontractors' agents or employees ·or any other persons (except his own employees and agents) at the Project site or otherwise performing any of the work of the Project.

SECTION 2—ADDITIONAL SERVICES OF ENGINEER

2.1. General.

If authorized in writing by OWNER, ENGINEER shall furnish or obtain from others Additional Services of the following types which are not considered normal or customary Basic Services; these will be paid for by OWNER as indicated in Section 5.

2.1.1. Preparation of applications and supporting documents for governmental grants, loans or advances in con-. nection with the Project; preparation or review of environmental assessments and impact statements; and assistance in obtaining approvals of authorities having jurisdiction over the anticipated environmental impact of the Project.

2.1.2. Services to make measured drawings of or to investigate existing conditions or facilities, or to verify the accuracy of drawings or, other information furnished by OWNER.

2.1.3. Services resulting from significant changes in general scope of the Project or its design including, but not limited to, changes in size, complexity, OWNER's schedule, or character of construction; and revising previously accepted studies, reports, design documents or Contract Documents when such revisions are due to causes beyond ENGINEER's control.

2.1.4. Providing renderings or models for OWNER's use.

2.1.5. Preparing documents for alternate bids requested by OWNER for work which is not executed or documents for out-of-sequence work.

2.1.6. Investigations involving detailed consideration of

operations, maintenance and overhead expenses; and the preparation of rate schedules, earnings and expense statements, feasibility studies, appraisals and valuations; detailed quantity surveys of material, equipment and labor; and audits or inventories required in connection with construction performed by OWNER.

2.1.7. Furnishing the services of special consultants for other than the normal civil, structural, mechanical and electrical engineering and normal architectural design incidental thereto, such as consultants for interior design, selection of furniture and furnishings, communications, acoustics, kitchens and landscaping.

2.1.8. Services resulting from the involvement of more separate prime contracts for construction or for equipment than are contemplated by paragraphs 5.1.1.2 or 5.1.1.4.

2.1.9. Services in connection with change orders to reflect changes requested by OWNER if the resulting change in compensation for Basic Services is not commensurate with the additional services rendered, and services resulting from significant delays, changes or price increases occurring as a direct or indirect result of material, equipment or energy shortages.

2.1.10. Services during out-of-town travel required of ENGINEER other than visits to the Project site as required by Section 1.

2.1.11. Preparing for OWNER, on request, a set of reproducible record prints of Drawings showing those changes made during the construction process, based on the marked-up prints, drawings and other data furnished by Contractor(s) to ENGINEER and which ENGINEER considers significant.

2.1.12. Additional or extended services during construction made necessary by (1) work damaged by fire or other cause during construction, (2) a significant amount of defective or neglected work of any Contractor, (3) prolongation of the contract time of any prime contract by more than sixty days, (4) acceleration of the work schedule involving services beyond normal working hours, and (5) default by any Contractor.

2.1.13. Preparation of operating and maintenance manuals; extensive assistance in the utilization of any equipment or system (such as initial start-up, testing, adjusting and balancing); and training personnel for operation and maintenance.

2.1.14. Services after completion of the Construction Phase, such as inspections during any guarantee period and reporting observed discrepancies under guarantees called for in any contract for the Project.

2.1.15. Preparing to serve or serving as a consultant or witness for OWNER in any litigation, public hearing or other legal or administrative proceeding involving the Project.

2.1.16. Additional services in connection with the Project, including services normally furnished by OWNER and services not otherwise provided for in this Agreement.

2.2. Resident Services During Construction.

2.2.1. If requested by OWNER or recommended by ENGINEER and agreed to in writing by the other, a Resident Project Representative and assistants will be furnished and will act as directed by ENGINEER in order to provide more extensive representation at the Project site during the Construction Phase. Such services will be paid for by OWNER as indicated in paragraph 5.1.2.4.

2.2.2. The duties and responsibilities and the limitations on the authority of the Resident Project Representative and assistants will be set forth in Exhibit A which is to be identified, attached to and made a part of this Agreement before such services begin.

2.2.3. Through more extensive on-site observations of the work in progress and field checks of materials and equipment by the Resident Project Representative (if furnished) and assistants, ENGINEER shall endeavor to provide further protection for OWNER against defects and deficiencies in the work, but the furnishing of such resident Project representation will not make ENGINEER responsible for construction means, methods, techniques, sequences or procedures or for safety precautions or programs, or for Contractor(s)' failure to perform the construction work in accordance with the Contract Documents.

SECTION 3—OWNER'S RESPONSIBILITIES

OWNER shall:

3.1. Provide full information as to his requirements for the Project.

3.2. Assist ENGINEER by placing at his disposal all available information pertinent to the Project including previous reports and any other data relative to design and construction of the Project.

3.3. Furnish to ENGINEER, as required by him for performance of his Basic Services, data prepared by or services of others, such as core borings, probings and subsurface explorations, hydrographic surveys, laboratory tests and inspections of samples, materials and equipment; appropriate professional interpretations of all of the foregoing; property, boundary, easement, right-of-way, topographic and utility surveys and property descriptions; zoning and deed restriction; and other special data or consultations not covered in paragraph 2.1; all of which ENGINEER may rely upon in performing his services.

3.4. Provide engineering surveys to enable Contractor(s) to proceed with their work.

3.5. Guarantee access to and make all provisions for ENGINEER to enter upon public and private property as required for ENGINEER to perform his services.

3.6. Examine all studies, reports, sketches, Drawings, Specifications, proposals and other documents presented by ENGINEER, obtain advice of an attorney, insurance counselor and other consultants as he deems appropriate for such examination and render in writing decisions pertaining thereto within a reasonable time so as not to delay the services of ENGINEER.

3.7. Pay all costs incident to obtaining bids or proposals from contractors.

3.8. Provide such legal, accounting, independent cost estimating and insurance counseling services as may be required for the Project, and such auditing service as OWNER may require to ascertain how or for what purpose any contractor has used the moneys paid to him under the construction contract.

3.9. Designate in writing a person to act as OWNER's representative with respect to the work to be performed under this Agreement. Such person shall have complete authority to transmit instructions, receive information, interpret and define OWNER's policies and decisions with respect to materials, equipment, elements and systems pertinent to ENGINEER's services.

3.10. Give prompt written notice to ENGINEER whenever OWNER observes or otherwise becomes aware of any defect in the Project.

3.11. Furnish approvals and permits from all governmental authorities having jurisdiction over the Project and such approvals and consents from others as may be necessary for completion of the Project.

3.12. Furnish, or direct ENGINEER to provide, necessary Additional Services as stipulated in Section 2 of this Agreement or other services as required.

3.13. Bear all costs incident to compliance with the requirements of this Section 3.

SECTION 4—PERIOD OF SERVICE

4.1. The provisions of 4.2 through 4.8, inclusive, and the various rates of compensation for ENGINEER's services provided for elsewhere in this Agreement have been agreed to in anticipation of the orderly and continuous progress of the Project through completion of the Construction Phase. ENGINEER's obligation to render services hereunder will extend for a period which may reasonably be required for the design, award of contracts and construction of the Project including extra work and required extensions thereto.

4.2. The services called for in the Study and Report Phase will be completed and the Report submitted within .... calendar days following the authorization to proceed with that phase of services.

4.3. After acceptance by OWNER of the Report, indicating any specific modifications or changes in scope desired by OWNER, and upon written authorization from OWNER, ENGINEER shall proceed with the performance of the services called for in the Preliminary Design Phase, and shall submit preliminary design documents and a revised opinion of probable Project Cost within .... calendar days following the authorization to proceed with that phase of services.

4.4. After acceptance by OWNER of the preliminary design documents and revised opinion of probable Project Cost, indicating any specific modifications or changes in scope desired by OWNER, and upon written authorization from OWNER, ENGINEER shall proceed with the performance of the services called for in the Final Design Phase, so as to deliver Contract Documents and a revised opinion of probable Project Cost for all authorized work on the Project within .... calendar days after the authorization to proceed with that phase of services.

4.5. ENGINEER's services under the Study and Report Phase, Preliminary Design Phase and Final Design Phase shall each be considered complete at the earlier of (1) the date when the submissions for that phase have been accepted by OWNER or (2) thirty days after the date when such submissions are delivered to OWNER for final acceptance, plus such additional time as may be considered reasonable for obtaining approval of governmental authorities having jurisdiction over design criteria applicable to the Project.

4.6. After acceptance by OWNER of the Contract Documents and ENGINEER's most recent opinion of probable

Page 4 of .... pages

Project Cost and upon written authorization to proceed, ENGINEER shall proceed with performance of the services called for in the Bidding or Negotiating Phase. This Phase shall terminate and the services to be rendered thereunder shall be considered complete upon commencement of the Construction Phase or upon cessation of negotiations with Contractor(s) (except as may be otherwise required to complete the services called for in paragraph 6.3.2.5).

4.7. The Construction Phase will commence with the execution of the first prime contract to be executed for the work of the Project or any part thereof, and will terminate upon written approval by ENGINEER of final payment on the last prime contract to be completed. Construction Phase services may be rendered at different times in respect of separate prime contracts if the Project involves more than one prime contract.

4.8. In the event that the work of the Project is to be performed under more than one prime contract, OWNER and ENGINEER shall, prior to commencement of the Final Design Phase, develop a schedule for performance of ENGINEER's services during the Final Design, Bidding or Negotiating and Construction Phases in order to sequence and coordinate properly such services as applicable to the work under such separate contracts. This schedule is to be prepared whether or not the work under such contracts is to proceed concurrently and is to be attached as an exhibit to and made a part of this Agreement, and the provisions of paragraphs 4.4 thru 4.6 inclusive, will be modified accordingly.

4.9. If OWNER has requested significant modifications or changes in the scope of the Project, the time of performance of ENGINEER's services shall be adjusted appropriately.

4.10. If OWNER fails to give prompt written authorization to proceed with any phase of services after completion of the immediately preceding phase, or if the Construction Phase has not commenced within .... calendar days (plus such additional time as may be required to complete the services called for under paragraph 6.3.2.5) after completion of the Final Design Phase, ENGINEER may, after giving seven days' written notice to OWNER, suspend services under this Agreement.

4.11. If ENGINEER's services for design or during construction of the Project are delayed or suspended in whole or in part by OWNER for more than three months for reasons beyond ENGINEER's control, ENGINEER shall on written demand to OWNER (but without termination of this Agreement) be paid as provided in paragraph 5.3.5 for the services delayed or suspended. If such delay or suspension extends for more than one year for reasons beyond ENGINEER's control, or if ENGINEER for any reason is required to render services more than one year after Substantial Completion, the various rates of compensation provided for elsewhere in this Agreement shall be subject to renegotiation.

SECTION 5—PAYMENTS TO ENGINEER

5.1. Methods of Payment for Services and Expenses of ENGINEER.

5.1.1. *Basic Services.* OWNER shall pay ENGINEER for Basic Services rendered under Section 1 on one of the following bases (except as otherwise provided in paragraph 5.1.1.6):

5.1.1.1. *Lump Sum.* If the work of the entire Project is awarded on the basis of one prime contract, a lump sum fee of $ .... ; but, if the prime contract contains cost-plus or incentive savings provisions for the Contractor's basic compensation, a lump sum fee of $.......................... .
or

5.1.1.2. *Lump Sum.* If the work of the Project is awarded on the basis of not more than a total of .... separate prime contracts for construction and for equipment, a lump sum fee of $.......................... ; but, if the prime contracts contain cost-plus or incentive savings provisions for the Contractor's basic compensation, a lump sum fee of $.......................... .
or

5.1.1.3. *Percentage.* If the work of the entire Project is awarded on the basis of one prime contract, ...% of the Construction Cost; but, if the prime contract contains cost-plus or incentive savings provisions for the Contractor's basic compensation, ...% of the Construction Cost.
or

5.1.1.4. *Percentage.* If the work of the Project is awarded on the basis of not more than a total of .... separate prime contracts for construction and for equipment, ... % of the Construction Cost; but, if the prime contracts contain cost-plus or incentive savings provisions for the Contractor's basic compensation, ...% of the Construction Cost.
or

5.1.1.5. *Payroll Cost Times a Factor.* An amount based on the payroll costs times a factor of .... for services rendered by principals and employees assigned to the Project.
or

5.1.1.6. *Other Method. (To be used in case none of the above methods of compensation is applicable.)*
(Refer to and attach schedule when applicable.)

5.1.2. *Additional Services.* OWNER shall pay ENGINEER for Additional Services rendered under Section 2 as follows:

5.1.2.1. *General.* For Additional Services rendered under paragraphs 2.1.1 through 2.1.16, inclusive (except services covered by paragraph 2.1.7 and services as a consultant or witness under paragraph 2.1.15), on the basis of payroll costs times a factor of .... for services rendered by principals and employees assigned to the Project.

5.1.2.2. *Special Consultants.* For services and reimbursable expenses of special consultants employed by ENGINEER pursuant to paragraphs 2.1.7 or 2.1.16, the amount billed to ENGINEER therefor times a factor of ....

5.1.2.3. *Serving as a Witness.* For the services of the principals and employees as consultants or witnesses in any litigation, hearing or proceeding in accordance with paragraph 2.1.15, at the rate of $............... per day or any portion thereof (but compensation for time spent in preparing to appear in any such litigation, hearing or proceeding will be on the basis provided in paragraph 5.1.2.1).

5.1.2.4. *Resident Project Services.* For resident services during construction furnished under paragraph 2.2.1, on the basis of payroll costs times a factor of .... for services rendered by principals and employees assigned to field offices in connection with resident Project representation.

[Delete inapplicable paragraphs and initial]

5.1.3. *Reimbursable Expenses.* In addition to payments provided for in paragraphs 5.1.1 and 5.1.2, OWNER shall pay ENGINEER the actual costs of all reimbursable expenses incurred in connection with all Basic and Additional Services.

5.1.4. As used in this paragraph 5.1, the terms "Construction Cost," "payroll costs" and "reimbursable expenses" will have the meanings assigned to them in paragraphs 5.3.1, 5.3.2 and 5.3.3.

5.2. Times of Payment.

5.2.1. ENGINEER shall submit monthly statements for Basic and Additional Services rendered and for reimbursable expenses incurred. When compensation is on the basis of a lump sum or percentage of construction cost the statements will be based upon ENGINEER's estimate of the proportion of the total services actually completed at the time of billing. Otherwise, these monthly statements will be based upon ENGINEER's payroll cost times a factor. OWNER shall make prompt monthly payments in response to ENGINEER's monthly statements.

5.2.2. Where compensation for Basic Services is on the basis of a lump sum or percentage of Construction Cost, OWNER shall, upon conclusion of each phase of Basic Services, pay such additional amount, if any, as may be necessary to bring total compensation paid on account of such phase to the following percentages of total compensation for all phases of Basic Services:

Phase	Suggested Range	Insert Actual Percentage and Initial in Margin
Study and Report	(5-30%)	.... %
Preliminary Design	(5-30%)	.... %
Final Design	(35-75%)	.... %
Bidding or Negotiating	(2-10%)	.... %
Construction	(10-20%)	.... %
TOTAL	X	100 %

5.2.3. Payments for Basic Services in accordance with paragraph 5.1.1.6 shall be made as follows:

(Refer to and attach schedule when applicable.)

5.3. General.

5.3.1. The construction cost of the entire Project (herein referred to as "Construction Cost") means the total cost of the entire Project to OWNER, but it will not include ENGINEER's compensation and expenses, the cost of land, rights-of-way, or compensation for or damages, to properties unless this Agreement so specifies, nor will it include OWNER's legal, accounting, insurance counseling or auditing services, or interest and financing charges incurred in connection with the Project. When Construction Cost is used as a basis for payment it will be based on one of the following sources with precedence in the order listed:

5.3.1.1. For completed construction the total cost of all work performed as designed or specified by ENGINEER.

5.3.1.2. For work not constructed, the lowest bona fide bid received from a qualified bidder for such work; or if the work is not bid, the lowest bona fide negotiated proposal for such work.

5.3.1.3. For work for which no such bid or proposal is received, ENGINEER's most recent opinion of probable Project Cost.

Labor furnished by OWNER for the Project will be included in the Construction Cost at current market rates

including a reasonable allowance for overhead and profit. Materials and equipment furnished by OWNER will be included at current market prices except used materials and equipment will be included as if purchased new for the Project. No deduction is to be made from ENGINEER's compensation on account of any penalty, liquidated damages, or other amounts withheld from payments to Contractor(s).

5.3.2. The payroll costs used as a basis for payment mean the salaries and wages paid to all personnel engaged directly on the Project, including, but not limited to, engineers, architects, surveymen, designers, draftsmen, specification writers, estimators, other technical personnel, stenographers, typists and clerks; plus the cost of customary and statutory benefits including, but not limited to, social security contributions, unemployment, excise and payroll taxes, workmen's compensation, health and retirement benefits, sick leave, vacation and holiday pay applicable thereto. For the purposes of this Agreement, the principals of ENGINEER and their hourly payroll costs are:

....
....

The amount of customary and statutory benefits of all other personnel will be considered equal to ....% of salaries and wages.

5.3.3. Reimbursable expenses mean the actual expenses incurred directly or indirectly in connection with the Project for: transportation and subsistence incidental thereto; obtaining bids or proposals from Contractor(s); furnishing and maintaining field office facilities; subsistence and transportation of Resident Project Representatives and their assistants; toll telephone calls and telegrams; reproduction of reports, Drawings and Specifications, and similar Project-related items in addition to those required under Section 1; computer time including an appropriate charge for previously established programs; and, if authorized in advance by OWNER, overtime work requiring higher than regular rates.

5.3.4. If OWNER fails to make any payment due ENGINEER for services and expenses within sixty days after receipt of ENGINEER's bill therefor, the amounts due ENGINEER shall include a charge at the rate of 1% per month from said sixtieth day, and in addition ENGINEER may, after giving seven days' written notice to OWNER, suspend services under this Agreement until he has been paid in full all amounts due him for services and expenses.

5.3.5. If this Agreement is terminated by OWNER upon the completion of any phase of the Basic Services, progress payments due ENGINEER for services rendered through such phase shall constitute total payment for such services. If this Agreement is terminated by OWNER during any phase of the Basic Services, ENGINEER will be paid for services rendered during that phase on the basis of payroll costs times a factor of .... for services rendered during that phase to date of termination by principals and employees assigned to the Project. In the event of any termination, ENGINEER will be paid for all unpaid Additional Services and unpaid reimbursable expenses, plus all termination expenses. Termination expenses mean reimbursable expenses directly attributable to termination, which shall include an amount computed as a percentage of total compensation for Basic Services earned by ENGINEER to the date of termination, as follows:

20% if termination occurs after commencement of the Preliminary Design Phase but prior to commencement of the Final Design Phase; or

10% if termination occurs after commencement of the Final Design Phase.

Page 6 of .... pages

SECTION 6—GENERAL CONSIDERATIONS

6.1. Termination

This Agreement may be terminated by either party upon seven days' written notice in the event of substantial failure by the other party to perform in accordance with the terms hereof through no fault of the terminating party.

6.2. Reuse of Documents.

All documents including Drawings and Specifications furnished by ENGINEER pursuant to this Agreement are instruments of his services in respect of the Project. They are not intended or represented to be suitable for reuse by OWNER or others on extensions of the Project or on any other project. Any reuse without specific written verification or adaptation by ENGINEER will be at OWNER's sole risk and without liability or legal exposure to ENGINEER, and OWNER shall indemnify and hold harmless ENGINEER from all claims, damages, losses and expenses including attorneys' fees arising out of or resulting therefrom. Any such verification or adaptation will entitle ENGINEER to further compensation at rates to be agreed upon by OWNER and ENGINEER.

6.3. Estimates of Cost.

6.3.1. Since ENGINEER has no control over the cost of labor, materials or equipment, or over the Contractor(s)' methods of determining prices, or over competitive bidding or market conditions, his opinions of probable Project Cost or Construction Cost provided for herein are to be made on the basis of his experience and qualifications and represent his best judgment as a design professional familiar with the construction industry, but ENGINEER cannot and does not guarantee that proposals, bids or the Construction Cost will not vary from opinions of probable cost prepared by him. If prior to the Bidding or Negotiating Phase OWNER wishes greater assurance as to the Construction Cost he shall employ an independent cost estimator as provided in paragraph 3.8.

6.3.2. If a Construction Cost limit is established as a condition to this Agreement, the following will apply:

6.3.2.1. The acceptance by OWNER at any time during the Basic Services of a revised opinion of probable Project Cost in excess of the then established cost limit will constitute a corresponding increase in the Construction Cost limit.

6.3.2.2. Any Construction Cost limit established by this Agreement will include a bidding contingency of ten percent unless another amount is agreed upon in writing.

6.3.2.3. ENGINEER will be permitted to determine what materials, equipment, component systems and types of construction are to be included in the Drawings and Specifications and to make reasonable adjustments in the scope of the Project to bring it within the cost limit.

6.3.2.4. If the Bidding or Negotiating Phase has not commenced within six months of the completion of the Final Design Phase, the established Construction Cost limit will not be effective or binding on ENGINEER, and OWNER shall consent to an adjustment in such cost limit commensurate with any applicable change in the general level of prices in the construction industry between the date of completion of the Final Design Phase and the date on which proposals or bids are sought.

6.3.2.5. If the lowest bona fide proposal or bid exceeds the established Construction Cost limit, OWNER shall (1) give written approval to increase such cost limit, (2) authorize negotiating or rebidding the Project within a reasonable time, or (3) cooperate in revising the Project scope or quality. In the case of (3), ENGINEER shall, without additional charge, modify the Contract Documents as necessary to bring the Construction Cost within the cost limit. The providing of such service will be the limit of ENGINEER's responsibility in this regard and, having done so, ENGINEER shall be entitled to payment for his services in accordance with this Agreement.

6.4. Arbitration

6.4.1. All claims, counter-claims, disputes and other matters in question between the parties hereto arising out of or relating to this Agreement or the breach thereof will be decided by arbitration in accordance with the Construction Industry Arbitration Rules of the American Arbitration Association then obtaining, subject to the limitations stated in paragraphs 6.4.3 and 6.4.4 below. This agreement so to arbitrate and any other agreement or consent to arbitrate entered into in accordance therewith as provided below, will be specifically enforceable under the prevailing law of any court having jurisdiction.

6.4.2. Notice of demand for arbitration must be filed in writing with the other parties to this Agreement and with the American Arbitration Association. The demand must be made within a reasonable time after the claim, dispute or other matter in question has arisen. In no event may the demand for arbitration be made after the time when institution of legal or equitable proceedings based on such claim, dispute or other matter in question would be barred by the applicable statute of limitations.

6.4.3. All demands for arbitration and all answering statements thereto which include any monetary claim must contain a statement that the total sum or value in controversy as alleged by the party making such demand or answering statement is not more than $200,000 (exclusive of interest and costs). The arbitrators will not have jurisdiction, power or authority to consider, or make findings (except in denial of their own jurisdiction) concerning any claim, counter-claim, dispute or other matter in question where the amount in controversy thereof is more than $200,000 (exclusive of interest and costs) or to render a monetary award in response thereto against any party which totals more than $200,000 (exclusive of interest and costs).

6.4.4. No arbitration arising out of, or relating to, this Agreement may include, by consolidation, joinder or in any other manner, any additional party not a party to this Agreement.

6.4.5. By written consent signed by all the parties to this Agreement and containing a specific reference hereto, the limitations and restrictions contained in paragraphs 6.4.3 and 6.4.4 may be waived in whole or in part as to any claim, counter-claim, dispute or other matter specifically described in such consent. No consent to arbitration in respect of a specifically described claim, counter-claim, dispute or other matter in question will constitute consent to arbitrate any other claim, counter-claim, dispute or other matter in question which is not specifically described in such consent or in which the sum or value in controversy exceeds $200,000 (exclusive of interest and costs) or which is with any party not specifically described therein.

6.4.6. The award rendered by the arbitrators will be final, not subject to appeal and judgment may be entered upon it in any court having jurisdiction thereof.

6.5. Successors and Assigns.

OWNER and ENGINEER each binds himself and his partners, successors, executors, administrators and assigns to the other party of this Agreement and to the partners, successors, executors, administrators and assigns of such other party, in respect to all covenants of this Agreement; except as above, neither OWNER nor ENGINEER shall assign, sublet or transfer his interest in this Agreement without the written consent of the other. Nothing herein shall be construed as creating any personal liability on the part of any officer or agent of any public body which may be a party hereto, nor shall it be construed as giving any rights or benefits hereunder to anyone other than OWNER and ENGINEER.

SECTION 7—SPECIAL PROVISIONS

OWNER and ENGINEER agree that this Agreement is subject to the following special provisions which together with the provisions hereof and the exhibits and schedules hereto represent the entire Agreement between OWNER and ENGINEER; they may only be altered, amended or repealed by a duly executed written instrument.

7.1.

IN WITNESS WHEREOF the parties hereto have made and executed this Agreement as of the day and year first above written.

OWNER: ENGINEER:

--   --

--   --

--   --

Page 8 of .... pages

Number:

Date:

TIME AND MATERIAL CONTRACT

TO:

PROPOSAL:
On the premises located at

We herewith propose to furnish, as specified below, and subject to the General Conditions herein, all labor and material, and perform all said work in a good and workmanlike manner, for the construction, installation and completion of

CONTRACT AMOUNT:
The work as described herein shall be performed for the sum of the following:
   a)   The actual cost, as hereinafter defined and qualified, of material, equipment and labor, plus _____ % to cover overhead, plus _____ % to cover profit.
   b)   The actual cost of all sub-contracts for labor and material plus _____ % to cover overhead.

PAYMENTS:
Terms of payment shall be as follows:

349

COSTS:

Costs shall be defined as follows:

a) Actual cost of all material allowed and equipment. Trade discounts are deducted from costs. Cash discounts are not excluded from cost of materials and equipment.

b) Cost of all labor incorporated, field supervision of general superintendent and travel expense in accord with Union Agreement.

c) Rigging, erecting, and machinery moving charges.

d) Cost of all permits, fees and charges for which the contractor is liable in conjunction with the work performed.

e) Cost of all royalties.

f) Federal, state, municipal or other taxes directly chargeable to the work which the contractor is legally obligated to pay and does pay in connection with this contract.

g) Cost of temporary field facilities including transportation thereof, at rates approved by contractor and buyer.

h) Cost of welding consumables such as oxygen, welding rod, solder, acetylene, demurrage on cylinders, etc.

i) Cost of electric power and gas and oil for contractor's equipment on job.

j) Rental charges for all rented tools and equipment.

k) Cost of scaffolding.

l) Cost of living expense for union men, in accordance with Area Agreement.

m) Traveling, transportation, and living expenses of contractor's expediters engaged in expediting the production and transportation of materials and equipment, if prior authorization has been given by Buyer, and all long distance telephone calls or telegraph service for this purpose.

n) Cost of "shop" and "as-built" drawings as required.

o) Salary of personnel employed at field office.

p) Cost of special insurance as required.

q) Cost of payment and performance bonds as required.

r) Service reserve.

s) Cost of testing and adjusting system.

t) · Freight and delivery charges, including cartage, use of contractor's trucks, at $_____ per trip and charges for truck rental.

u) Use of contractor-owned equipment at the following rates per item:

Items (list):	Daily Rental	Weekly Rental	Monthly Rental

v) Cost of expendable tools, at the rate of___ % of total labor cost.

w) A sum equal to___ % of item (b) to cover the following:

  1. Workmen's Compensation.
  2. Occupational Diseases.
  3. Public Liability.
  4. Property Damage.
  5. State Unemployment.
  6. Federal Unemployment.
  7. Old Age Benefit.
  8. Pipefitters Welfare and Retirement Fund payments.
  9. Education, Apprenticeship, Association and Industry Fund payments.

x) Costs of protection of equipment, materials, and tools against loss or damage.

y) Costs of all sub-contracts.

All terms and conditions as set forth on reverse side are a part of this proposal.

This proposal is subject to acceptance within _____ days from date hereof by the Buyer, otherwise at our option it becomes null and void.

ACCEPTED: _____

_____
Company Name

By: _____
Title

By: _____
Title

Date: _____

_____

351

GENERAL CONDITIONS

1. All materials, equipment and workmanship furnished under this contract shall be guaranteed by the Seller against defect and Seller agrees to replace or repair any defective material or equipment, and any defective workmanship not caused by ordinary wear and tear or to improper use or maintenance within one (1) year from date of completion of the contract. Seller further agrees to replace any refrigerant lost during that period, caused by defects in the installation, and not due to improper use or maintenance. In no event shall Seller be liable for consequential damages.

2. The work to be furnished under this contract shall be guaranteed by Seller to produce capacities, meet design limitations and to function (1) as called for in plans, specifications and addenda (2) as herein set forth or (3) as published by the manufacturer for equipment involved. In the event the foregoing requirements are not met, Seller's liability shall be limited to remedying any deficiency without expense to the Buyer. This guarantee shall not apply in any case in which the Buyer specifies the type of equipment to be used.

3. Seller shall not be held responsible or liable for any loss, damage, detention or delay caused by accidents, strikes, lockouts or by any other cause which is unavoidable or beyond Seller's control.

4. Buyer shall provide adequate fire insurance to protect the interest of the Seller against loss or damage to equipment, materials, and tools on the job site.

5. Seller shall have the right to bill the Buyer for the amount of material when delivered on the job site, even though not actually installed. In the event that material is ready for delivery and installation, but Buyer is unable to receive same, Seller shall have the right to bill Buyer for the amount of the material, including storage and insurance costs incurred by the Seller, which Buyer agrees to pay.

6. The Buyer shall be responsible for identifying and disclosing to the Seller all concealed piping, fixtures, wiring or other equipment or conditions which might be damaged, cause damage or otherwise affect or be affected by the work. In the event of damage, or a claim of damage without disclosure being given, the Buyer shall waive and hold the Seller harmless against all claims, suits, judgments and awards resulting therefrom.

7. Seller will furnish all necessary lien waivers, affidavits and other documents required to keep the Buyer's premises from liens or claims for liens of all materialmen, subcontractors or laborers as payments are made under this contract.

8. The price or prices set forth in this contract shall be increased in an amount or amounts equal to the tax or taxes which may be assessed on the equipment or materials supplied hereunder or which may be due or become due, or which may be required to be paid with respect to this contract as a result of any excise, sales, use, occupation or similar tax not now in effect but hereafter imposed or made effective by the United States Government or any state or local government.

9. Buyer shall be responsible for structural ability of the premises to contain the equipment in the manner and location specified in the contract or shown on drawings, and Seller shall not be liable for any failure, or damage resulting from such failure, of premises, due to such structural deficiency.

10. This proposal, when signed and accepted by the Buyer, and approved by an authorized representative of the Seller, shall constitute exclusively the contract between the parties, and all prior representations or agreements, whether written or verbal, not incorporated herein, are superseded.

11. All changes, alterations or omissions to be made in the work as specified shall be performed in accordance with a written agreement between Buyer and Seller, which shall define the amount of any increase or credit in price adjustment. Neither party to this contract shall assign this contract or monies due hereunder without the prior written consent of the other.

352

TIME AND MATERIAL CONTRACT WITH MAXIMUM PRICE

TO:

PROPOSAL:
On the premises located at

We herewith propose to furnish, as specified below, and subject to the General Conditions herein, all labor and material, and perform all said work in a good and workmanlike manner, for the construction, installation and completion of

CONTRACT AMOUNT:
The work described above shall be performed for the sum of $_____designated the maximum price.
The cost of the work shall be the sum of the following:

    a)   The actual cost of material, equipment, labor, and such other cost items applicable to the work, as hereinafter defined and qualified, plus _____% to cover overhead, plus _____% of the aggregate to cover profit.

    b)   The actual cost of all sub-contracts, plus_____% to cover overhead.

In the event the actual cost of the work is less than the maximum price, the Buyer will pay to the contractor _____% of the difference between the maximum price and the actual cost, in addition to the costs as defined in (a) and (b) above, as additional compensation.

In the event the actual cost of the work exceeds the maximum price, such excess cost shall be borne by the contractor.

PAYMENTS:
Terms of payment shall be as follows:

COSTS:

Costs shall be defined as follows:

a) Actual cost of all material allowed and equipment. Trade discounts are deducted from costs. Cash discounts are not excluded from cost of materials and equipment.

b) Cost of all labor incorporated, field supervision of general superintendent and travel expense in accord with Union Agreement.

c) Rigging, erecting, and machinery moving charges.

d) Cost of all permits, fees and charges for which the contractor is liable in conjunction with the work performed.

e) Cost of all royalties.

f) Federal, state, municipal or other taxes directly chargeable to the work which the contractor is legally obligated to pay and does pay in connection with this contract.

g) Cost of temporary field facilities including transportation thereof, at rates approved by contractor and buyer.

h) Cost of welding consumables such as oxygen, welding rod, solder, acetylene, demurrage on cylinders, etc.

i) Cost of electric power and gas and oil for contractor's equipment on job.

j) Rental charges for all rented tools and equipment.

k) Cost of scaffolding.

l) Cost of living expense for union men, in accordance with Area Agreement.

m) Traveling, transportation, and living expenses of contractor's expediters engaged in expediting the production and transportation of materials and equipment, if prior authorization has been given by Buyer, and all long distance telephone calls or telegraph service for this purpose.

n) Cost of "shop" and "as-built" drawings as required.

o) Salary of personnel employed at field office.

p) Cost of special insurance as required.

q) Cost of payment and performance bonds as required.

r) Service reserve.

s) Cost of testing and adjusting system.

354

t)   Freight and delivery charges, including cartage, use of contractor's trucks, at $_____ per trip and charges for truck rental.

u)   Use of contractor-owned equipment at the following rates per item:

Items (list):	Daily Rental	Weekly Rental	Monthly Rental

v)   Cost of expendable tools, at the rate of ____ % of total labor cost.

w)   A sum equal to ____ % of item (b) to cover the following:

1. Workmen's Compensation.
2. Occupational Diseases.
3. Public Liability.
4. Property Damage.
5. State Unemployment.
6. Federal Unemployment.
7. Old Age Benefit.
8. Pipefitters Welfare and Retirement Fund payments.
9. Education, Apprenticeship, Association and Industry Fund payments.

x)   Costs of protection of equipment, materials, and tools against loss or damage.

y)   Costs of all sub-contracts.

All terms and conditions as set forth on reverse side are a part of this proposal.

This proposal is subject to acceptance within ____ days from date hereof by the Buyer, otherwise at our option it becomes null and void.

ACCEPTED: _____

By: _____
                                    Title

Date: _____

_____
                              Company Name

By: _____
                                    Title

_____

355

GENERAL CONDITIONS

1. All materials, equipment and workmanship furnished under this contract shall be guaranteed by the Seller against defects and Seller agrees to replace or repair any defective material or equipment, and any defective workmanship not caused by ordinary wear and tear or to improper use or maintenance within one (1) year from date of completion of the contract. Seller further agrees to replace any refrigerant lost during that period, caused by defects in the installation, and not due to improper use or maintenance. In no event shall Seller be liable for consequential damages.

2. The work to be furnished under this contract shall be guaranteed by Seller to produce capacities, meet design limitations and to function (1) as called for in plans, specifications and addenda (2) as herein set forth or (3) as published by the manufacturer for equipment involved. In the event the foregoing requirements are not met, Seller's liability shall be limited to remedying any deficiency without expense to the Buyer. This guarantee shall not apply in any case in which the Buyer specifies the type of equipment to be used.

3. Seller shall not be held responsible or liable for any loss, damage, detention or delay caused by accidents, strikes, lockouts or by any other cause which is unavoidable or beyond Seller's control.

4. Buyer shall provide adequate fire insurance to protect the interest of the Seller against loss or damage to equipment, materials, and tools on the job site.

5. Seller shall have the right to bill the Buyer for the amount of material when delivered on the job site, even though not actually installed. In the event that material is ready for delivery and installation, but Buyer is unable to receive same, Seller shall have the right to bill Buyer for the amount of the material, including storage and insurance costs incurred by the Seller, which Buyer agrees to pay.

6. The Buyer shall be responsible for identifying and disclosing to the Seller all concealed piping, fixtures, wiring or other equipment or conditions which might be damaged, cause damage or otherwise affect or be affected by the work. In the event of damage, or a claim of damage without disclosure being given, the Buyer shall waive and hold the Seller harmless against all claims, suits, judgments and awards resulting therefrom.

7. Seller will furnish all necessary lien waivers, affidavits and other documents required to keep the Buyer's premises free from liens or claims for liens of all materialmen, subcontractors or laborers as payments are made under this contract.

8. The price or prices set forth in this contract shall be increased in an amount or amounts equal to the tax or taxes which may be assessed on the equipment or materials supplied hereunder or which may be due or become due, or which may be required to be paid with respect to this contract as a result of any excise, sales, use, occupation or similar tax not now in effect but hereafter imposed or made effective by the United States Government or any state or local government.

9. Buyer shall be responsible for structural ability of the premises to contain the equipment in the manner and location specified in the contract or shown on drawings, and Seller shall not be liable for any failure, or damage resulting from such failure, of premises, due to such structural deficiency.

10. This proposal, when signed and accepted by the Buyer, and approved by an authorized representative of the Seller, shall constitute exclusively the contract between the parties, and all prior representations or agreements, whether written or verbal, not incorporated herein, are superseded.

11. All changes, alterations or omissions to be made in the work as specified shall be performed in accordance with a written agreement between Buyer and Seller, which shall define the amount of any increase or credit in price adjustment. Neither party to this contract shall assign this contract or monies due hereunder without the prior written consent of the other.

12. Unless otherwise agreed, this contract shall be performed during the regular working days consisting of____hours per day,____days per week. Should the Buyer request overtime, then this contract shall be increased to compensate for such work at the standard overtime charges, including insurance, taxes, overhead and profit, and the loss of efficiency factor as described in Mechanical Contractors Association of America, Inc. Management Methods Committee Bulletin No. 18A "How Much Does Overtime Really Cost?" and Bulletin No. 20 "Tables for Calculation of Premium Time and Efficiency on Overtime Work".

13. If the scope of work contemplated herein is changed by____percent or more, compensation provided herein shall be adjusted in accordance with Mechanical Contractors Association of America, Inc. Management Methods Committee Bulletin No. 32 "Change Orders in the Mechanical Contracting Industry" (Revised 1972).

COST PLUS FIXED FEE CONTRACT

TO:

PROPOSAL:
On the premises located at

We herewith propose to furnish, as specified below, and subject to the General Conditions herein, all labor and material, and perform all said work in a good and workmanlike manner, for the construction, installation and completion of

CONTRACT AMOUNT:
The work as described herein shall be performed for the sum of the following:

 a) The actual cost, as hereinafter defined and qualified, of material, equipment, labor, sub-contracts, and such other cost items applicable to the work.

 b) A fixed fee of $_____to represent overhead and profit.

PAYMENTS:
Terms of payment shall be as follows:

357

COSTS:

Costs shall be defined as follows:

a)   Actual cost of all material allowed and equipment. Trade discounts are deducted from costs. Cash discounts are not excluded from cost of materials and equipment.

b)   Cost of all labor incorporated, field supervision of general superintendent and travel expense in accord with Union Agreement.

c)   Rigging, erecting, and machinery moving charges.

d)   Cost of all permits, fees and charges for which the contractor is liable in conjunction with the work performed.

e)   Cost of all royalties.

f)   Federal, state, municipal or other taxes directly chargeable to the work which the contractor is legally obligated to pay and does pay in connection with this contract.

g)   Cost of temporary field facilities including transportation thereof, at rates approved by contractor and buyer.

h)   Cost of welding consumables such as oxygen, welding rod, solder, acetylene, demurrage on cylinders, etc.

i)   Cost of electric power and gas and oil for contractor's equipment on job.

j)   Rental charges for all rented tools and equipment.

k)   Cost of scaffolding.

l)   Cost of living expense for union men, in accordance with Area Agreement.

m)   Traveling, transportation, and living expenses of contractor's expediters engaged in expediting the production and transportation of materials and equipment, if prior authorization has been given by Buyer, and all long distance telephone calls or telegraph service for this purpose.

n)   Cost of "shop" and "as-built" drawings as required.

o)   Salary of personnel employed at field office.

p)   Cost of special insurance as required.

q)   Cost of payment and performance bonds as required.

r)   Service reserve.

s)   Cost of testing and adjusting system.

358

t)   Freight and delivery charges, including cartage, use of contractor's trucks, at $_____ per trip and charges for truck rental.

u)   Use of contractor-owned equipment at the following rates per item:

Items (list):

	Daily Rental	Weekly Rental	Monthly Rental

v)   Cost of expendable tools, at the rate of ____ % of total labor cost.

w)   A sum equal to ____ % of item (b) to cover the following:

1. Workmen's Compensation.
2. Occupational Diseases.
3. Public Liability.
4. Property Damage.
5. State Unemployment.
6. Federal Unemployment.
7. Old Age Benefit.
8. Pipefitters Welfare and Retirement Fund payments.
9. Education, Apprenticeship, Association and Industry Fund payments.

x)   Costs of protection of equipment, materials, and tools against loss or damage.

y)   Costs of all sub-contracts.

All terms and conditions as set forth on reverse side are a part of this proposal.

This proposal is subject to acceptance within ____ days from date hereof by the Buyer, otherwise at our option it becomes null and void.

ACCEPTED: _____   _____
                                                                                      Company Name

By: _____   By: _____
                                   Title   Title

Date: _____   _____

359

GENERAL CONDITIONS

1. All materials, equipment and workmanship furnished under this contract shall be guaranteed by the Seller against defects and Seller agrees to replace or repair any defective material or equipment, and any defective workmanship not caused by ordinary wear and tear or to improper use or maintenance within one (1) year from date of completion of the contract. Seller further agrees to replace any refrigerant lost during that period, caused by defects in the installation, and not due to improper use or maintenance. In no event shall Seller be liable for consequential damages.

2. The work to be furnished under this contract shall be guaranteed by Seller to produce capacities, meet design limitations and to function (1) as called for in plans, specifications and addenda (2) as herein set forth or (3) as published by the manufacturer for equipment involved. In the event the foregoing requirements are not met, Seller's liability shall be limited to remedying any deficiency without expense to the Buyer. This guarantee shall not apply in any case in which the Buyer specifies the type of equipment to be used.

3. Seller shall not be held responsible or liable for any loss, damage, detention or delay caused by accidents, strikes, lockouts or by any other cause which is unavoidable or beyond Seller's control.

4. Buyer shall provide adequate fire insurance to protect the interest of the Seller against loss or damage to equipment, materials, and tools on the job site.

5. Seller shall have the right to bill the Buyer for the amount of material when delivered on the job site, even though not actually installed. In the event that material is ready for delivery and installation, but Buyer is unable to receive same, Seller shall have the right to bill Buyer for the amount of the material, including storage and insurance costs incurred by the Seller, which Buyer agrees to pay.

6. The Buyer shall be responsible for identifying and disclosing to the Seller all concealed piping, fixtures, wiring or other equipment or conditions which might be damaged, cause damage or otherwise affect or be affected by the work. In the event of damage, or a claim of damage without disclosure being given, the Buyer shall waive and hold the Seller harmless against all claims, suits, judgments and awards resulting therefrom.

7. Seller will furnish all necessary lien waivers, affidavits and other documents required to keep the Buyer's premises free from liens or claims for liens of all materialmen, subcontractors or laborers as payments are made under this contract.

8. The price or prices set forth in this contract shall be increased in an amount or amounts equal to the tax or taxes which may be assessed on the equipment or materials supplied hereunder or which may be due or become due, or which may be required to be paid with respect to this contract as a result of any excise, sales, use, occupation or similar tax not now in effect but hereafter imposed or made effective by the United States Government or any state or local government.

9. Buyer shall be responsible for structural ability of the premises to contain the equipment in the manner and location specified in the contract or shown on drawings, and Seller shall not be liable for any failure, or damage resulting from such failure, of premises, due to such structural deficiency.

10. This proposal, when signed and accepted by the Buyer, and approved by an authorized representative of the Seller, shall constitute exclusively the contract between the parties, and all prior representations or agreements, whether written or verbal, not incorporated herein, are superseded.

11. All changes, alterations or omissions to be made in the work as specified shall be performed in accordance with a written agreement between Buyer and Seller, which shall define the amount of any increase or credit in price adjustment. Neither party to this contract shall assign this contract or monies due hereunder without the prior written consent of the other.

GENERAL CONDITIONS

The following General Conditions shall be considered a part of this contract:

1. Seller guarantees the equipment and workmanship of the apparatus furnished under this contract, and will replace or repair any defects, not due to ordinary wear and tear, or to improper use or maintenance, which may develop within one year from date of completion. Seller further agrees to replace any refrigerant lost during that period, caused by defects in the installation, and not due to improper use or maintenance.

2. Seller's liability resulting from design, manufacture, and erection of equipment, whether on warranties or otherwise, shall be limited to the cost of correcting defects in the installation, as further provided herein. In no event shall Seller be liable for consequential damages.

3. Unless otherwise agreed, it is understood that work will be performed during regular working hours. If overtime work is mutually agreed upon and performed, the additional price, at Seller's usual rates for such work, shall be added to the contract price.

4. Buyer shall provide Seller's workmen a safe place in which to work, and Seller shall have the right to discontinue work when, in Seller's opinion, this clause is being violated. Seller shall not be liable for any delay, loss, or damage caused by such delay.

5. On delivery of equipment by Seller, or any part thereof, to the premises of Buyer, Buyer shall assume risk of loss or damage to such equipment and shall cause same to be insured in all respects against loss or damage in an amount to protect interest of Seller. Cost of insurance is to be paid by Buyer.

6. In the event that the material incorporated in this contract is ready for delivery and installation, and Buyer is unable to receive same, Seller shall have the right to bill Buyer for the amount of the material in accordance with the terms of the contract and also to provide suitable storage and insurance at the Buyer's risk and expense.

7. Seller shall not be held responsible or liable for any loss, damage, detention, or delay caused by accidents, strikes, lockouts, or by any other cause which is unavoidable or beyond Seller's control.

8. Title to the equipment remains in Seller until payment of the entire purchase price and all sums due Seller under this contract are fully made. All equipment, whether affixed to the realty or not, shall remain personal property and be deemed severable without injury to the freehold. Buyer shall do whatever may be required to maintain Seller's title.

   In the event of default of payment of any installment or failure to perform any terms or conditions of this contract, or in the event that a proceeding in bankruptcy or insolvency be instituted by or against Buyer, or if equipment is misused, illegally used, or imperiled, then at Seller's option the entire unpaid balance shall become immediately due and payable without notice or demand and in such case Seller may enter the premises and retake, remove and hold or resell the equipment or any part thereof at either private or public sale. If the unpaid balance plus interest is not satisfied by the proceeds of such sale after deducting the expenses of retaking, repairs necessary to place the equipment in saleable condition, storing, taxes, liens, attorneys' and collection agency's fees and other expenses in connection therewith, Buyer shall pay any deficiency as liquidated damages for breach of this contract. Seller shall retain all lien rights upon premises on which the installation is made, to the extent of the unpaid balance, until final payment is made.

9. Should Seller be delayed by reason of any default on the part of Buyer of the terms and conditions of this contract, the entire contract price, less payments theretofore made, shall become due, and shall bear interest at the full legal rate from the date of billing.

10. Buyer shall be responsible for structural ability of the premises to contain the equipment in the manner and location specified in the contract or shown on drawings, and Seller shall not be liable for any failure, or damage resulting from such failure, of premises, due to such structural deficiency.

11. Buyer shall keep equipment free of taxes and encumbrances, shall not remove said equipment from the premises without written permission of Seller, and shall not transfer any interest in said equipment or in this contract without written consent of Seller until all payments due hereunder have been made.

12. Any price or prices herein set forth shall be increased in an amount or amounts equal to the tax or taxes which may be assessed on the equipment supplied hereunder, or which may be due or become due from Seller, or which Seller may be required to pay with respect to this contract as a result of any excise, sales, use, occupation, or similar tax not now in effect but hereafter imposed or made effective by the United States Government or any state or local government.

13. Upon completion of the installation, Seller shall fully instruct Buyer in regard to operation and maintenance. If for a period of eight months immediately after the equipment supplied hereunder is installed, Buyer fails to notify Seller in writing of any claim that the said equipment as supplied does not fulfill the terms and conditions of this contract, specifying in what particulars it fails, this shall be an acknowledgment by Buyer that said equipment as supplied does fulfill said terms and conditions, and shall constitute a complete acceptance of the installation.

   If Buyer claims that the plant does not fulfill the terms and conditions of the contract, he shall notify Seller in writing to this effect, specifying in what particular it fails. A reasonable length of time shall then be allowed Seller to remedy any defects or deficiencies that may exist, or to demonstrate to Buyer the capability of the plant to fulfill the terms and conditions. If the plant then fails to fulfill the terms and conditions of the contract, Seller may then remove the equipment upon refunding all moneys paid therefor, and thereafter no liability shall exist whatsoever in favor of either party as against the other and this contract shall thereupon be terminated.

14. Buyer shall not assign this contract or any rights thereunder without Seller's written consent.

15. This proposal, when signed and accepted by the Buyer, and approved by an authorized representative of Advance Heating and Air Conditioning Corporation, shall constitute exclusively the contract between the parties, and all prior representations or agreements, whether written or verbal, not incorporated herein, are superseded.

16. This contract is not valid unless approved by a duly authorized representative of Advance Heating and Air Conditioning Corporation.

APPENDIX C

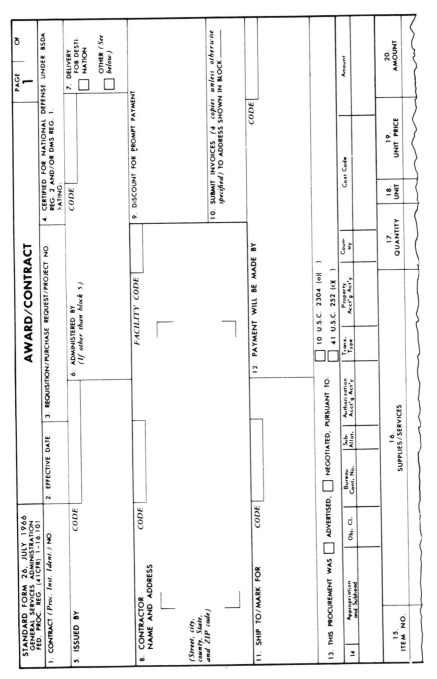

21.

TOTAL AMOUNT OF CONTRACT $

CONTRACTING OFFICER WILL COMPLETE BLOCK 22 OR 26 AS APPLICABLE

22. ☐ CONTRACTOR'S NEGOTIATED AGREEMENT (Contractor is required to sign this document and return _____ copies to issuing office.) Contractor agrees to furnish and deliver all items or perform all the services set forth or otherwise identified above and on any continuation sheets for the consideration stated herein. The rights and obligations of the parties to this contract shall be subject to and governed by the following documents: (a) this award/contract, (b) the solicitation, if any, and (c) such provisions, representations, certifications, and specifications, as are attached or incorporated by reference herein. (Attachments are listed herein.)

26. ☐ AWARD (Contractor is not required to sign this document.) Your offer on Solicitation Number _____, including the additions or changes made by you which additions or changes are set forth in full above, is hereby accepted as to the items listed above and on any continuation sheets. This award consummates the contract which consists of the following documents: (a) the Government's solicitation and your offer, and (b) this award/contract. No further contractual document is necessary.

23. NAME OF CONTRACTOR

BY _____
(Signature of person authorized to sign)

24. NAME AND TITLE OF SIGNER (Type or print)

25. DATE SIGNED

27. UNITED STATES OF AMERICA

BY _____
(Signature of Contracting Officer)

28. NAME OF CONTRACTING OFFICER (Type or print)

29. DATE SIGNED

NAVFAC 4-4330/11(10-67)
Supersedes NavDocks 424 (Rev. 4–65)

SCHEDULE

It has been determined that the execution of this contract is advantageous to the national defense and that the existing facilities of the Naval Establishment are inadequate.

1. DESCRIPTION OF THE PROJECT. The Architect-Engineer shall perform all the services required under this contract for the project generally described as follows: (hereinafter refered to as the "Project")

_____

_____
_____

having an estimated construction cost of $_____, and more specifically described in APPENDIX "A" which is attached hereto and made a part hereof.

2. STATEMENT OF ARCHITECT-ENGINEER SERVICES. The Architect-Engineer shall prepare and furnish to the Government, complete and ready for use, all necessary studies, preliminary sketches, estimates, working records and other drawings (including large scale details as required), and specifications; shall check shop drawings furnished by the construction contractor; shall furnish consultation and advice as requested by the Government during the construction (but not including the supervision of the construction work); and shall furnish all other architectural and engineering services, including without limitations those specified hereinafter and required in connection with the accomplishment of Naval public works and/or utilities projects.

3. PERIOD OF SERVICE. The Architect-Engineer shall complete all work and services within _____ months after receipt of notice to proceed.

4. DESIGN WITHIN COST LIMITATION. The Architect-Engineer agrees to design a facility which can be constructed (under normal Bureau of Yards and Docks procedures) within the estimated net construction cost, with each item of the facility to be designed so as to be constructed within applicable statutory cost limitations indicated below.

If, after receipt of competitive bids, it is found that a construction contract cannot be awarded within the aforementioned estimated construction cost and within any applicable statutory cost limitations, the Architect-Engineer will, as a part of this contract, and at no additional expense to the Government, perform such redesign, re-estimating and other services as may, in the opinion of the Officer in Charge, be required to produce a usable facility within the estimated construction cost and/or within the statutory cost limitations. In connection with the foregoing, the Architect-Engineer shall be obligated to perform such additional services at no increase in contract price only when the Navy has received competitive bids within six (6) months from the date of final approval of the plans and specifications.

STATUTORY COST LIMITS

5. ARCHITECTURAL DESIGNS AND DATA–GOVERNMENT RIGHTS (UN-LIMITED) (APR.1966). The Government shall have unlimited rights, for the benefit of the Government, in all drawings, designs, specifications, architectural designs of buildings and structures, notes and other architect-engineer work produced in the performance of this contract, or in contemplation thereof, and all as-built drawings produced after completion of the work, including the right to use same on any other Government work without additional cost to the Government; and with respect thereto the Architect-Engineer agrees to and does hereby grant to the Government a royalty-free license to all such data which he may cover by copyright and to all architectural designs as to which he may assert any rights or establish any claim under the design patent or copyright laws. The Architect-Engineer for a period of three (3) years after completion of the project agrees to furnish and to provide access to the originals or copies of all such materials on the request of the Contracting Officer.

6. ADDITIONAL STATEMENT OF ARCHITECT-ENGINEER SERVICES. It is agreed, without limiting the generality of the foregoing, that (a) The Architect-Engineer shall, if necessary, visit the site and shall hold such conferences with representatives of the Government and take such other action as may be necessary to obtain the data upon which to develop the design and preliminary sketches showing the contemplated project, (b) The preliminary sketches shall include plans, elevations and sections developed in such detail and with such descriptive specifications as will clearly indicate the scope of the work, and make possible a reasonable estimate of the cost, (c) Preliminary sketches together with an estimate of the cost of the project shown on the sketches shall be submitted for the approval of the Officer in Charge, (d) The Architect-Engineer shall change the preliminary sketches for the extent necessary to meet the requirements of the Government, and after approval by the Officer in Charge the Architect-Engineer shall furnish necessary prints of the approved preliminary sketches to the Officer in Charge, (e) After the preliminary sketches and estimates have been approved, the Architect-Engineer shall proceed with the preparation of complete working drawings and specifications as required by the Officer in Charge in connection with the construction of the said project. Working drawings, specifications and estimates shall be delivered to the Officer in Charge in such sequence and at such times as required by the Government and as will insure that the construction work can be initiated promptly, procurement of materials made without delay, and continuous prosecution of the work promoted. Working drawings and specifications shall be revised as necessary and as required by the Officer in Charge. After working drawings and specifications have been approved by the Officer in Charge, the Architect-Engineer shall furnish such number of sets of prints of the approved working drawings and such number of sets of the approved specifications, as may be required by the Officer in Charge (f) Upon approval of final plans the Architect-Engineer shall deliver to the Government one set of tracings, in such medium and on such materials, as may be required by the Officer in Charge, suitable for blue-printing, showing complete approved construction requirements (not of "as-built" construction unless otherwise stipulated), provided, however, that should this contract be terminated by the Government, the Architect-Engineer shall deliver to the Government one set of such tracings as may be required by the Officer in Charge. Such tracings as are delivered shall be signed by the Officer in Charge as an indication of approval thereof, and shall become and remain the property of the Government. (g) The Architect-Engineer shall perform all necessary architectural-engineering services of every kind required in connection with the studies, designs and the preparation of drawings and specifications, but said services shall not, unless otherwise stipulated, include borings, test piles and pits, or supervision of construction work executed from the drawings and specifications, provided, however, that the Architect-Engineer shall furnish upon request and without additional compensation, such

amplifications and explanations and attend such conferences as may, in the opinion of the Officer-in-Charge, be necessary to clarify the intent of the drawings and specifications and shall afford the benefit of his advice on questions that may arise in connection with the construction of the project. (h) The Architect-Engineer shall without additional fee correct or revise the drawings, specifications, or other materials furnished under this contract, if the Officer in Charge finds that such revision is necessary to correct errors or deficiencies for which the Architect-Engineer is responsible.

7. DEFINITIONS (JUNE 1964). (a) The term "head of the agency" or "Secretary" as used herein means the Secretary, the Under Secretary, and Assistant Secretary, or any other head or assistant head of the executive or military department or other Federal agency; and the term "his duly authorized representative" means any person or persons or board (other than the Contracting Officer) authorized to act for the head of the agency or the Secretary.

(b) The term "Contracting Officer" as used herein means the person executing this contract on behalf of the Government and includes a duly appointed successor or authorized representative. (See "Government Representatives" clause).

(c) Except as otherwise provided in this contract, the term "subcontracts" includes purchase orders under this contract.

8. GOVERNMENT REPRESENTATIVES. The work will be under the general direction of the Contracting Officer, The Commander of Naval Facilities Engineering Command, who shall designate an officer of the Civil Engineer Corps, United States Navy, or other officer or representative of the Government, as Officer in Charge, hereinafter referred to as the "OIC" who, except in connection with the Disputes Clause shall be the authorized representative of the Contracting Officer and under the direction of the Contracting Officer have complete charge of the work, and shall exercise full supervision and general direction of the work, so far as it affects the interest of the Government. For the purposes of the "Disputes" clause the Contracting Officer shall mean the Commander, Naval Facilities Engineering Command, the Acting Commander, their successors, or their representatives specially designated for this purpose.

9. CONTRACTING OFFICER'S DECISIONS (JAN. 1965). The extent and character of the work to be done by the Architect-Engineer shall be subject to the general supervision, direction, control and approval of the Contracting Officer.

10. TECHNICAL ADEQUACY. Approval by the Officer in Charge of drawings, designs, specifications and other incidental architectural engineering work or materials furnished hereunder shall not in any way relieve the Architect-Engineer of responsibility for the technical adequacy of the work.

11. COMPENSATION. The Architect-Engineer shall be paid the lump sum of $ as full compensation for all services, labor and material required hereby, and for all expenditures which may be made and expenses incurred except as are otherwise expressly provided herein.

12. METHOD OF PAYMENT (JAN. 1965). (a) Estimates shall be made monthly of the amount and value of the work and services performed by the Architect-Engineer under this contract, such estimates to be prepared by the Architect-Engineer and accompanied by such supporting data as may be required by the Contracting Officer.

(b) Upon approval of such estimate by the Contracting Officer payment upon properly certified vouchers shall be made to the Architect-Engineer as soon as practicable of

90% of the amount as determined above, less all previous payments; provided, however, that if the Contracting Officer determines that the work is substantially complete and that the amount of retained percentages is in excess of the amount considered by him to be adequate for the protection of the Government, he may at his discretion release to the Architect-Engineer such excess amount.

(c) Upon satisfactory completion by the Architect-Engineer and acceptance by the Contracting Officer of the work done by the Architect-Engineer in accordance with the foregoing "Statement of Architect-Engineer Services," the Architect-Engineer will be paid the unpaid balance of any money due for work under said statement, including retained percentages relating to this portion of the work.

(d) Upon satisfactory completion of the construction work and its final acceptance, the Architect-Engineer shall be paid the unpaid balance of any money due hereunder. Prior to such final payment under this contract, or prior settlement upon termination of the contract, and as a condition precedent thereto, the Architect-Engineer shall execute and deliver to the Contracting Officer a release of all claims against the Government arising under or by virtue of this contract, other than such claims, if any, as may be specifically excepted by the Architect-Engineer from the operation of the release in stated amounts to be set forth therein.

13. CHANGES (JAN. 1965). The Contracting Officer may at any time, by written order, and without notice to the sureties, make changes within the general scope of the contract in the work and service to be performed. If any such changes cause an increase or decrease in the Architect-Engineer's cost of, or the time required for, performance of this contract, an equitable adjustment shall be made and the contract shall be modified in writing accordingly. Any claim by the Architect-Engineer for adjustment under this clause must be asserted in writing within 30 days from the date of receipt by the Architect-Engineer of the notification of change unless the Contracting Officer grants a further period of time before the date of final payment under the contract. If the parties fail to agree upon the adjustment to be made, the dispute shall be determined as provided in the "Disputes" clause of this contract; but nothing provided in this clause shall excuse the Architect-Engineer from proceeding with the prosecution of the work as changed. Except as otherwise provided in this contract, no charge for any extra work or material will be allowed.

14. FEDERAL, STATE, AND LOCAL TAXES (JUL. 1960). (a) As used throughout this clause, the term "contract date" means the date of this contract. As to additional supplies or services procured by modification of this contract, the term "contract date" means the date of such modification.

(b) Except as may be otherwise provided in this contract, the contract price includes, to the extent allocable to this contract, all Federal, State, and local taxes which, on the contract date:

(i) by Constitution, statute, or ordinance, are applicable to this contract, or to the transactions covered by this contract, or to property or interests in property; or

(ii) pursuant to written ruling or regulation, the authority charged with administering any such tax is assessing or applying to, and is not granting or honoring an exemption for, a contractor under this kind of contract, or the transactions covered by this contract, or property or interests in property.

(c) Except as may be otherwise provided in this contract, duties in effect on the contract date are included in the contract price, to the extent allocable to this contract.

(d) (1) If the Architect-Engineer is required to pay or bear the burden—

(i) of any tax or duty which either was not to be included in the contract price pursuant to the requirements of paragraphs (b) and (c), or of a tax or duty specifically excluded from the contract price by a provision of this contract; or

(ii) of an increase in rate of any tax or duty, whether or not such tax or duty was excluded from the contract price; or

(iii) of any interest or penalty on any tax or duty referred to in (i) or (ii) above; the contract price shall be increased by the amount of such tax, duty, interest, or penalty allocable to this contract; *provided,* that the Architect-Engineer warrants in writing that no amount of such tax, duty, or rate increase was included in the contract price as a contingency reserve or otherwise; and *provided further,* that liability for such tax, duty, rate increase, interest, or penalty was not incurred through the fault or negligence of the Architect-Engineer or his failure to follow instructions of the Contracting Officer.

(2) If the Architect-Engineer is not required to pay or bear the burden, or obtains a refund or drawback, in whole or in part, of any tax, duty, interest, or penalty which:

(i) was to be included in the contract price pursuant to the requirements of paragraphs (b) and (c);

(ii) was included in the contract price; or

(iii) was the basis of an increase in the contract price; the contract price shall be decreased by the amount of such relief, refund, or drawback allocable to this contract, or the allocable amount of such relief, refund, or drawback shall be paid to the Government, as directed by the Contracting Officer. The contract price also shall be similarly decreased if the Contractor, through his fault or negligence or his failure to follow instructions of the Contracting Officer, is required to pay or bear the burden, or does not obtain a refund or drawback of any such tax, duty, interest, or penalty. Interest paid or credited to the Architect-Engineer incident to a refund of taxes shall inure to the benefit of the Government to the extent that such interest was earned after the Architect-Engineer was paid or reimbursed by the Government for such taxes.

(3) Invoices or vouchers covering any adjustment of the contract price pursuant to this paragraph (d) shall set forth the amount thereof as a separate item and shall identify the particular tax or duty involved.

(4) This paragraph (d) shall not be applicable to social security taxes; income and franchise taxes, other than those levied on or measured by (i) sales or receipts from sales, or (ii) the Architect-Engineer's possession of, interest in, or use of property, title to which is in the Government; excess profits taxes; capital stock taxes; unemployment compensation taxes; or property taxes, other than such property taxes, allocable to this contract, as are assessed either on completed supplies covered by this contract, or on the Architect-Engineer's possession of, interest in, or use of property, title to which is in the Government.

(5) No adjustment of less than $100 is required to be made in the contract price pursuant to this paragraph (d).

(e) Unless there does not exist any reasonable basis to sustain an exemption, the Government upon request of the Architect-Engineer, without further liability, agrees, except as otherwise provided in this contract, to furnish evidence appropriate to establish exemption from any tax which the Architect-Engineer warrants in writing was excluded from the contract price. In addition, the Contracting Officer may furnish evidence appropriate to establish exemption from any tax that may, pursuant to this clause, give rise to either an increase or decrease in the contract price. Except as otherwise provided in this contract, evidence appropriate to establish exemption from duties will be furnished only at the discretion of the Contracting Officer.

(f) (1) The Architect-Engineer shall promptly notify the Contracting Officer on all matters pertaining to Federal, State, and local taxes, and duties, that reasonably may be expected to result in either an increase or decrease in the contract price.

(2) Whenever an increase or decrease in the contract price may be required under this clause, the Architect-Engineer shall take action as directed by the Contracting Officer, and the contract price shall be equitably adjusted to cover the costs of such action, including any interest, penalty, and reasonable attorneys' fees.

15. RESPONSIBILITY OF THE ARCHITECT-ENGINEER. (a) The Architect-Engineer shall be responsible for the professional and technical accuracy and the coordination of all designs, drawings, specifications and other work or materials furnished by the Architect-Engineer under this contract. The Architect-Engineer shall without additional cost or fee to the Government correct or revise any errors or deficiencies in his performance.

(b) Neither the Government's review, approval or acceptance of, nor payment for, any of the services required under this contract shall be construed to operate as a waiver of any rights under this contract or of any cause of action arising out of the performance of this contract, and the Architect-Engineer shall be and remain liable to the Government for all costs of any kind which were incurred by the Government as a result of the Architect-Engineer's negligent performance of and of the services furnished under this contract.

(c) The rights and remedies of the Government provided for under this contract are in addition to any other rights and remedies provided by law; the Government may assert its right of recovery by any appropriate means including, but not limited to, set-off, suit, withholding, recoupment, or counterclaim, either during or after performance of this contract.

16. MILITARY SECURITY REQUIREMENTS (APR.1966). (a) The provisions of this clause shall apply to the extent that this contract involves access to information classified "Confidential" including "Confidential–Modified Handling Authorized" or higher.

(b) The Government shall notify the Architect-Engineer of the security classification of this contract and the elements thereof, and of any subsequent revisions in such security classification, by the use of a Security Requirements Check List (DD Form 254), or other written notification.

(c) To the extent the Government has indicated as of the date of this contract or thereafter indicates security classification under this contract as provided in paragraph (b) above, the Architect-Engineer shall safeguard all classified elements of this contract and shall provide and maintain a system of security controls within its own organization in accordance with the requirements of:

(i) the Security Agreement (DD Form 441), including the Department of Defense Industrial Security Manual for Safeguarding Classified Information as in effect on the date of this contract, and any modification to the Security Agreement for the purpose of adapting the Manual to the Architect-Engineer's business; and

(ii) any amendments to said Manual made after the date of this contract, notice of which has been furnished to the Architect-Engineer by the Security Office of the Military Department having security cognizance over the facility.

(d) Representatives of the Military Department having security cognizance over the facility and representatives of the contracting Military Department shall have the right to inspect at reasonable intervals the procedures, methods, and facilities utilized by the Architect-Engineer in complying with the security requirements under this contract. Should the Government, through these representatives, determine that the Architect-Engineer is not complying with the Security requirements of this contract the Architect-Engineer shall be informed in writing by the Security Office of the cognizant Military Department of the proper action to be taken in order to effect compliance with such requirements.

(e) If, subsequent to the date of this contract, the security classifications or security requirements under this contract are changed by the Government as provided in this clause and the security costs or time required for delivery under this contract are thereby increased or decreased, the contract price, delivery schedule, or both and any other provision of the contract that may be affected shall be subject to an equitable adjustment by reason of such increased or decreased costs. Any equitable adjustment shall be accomplished in the same manner as if such changes were directed under the "Changes" clause in this contract.

(f) The Architect-Engineer agrees to insert, in all subcontracts hereunder which involve access to classified information, provisions which shall conform substantially to the language of this clause, including this paragraph (f) but excluding the last sentence of paragraph (e) of this clause.

(g) The Architect-Engineer also agrees that it shall determine that any subcontractor proposed by it for the furnishing of supplies and services which involve access to classified information in the Architect-Engineer's custody has been granted an appropriate facility security clearance which is still in effect, prior to being accorded access to such classified information.

17. TERMINATION FOR DEFAULT (JAN. 1965). (a) The performance of work under this contract may be terminated by the Government in accordance with this clause in whole, or from time to time in part, whenever the Architect-Engineer shall default in performance of this contract in accordance with its terms (including in the term "default" any such failure by the Architect-Engineer to make progress in the prosecution of the work hereunder as endangers such performance), and shall fail to cure such default within a period of ten days (or such longer periods as the Contracting Officer may allow) after receipt by the Architect-Engineer of a notice specifying the default.

(b) If the contract is so terminated, the Government may take over the work and services and prosecute the same to completion by contract or otherwise, and the Architect-Engineer shall be liable to the Government for any excess cost occasioned to the Government thereby.

(c) The contract may not be so terminated if the failure to perform arises from unforseeable causes beyond the control and without the fault or negligence of the Architect-Engineer. Such causes may include, but are not restricted to acts of God, acts of the public enemy, acts of the Government in either its sovereign or contractural capacity, fires, floods, epidemics, quarantine restrictions, strikes and unusually severe weather; but in every case the failure to perform must be beyond the control and without the fault or negligence of the Architect-Engineer; and the Architect-Engineer, within ten days from the beginning of any such delay (unless the Contracting Officer grants a further period of time before the date of final payment under the contract) notifies the Contracting Officer in writing of the causes of delay. The Contracting Office shall ascertain the facts and the extent of delay and extend the time for completing the work when, in his judgment, the findings of fact justify such an extension, and his finding of fact shall be final and conclusive on the parties, subject only to appeal as provided in the clause of this contract entitled "Disputes."

(d) If, after Notice of Termination of the contract under the provisions of this clause, it is determined for any reason that the Architect-Engineer was not in default under the provisions of this clause or that the default was excusable under the provisions of this clause, the rights and obligations of the parties shall be the same as if the Notice of Termination had been issued pursuant to the clause of this contract entitled "Termination for the Convenience of the Government."

(e) The rights and remedies of the Government provided in this clause are in addition to any other rights and remedies provided by law or under this contract.

18. DISPUTES (JUNE 1964). (a) Except as otherwise provided in this contract, any dispute concerning a question of fact arising under this contract which is not disposed of by agreement shall be decided by the Contracting Officer, who shall reduce his decision to writing and mail or otherwise furnish a copy thereof to the Architect-Engineer. The decision of the Contracting Officer shall be final and conclusive unless, within 30 days from the date of receipt of such copy, the Architect-Engineer mails or otherwise furnishes to the Contracting Officer a written appeal addressed to the head of the agency involved. The decision of the head of the agency or his duly authorized representative for the determination of such appeals shall be final and conclusive. This provision shall not be pleaded in any suit involving a question of fact arising under this contract as limiting judicial review of any such decision to cases where fraud by such official or his representative or board is alleged: Provided, however, that any such decision shall be final and conclusive unless the same is fraudulent or capricious or arbitrary or so grossly erroneous as necessarily to imply bad faith or is not supported by substantial evidence. In connection with any appeal proceeding under this clause, the Architect-Engineer shall be afforded an opportunity to be heard and to offer evidence in support of his appeal. Pending final decision of a dispute hereunder, the Architect-Engineer shall proceed diligently with the performance of the contract and in accordance with the Contracting Officer's decision.

(b) This Disputes clause does not preclude consideration of questions of law in connection with decisions provided for in paragraph (a) above. Nothing in this contract, however, shall be construed as making final the decision of any administrative official, representative, or board on a question of law.

19. OFFICIALS NOT TO BENEFIT (JUNE 1964). No member of or delegate to Congress, or resident commissioner, shall be admitted to any share or part of this contract, or to any benefit that may arise therefrom; but this provision shall not be construed to extend to this contract if made with a corporation for its general benefit.

20. COVENANT AGAINST CONTINGENT FEES (JAN. 1958). The Architect-Engineer warrants that no person or selling agency has been employed or retained to solicit or secure this contract upon an agreement or understanding for a commission, percentage, brokerage, or contingent fee, excepting bona fide employees or bona fide established commercial or selling agencies maintained for the Architect-Engineer for the purpose of securing business. For breach or violation of this warranty the Government shall have the right to annul this contract without liability or in its discretion to deduct from the contract price or consideration the full amount of such commission, percentage, brokerage, or contingent fee.

21. ASSIGNMENT OF CLAIMS (JUNE 1964). (a) Pursuant to the provisions of the Assignment of Claims Act of 1940 as amended (31 U.S. Code 203, 41 U.S. Code 15), if this

contract provides for payments aggregating $1,000 or more, claims for moneys due or to become due the Architect-Engineer from the Government under this contract may be assigned to a bank, trust company or other financing institution, including any Federal lending agency, and may thereafter be further assigned and reassigned to any such institution. Any such assignment or reassignment shall cover all amounts payable under this contract and not already paid, and shall not be made to more than one party, except that any such assignment or reassignment may be made to one party as agent or trustee for two or more parties participating in such financing. Unless otherwise provided in this contract, payments to an assignee of any moneys due or to become due under this contract shall not, to the extent provided in said Act as amended, be subject to reduction or set-off.

(b) In no event shall copies of this contract or of any plans, specifications, or other similar documents relating to work under this contract, if marked "Top Secret," "Secret," or "Confidential" be furnished to any assignee of any claim arising under this contract or to any other person not entitled to receive the same. However a copy of any part or all of this contract so marked may be furnished, or any information contained therein may be disclosed, to such assignee upon the prior written authorization of the Contracting Officer.

22. CONTRACT WORK HOURS STANDARDS ACT–OVERTIME COMPENSATION (40 U.S.C. 327-330) (APR. 1965). *(This clause is only applicable to contracts exceeding $2,500 and is required by 40 USC 327-330).* (a) The Architect-Engineer shall not require or permit any laborer or mechanic in any workweek in which he is employed on any work under this contract to work in excess of eight (8) hours in any calendar day or in excess of forty (40) hours in such workweek on work subject to the provisions of the Contract Work Hours Standards Act unless such laborer or mechanic receives compensation at a rate not less than one and one-half times his basic rate of pay for all such hours worked in excess of eight (8) hours in any calendar day or in excess of forty (40) hours in such workweeks, whichever is the greater number of overtime hours. The "basic rate of pay," as used in this clause, shall be the amount paid per hour, exclusive of the Architect-Engineer contribution or cost for fringe benefits and any cash payment made in lieu of providing fringe benefits, or the basic hourly rate contained in the wage determination, whichever is greater.

(b) In the event of any violation of the provisions of paragraph (a), the Architect-Engineer shall be liable to any affected employee for any amounts due, and to the United States for liquidated damages. Such liquidated damages shall be computed with respect to each individual laborer or mechanic employed in violation of the provisions of paragraph (a) in the sum of $10 for each calendar day in which such employee was required or permitted to be employed on such work in excess of eight (8) hours or in excess of the standard workweek of forty (40) hours without payment of the overtime wages required by paragraph (a).

23. CONVICT LABOR (MAR 1949). In connection with the performance of work under this contract, the Architect-Engineer agrees not to employ any person undergoing sentence of imprisonment at hard labor.

24. EQUAL OPPORTUNITY (APRIL 1964). (The following clause is applicable unless this contract is exempt under the rules and regulations of the President's Committee on Equal Employment Opportunity (41 C.F.R. Chapter 60). Exemptions include contracts and subcontracts (i) not exceeding $10,000, (ii) not exceeding $100,000 for standard commercial supplies or raw materials, and (iii) under which work is performed outside the United States and no recruitment of workers within the United States is involved).

During the performance of this contract, the Architect-Engineer agrees as follows:

(a) The Architect-Engineer will not discriminate against any employee or applicant for employment because of race, creed, color, or national origin. The Architect-Engineer will take affirmative action to ensure that applicants are employed, and that employees are treated during employment, without regard to their race, creed, color, or national origin. Such action shall include, but not be limited to, the following: employment, upgrading, demotion or transfer; recruitment or recruitment advertising; layoff or termination; rates of pay or other forms of compensation; and selection for training, including apprenticeship. The Architect-Engineer agrees to post in conspicuous places, available to employees and applicants for employment, notices to be provided by the Contracting Officer setting forth the provisions of this nondiscrimination clause.

(b) The Architect-Engineer will, in all solicitations or advertisements for employees placed by or on behalf of the Architect-Engineer, state that all qualified applicants will receive consideration for employment without regard to race, creed, color, or national origin.

(c) The Architect-Engineer will send to each labor union or representative of workers with which he has a collective bargaining agreement or other contract or understanding, a notice, to be provided by the Agency Contracting Officer, advising the said labor union of workers' representative of the Architect-Engineer's commitments under this nondiscrimination clause, and shall post copies of the notice in conspicuous places available to employees and applicants for employment.

(d) The Architect-Engineer will comply with all provisions of Executive Order No. 10925 of March 6, 1961, as amended, and of the rules, regulations, and relevant orders of the President's Committee on Equal Employment Opportunity created thereby.

(e) The Architect-Engineer will furnish all information and reports required by Executive Order No. 10925 of March 6, 1961, as amended, and by the rules, regulations, and orders of the said Committee, or pursuant thereto, and will permit access to his books, records, and accounts by the contracting agency and the Committee for purposes of investigation to ascertain compliance with such rules, regulations, and orders.

(f) In the event of the Architect-Engineer's noncompliance with the nondiscrimination clause of this contract or with any of the said rules, regulations, or orders, this contract may be cancelled, terminated, or suspended in whole or in part and the Architect-Engineer may be declared ineligible for further Government contracts in accordance with procedures authorized in Executive Order No. 10925 of March 6, 1961, as amended, and such other sanctions may be imposed and remedies invoked as provided in the said Executive Order or by rule regulation, or order of the President's Committee on Equal Employment Opportunity, or as otherwise provided by law.

(g) The Architect-Engineer will include the provisions of paragraphs (a) through (g) in every subcontract or purchase order unless exempted by rules, regulations, or orders of the President's Committee on Equal Employment Opportunity issued pursuant to Section 303 of Executive Order No. 10925 of March 6, 1961, as amended, so that such provisions will be binding upon each subcontractor or vendor.* The Architect-Engineer will take such action with respect to any subcontract or purchase order as the contracting

*Unless otherwise provided, the "Equal Opportunity" clause is not required to be inserted in subcontracts below the second tier, except for subcontracts involving the performance of "construction work" as the "site of construction" (as those terms are defined in the Committee's rules and regulations) in which case the clause must be inserted in all such contracts. Subcontracts may incorporate by reference the "Equal Opportunity" clause.

agency may direct as a means of enforcing such provisions, including sanctions for non-compliance: *Provided, however,* that in the event the Architect-Engineer becomes involved in, or is threatened with, litigation with a subcontractor or vendor as a result of such direction by the contracting agency, the Architect-Engineer may request the United States to enter into such litigation to protect the interest of the United States.

25. RENEGOTIATION (OCT. 1959). (a) To the extent required by law, this contract is subject to the Renegotiation Act of 1951 (50 U.S.C. App. 1211, et seq.), as amended, and to any subsequent act of Congress providing for the renegotiation of Contracts. Nothing contained in this clause shall impose any renegotiation obligation with respect to this contract or any subcontract hereunder which is not imposed by an act of Congress heretofore or hereafter enacted. Subject to the foregoing this contract shall be deemed to contain all the provisions required by Section 104 of the Renegotiation Act of 1951, and by any such other act without subsequent contract amendment specifically incorporating such provisions.

(b) The Architect-Engineer agrees to insert the provisions of this clause, including this paragraph (b), in all subcontracts, as that term is defined in Section 103g of the Renegotiation Act of 1951, as amended.

26. EXAMINATION OF RECORDS (FEB. 1962). (a) The Architect-Engineer agrees that the Comptroller General of the United States or any of his duly authorized representatives shall, until the expiration of three years after final payment under this contract, have access to and the right to examine any directly pertinent books, documents, papers, and records of the Architect-Engineer involving transactions related to this contract.

(b) The Architect-Engineer further agrees to include in all his subcontracts hereunder a provision to the effect that the subcontractor agrees that the Comptroller General of the United States or any of his duly authorized representative shall, until the expiration of three years after final payment under the subcontract, have access to and the right to examine any directly pertinent books, documents, papers, and records of such subcontractor involving transactions related to the subcontract. The term "subcontract" as used in this clause excludes (i) purchase orders not exceeding $2,500 and (ii) subcontracts or purchase orders for public utility services at rates established for uniform applicability to the general public.

27. GRATUITIES (MAR. 1952). (a) The Government may, by written notice to the Architect-Engineer, terminate the right of the Architect-Engineer to proceed under this contract if it is found, after notice and hearing, by the Secretary or his duly authorized representative, that gratuities (in the form of entertainment, gifts, or otherwise) were offered or given by the Architect-Engineer, or any agent or representative of the Architect-Engineer, to any officer or employee of the Government with a view toward securing a contract or securing favorable treatment with respect to the awarding or amending, or the making of any determinations with respect to the performing, of such contract; provided, that the existence of the facts upon which the Secretary or his duly authorized representatives makes such finding shall be in issue and may be reviewed in any competent court.

(b) In the event this contract is terminated as provided in paragraph (a) hereof, the Government shall be entitled (i) to pursue the same remedies against the Architect-Engineer as it could pursue in the event of a breach of the contract by the Architect-Engineer, and (ii) as a penalty in addition to any other damages to which it may be entitled by law, to exemplary damages in an amount (as determined by the Secretary or his duly authorized representatives) which shall be not less than three nor more than ten times the costs incurred by the Architect-Engineer in providing any such gratuities to any such officer or employee.

(c) The rights and remedies of the Government provided in this clause shall not be exclusive and are in addition to any other rights and remedies provided by law or under this contract.

28. INTEREST (MAY 1963). Notwithstanding any other provision of this contract, unless paid within 30 days all amounts that become payable by the Architect-Engineer to the Government under this contract (net of any applicable tax credit under the Internal Revenue Code) shall bear interest at the rate of six percent per annum from the date due until paid and shall be subject to adjustments as provided by Part 6 of Appendix E of the Armed Services Procurement Regulation, as in effect on the date of this contract. Amounts shall be due upon the earliest one of (i) the date fixed pursuant to this contract, (ii) the date of the first written demand for payment, consistent with this contract, (iii) the date of transmittal by the Government to the Architect-Engineer of a proposed supplemental agreement to confirm completed negotiations fixing the amount of (iv) if this contract provided for revision of prices, the date of written notice to the Architect-Engineer stating the amount of refund payable in connection with a pricing proposal or in connection with a negotiated pricing agreement not confirmed by contract supplement.

29. ACCIDENT PREVENTION (JAN 1965). (a) In order to provide safety controls for protection to the life and health of employees and other persons; for prevention of damage to property, materials, supplies, and equipment; and for avoidance of work interruptions in the performance of this contract, the Architect-Engineer shall comply with all pertinent provisions of Corps of Engineers Mannual, EM385-1-1, dated 13 March 1958, entitled "General Safety Requirements," as amended, and will also take or cause to be taken such additional measures as the Contracting Officer may determine to be reasonably necessary for the purpose.

(b) The Architect-Engineer will maintain an accurate record of, and will report to the Contracting Officer in the manner and on the forms prescribed by the Contracting Officer, exposure data and all accidents resulting in death, traumatic injury, occupantional disease, and damage to property, materials, supplies and equipment incident to work performed under this contract.

(c) The Contracting Officer will notify the Architect-Engineer of any noncompliance with the foregoing provisions and the action to be taken. The Architect-Engineer shall, after receipt of such notice, immediately take corrective action. Such notice, when delivered to the Architect-Engineer or his representative at the site of the work, shall be deemed sufficient for the purpose. If the Architect-Engineer fails or refuses to comply promptly the Contracting Officer may issue an order stopping all or part of the work until satisfactory corrective action has been taken. No part of the time lost due to any such stop orders shall be made the subject of claim for extension of time or for excess costs or damages by the Architect-Engineer.

(d) Compliance with the provisions of this article by subcontractors will be the responsibility of the Architect-Engineer.

30. TERMINATION FOR CONVENIENCE OF THE GOVERNMENT (JAN. 1960). *(This clause is applicable only to contracts $10,000 and under)*. The Contracting Officer, by written notice, may terminate this contract, in whole or in part, when it is in the interest of the Government. If this contract is so terminated, the rights, duties and obligations of the parties hereto shall be in accordance with the applicable Sections of the Armed Services Procurement Regulation in effect on the date of this contract.

31. TERMINATION FOR CONVENIENCE OF THE GOVERNMENT (APR. 1966). *(This clause is applicable only to contracts in excess of $10,000)*. (a) The performance of

work under this contract may be terminated by the Government in accordance with this clause in whole, or from time to time in part, whenever the Contracting Officer shall determine that such termination is in the best interest of the Government. Any such termination shall be effected by delivery to the Architect-Engineer of a Notice of Termination specifying the extent to which performance of work under the contract is terminated, and the date upon which such termination becomes effective.

(b) After receipt of a Notice of Termination, and except as otherwise directed by the Contracting Officer, the Architect-Engineer shall:

(i) stop work under the contract on the date and to the extent specified in the Notice of Termination;

(ii) Place no further orders or subcontracts for materials, services or facilities, except as may be necessary for completion of such portion of the work under the contract as is not terminated;

(iii) terminate all orders and subcontracts to the extent that they relate to the performance of work terminated by the Notice of Termination;

(iv) assign to the Government, in the manner, at the times and to the extent directed by the Contracting Officer, all of the right, title, and interest of the Architect-Engineer under the orders and subcontracts so terminated, in which case the Government shall have the right, in its discretion, to settle or pay any or all claims arising out of the termination of such orders and subcontracts;

(v) settle all outstanding liabilities and all claims arising out of such termination of orders and subcontracts, with the approval or ratification of the Contracting Officer, to the extent he may require which approval or ratification shall be final for all the purposes of this clause;

(vi) transfer title and deliver to the Government, in the manner, at the times, and to the extent, if any, directed by the Contracting Officer, (A) the fabricated or unfabricated parts, work in process, completed work, supplies, and other material produced as a part of, or acquired in connection with the performance of, the work terminated by the Notice of Termination, and (B) the completed or partially completed plans, drawings, information, and other property which, if the contract had been completed, would have been required to be furnished to the Government.

(vii) use his best efforts to sell, in the manner, at the times, to the extent, and at the price or prices directed or authorized by the Contracting Officer, any property of the types referred to in (vi) above; provided, however, that the Architect-Engineer (A) shall not be required to extend credit to any purchaser, and (B) may acquire any such property under the conditions prescribed by and at a price or prices approved by the Contracting Officer; and provided further that the proceeds of any such transfer or disposition shall be applied in reduction of any payments to be made by the Government to the Architect-Engineer under this contract or shall otherwise be credited to the price of cost of the work covered by this contract or paid in such other manner as the Contracting Officer may direct;

(viii) complete performance of such part of the work as shall not have been terminated by the Notice of Termination; and

(ix) take such action as may be necessary, or as the Contracting Officer may direct, for the protection and preservation of the property related to this contract which is in the possession of the Architect-Engineer and in which the Government has or may acquire an interest.

At any time after expiration of the plant clearance period, as defined in Section VIII, Armed Services Procurement Regulation, as it may be amended from time to time, the Architect-Engineer may submit to the Contracting Officer a list, certified as to quantity and quality, of any or all items of termination inventory not previously disposed of, exclusive of items the disposition of which has been directed or authorized by the Contracting Officer, and may request the Government to remove such items or enter into a storage agreement covering them. Not later than fifteen (15) days thereafter, the Government will accept title to such items and remove them or enter into a storage agreement covering the same; provided, that the list submitted shall be subject to verification by the Contracting Officer upon removal of the items, or if the items are stored, within forty-five (45) days from the date of submission of the list, and any necessary adjustment to correct the list as submitted shall be made prior to final settlement.

(c) After receipt of a Notice of Termination, the Architect-Engineer shall submit to the Contracting Officer his termination claim, in the form and with certification prescribed by the Contracting Officer. Such claim shall be submitted promptly but in no event later than one year from the effective date of termination, unless one or more extensions in writing are granted by the Contracting Officer, upon request of the Architect-Engineer made in writing within such one year period or authorized extension thereof. However, if the Contracting Officer determines that the facts justify such action, he may receive and act upon any such termination claim at any time after such one year period or any extension thereof. Upon failure of the Architect-Engineer to submit his termination claim within the time allowed, the Contracting Officer may, subject to any Settlement Review Board approvals required by Section VIII of the Armed Services Procurement Regulation in effect as of the date of execution of this contract, determine on the basis of information available to him, the amount, if any, due to the Architect-Engineer by reason of the termination and shall thereupon pay to the Architect-Engineer the amount so determined.;

(d) Subject to the provisions of paragraph (c) and subject to any Settlement Review Board approvals required by Section VIII of the Armed Services Procurement Regulation in effect as of the date of execution of this contract, the Architect-Engineer and the Contracting Officer may agree upon the whole or any part of the amount or amounts to be paid to the Architect-Engineer by reason of the total or partial termination of work pursuant to this clause, which amount or amounts may include a reasonable allowance for profit on work done; provided, that such agreed amount or amounts exclusive of settlement costs, shall not exceed the total contract price as reduced by the amount of payments otherwise made and as further reduced by the contract price of work not terminated. The contract shall be amended accordingly, and the Architect-Engineer shall be paid the agreed amount. Nothing in paragraph (e) of this clause prescribing the amount to be paid to the Architect-Engineer in the event of failure of the Architect-Engineer and the Contracting Officer to agree upon the whole amount to be paid to the Architect-Engineer by reason of the termination of work pursuant to this clause, shall be deemed to limit, restrict, or otherwise determine or affect the amount or amounts which may be agreed upon to be paid to the Architect-Engineer pursuant to this paragraph (d).

(e) In the event of the failure of the Architect-Engineer and the Contracting Officer to agree as provided in paragraph (d), upon the whole amount to be paid to the Architect-Engineer by reason of the termination of work pursuant to this clause, the Contracting Officer shall, subject to any Settlement Review Board approvals required by Section VIII of the Armed Services Procurement Regulation in effect as of the date of execution of this contract, pay to the Architect-Engineer the amounts determined by the Contracting Officer as follows, but without duplication of any amounts agreed upon in accordance with paragraph (d):

(i) for completed work and services accepted by the Government, the price or prices specified in the contract for such work, less any payments previously made;

(ii) the total of—

(A) the costs incurred in the performance of the work and services terminated, including initial costs and preparatory expenses allocable thereto, but exclusive of any costs attributable to the work and services paid or to be paid for under paragraph (e) (i) hereof;

(B) the cost of settling and paying claims arising out of the termination of work or services under subcontracts or orders as provided in paragraph (b) (v) above, which are properly chargeable to the terminated portion of the contract (exclusive of amounts paid or payable on account of work or services delivered or furnished by subcontractors prior to the effective date of termination, which amounts shall be included in the costs payable under (A) above); and

(C) a sum, as profit on (A) above, determined by the Contracting Officer pursuant to 8–303 of the Armed Services Procurement Regulation, in effect as of the date of execution of this contract, to be fair and reasonable; *provided,* that if it appears that the Architect-Engineer would have sustained a loss on the entire contract had it been completed, no profit shall be included or allowed under this subdivision (C) and an appropriate adjustment shall be made reducing the amount of settlement to reflect the indicated rate of loss; and

(iii) the reasonable cost of the preservation and protection of property incurred pursuant to paragraph (b) (ix); and any other reasonable cost incidental to the termination of work under this contract, including expense incidental to the determination of the amount due to the Architect-Engineer as a result of the termination of work under this contract.

The total sum to be paid to the Architect-Engineer under (i) and (ii) above shall not exceed the total contract price as reduced by the amount of payments otherwise made and as further reduced by the contract price of work not terminated. Except for normal spoilage, and except to the extent that the Government shall have otherwise expressly assumed the risk of loss, there shall be excluded from the amounts payable to the Architect-Engineer under (ii) above, the fair value as determined by the Contracting Officer of property which is destroyed, lost, stolen, or damaged so as to become undeliverable to the Government, or a buyer pursuant to paragraph (b) (vii).

(f) Any determination of costs under paragraph (c) or (e) hereof shall be governed by the principles for consideration of costs set forth in Section XV, Part 2, of the Armed Services Procurement Regulation, as in effect on the date of this contract.

(g) The Architect-Engineer shall have the right of appeal, under the clause of this contract entitled "Disputes," from any determination made by the Contracting Officer under paragraph (c) or (e) above, except that if the Architect-Engineer has failed to submit his claim within the time provided in paragraph (c) above and has failed to request extension of such time, he shall have no such right of appeal. In any case where the Contracting Officer has made a determination of the amount due under paragraph (c) or (e) above, the Government shall pay to the Architect-Engineer the following: (i) if there is no right of appeal hereunder or if no timely appeal has been taken, the amount so determined by the Contracting Officer, or (ii) if an appeal has been taken, the amount finally determined on such appeal.

(h) In arriving at the amount due the Architect-Engineer under this clause there shall be deducted (i) all unliquidated advance or other payments on account theretofore made to the Architect-Engineer, applicable to the terminated portion of this contract, (ii) any claim which the Government may have against the Architect-Engineer in connection with his contract, and (iii) the agreed price for, or the proceeds of sale of, any materials, supplies, or other things acquired by the Architect-Engineer or sold, pursuant to the provisions of this clause, and not otherwise recovered by or credited to the Government.

(i) If the termination hereunder be partial, prior to the settlement of the terminated portion of this contract, the Architect-Engineer may file with the Contracting Officer a request in writing for an equitable adjustment of the price or prices specified in the contract relating to the contained portion of the contract (the portion not terminated by the Notice of Termination), and such equitable adjustment as may be agreed upon shall be made in such price or prices.

(j) The Government may from time to time, under such terms and conditions as it may prescribe, make partial payments and payments on account against costs incurred by the Architect-Engineer in connection with the terminated portion of this contract whenever in the opinion of the Contracting Officer the aggregate of such payments shall be within the amount to which the Architect-Engineer will be entitled hereunder. If the total of such payments is in excess of the amount finally agreed or determined to be due under this clause, such excess shall be payable by the Architect-Engineer to the Government upon demand, together with interest computed at the rate of 6 percent per annum, for the period from the date such excess payment is received by the Architect-Engineer to the date on which such excess is repaid to the Government; provided, however, that no interest shall be charged with respect to any such excess payment attributable to a reduction in the Architect-Engineer's claim by reason of retention or other disposition of termination inventory until ten days after the date of such retention or disposition, or such later date as determined by the Contracting Officer by reason of the circumstances.

(k) Unless otherwise provided for in this contract, or by applicable statute, the Architect-Engineer from the effective date of termination and for a period of three years after final settlement under this contract, shall preserve and make available to the Government at all reasonable times at the office of the Architect-Engineer but without direct charge to the Government, all his books, records, documents, and other evidence bearing on the costs and expenses of the Architect-Engineer under this contract and relating to the work terminated hereunder, or, to the extent approved by the Contracting Officer, photographs, micro-photographs, or other authentic reproductions thereof.

32. SUBCONTRACTORS AND OUTSIDE ASSOCIATES AND CONSULTANTS (JAN. 1965). Any subcontractors and outside associates or consultant required by the Architect-Engineer in connection with the services covered by the contract will be limited to such individuals or firms as were specifically identified and agreed to during negotiations. Any substitution in such subcontractors, associates, or consultants will be subject to the prior approval of the Contracting Officer.

33. COMPOSITION OF ARCHITECT-ENGINEER (JAN. 1965). If the Architect-Engineer hereunder is comprised of more than one legal entity, each such entity shall be jointly and severally liable hereunder.

34. AUDIT (SEP. 1964). (a) For purposes of verifying that cost or pricing data submitted, in conjunction with the negotiation of this contract or any contract change or other modification involving an amount in excess of $100,000 are accurate, complete, and current, the Contracting Officer, or his authorized representatives, shall–until the expiration of three years from the date of final payment under this contract–have the right to examine those books, records, documents and other supporting data which will permit adequate evaluation of the cost or pricing data submitted, along with the computations and projections used therein, which were available to the Architect-Engineer as of the date of execution of the Architect-Engineer's Certificate of Current Cost or Pricing Data.

(b) The Architect-Engineer agrees to insert the substance of this clause including this paragraph (b) in all subcontracts hereunder in excess of $100,000 so as to apply until three years after final payment under the subcontract, unless the price is based on adequate price competition, established catalog or market prices of commercial items sold in substantial quantities to the general public, or prices set by law or regulation. In each such excepted subcontract hereunder in excess of $100,000, the Architect-Engineer shall insert the substance of the following clause to apply until three years after final payment under the subcontract.

AUDIT-PRICE ADJUSTMENTS. (a) This clause shall become operative only with respect to any change or other modification made pursuant to one or more provisions of this contract which involves a price adjustment in excess of $100,000 unless the price adjustment is based on adequate price competition, established catalog or market prices of commercial items sold in substantial quantities to the general public, or prices set by law or regulation and further provided that such change or other modification to this contract must result from a change or other modification to the Government prime contract.

(b) For purposes of verifying that cost or pricing data submitted in conjunction with a contract change or modification involving an amount in excess of $100,000 are accurate, complete and current, the Contracting Officer or his authorized representative shall–until the expiration of three years from the data of final payment under this contract–have the right to examine those books, records, documents, and other supporting data which will permit adequate evaluation of the cost or pricing data submitted, along with the computations and projections used therein, which were available to the Contractor as of the date of execution of the Contractor's Certificate of Current Cost or Pricing Data.

(c) The Contractor agrees to insert the substance of this clause, including this paragraph (c), in all subcontracts hereunder in excess of $100,000 so as to apply until three years after final payment of the subcontract.

35. SUBCONTRACTOR COST AND PRICING DATA (SEP. 1964). (a) The Architect-Engineer shall require subcontractors hereunder to submit cost or pricing data under the following circumstances: (i) prior to award of any cost-reimbursement type, incentive, or price redeterminable subcontract; (ii) prior to the award of any subcontract the price of which is expected to exceed $100,000; (iii) prior to the pricing of any subcontract change or other modification for which the price adjustment is expected to exceed $100,000; except in the case of (ii) or (iii) where the price is based on adequate price competition, established catalog or market prices of commercial items sold in substantial quantities to the general public, or prices set by law or regulation.

(b) The Architect-Engineer shall require subcontractors to certify, in substantially the same form as that used in the certificate by the Architect-Engineer to the Government, that to the best of their knowledge and belief, the cost and pricing data submitted under (a) above is accurate, complete, and current as of the date of execution, which date shall be as close as possible to the date of agreement on the negotiated price of the subcontract.

(c) The Architect-Engineer shall insert the substance of this clause including this paragraph (c) in each of his cost-reimbursement type, price redeterminable, or incentive subcontracts hereunder, and in any other subcontract hereunder which exceeds $100,000 except where the price thereof is based on adequate price competition, established catalog or market prices of commercial items sold in substantial quantities to the general public, or prices set by law or regulation. In each such excepted subcontract hereunder in excess of $100,000, the Architect-Engineer shall insert the substance of the following clause:

SUBCONTRACTOR COST AND PRICING DATA-PRICE ADJUSTMENTS. (a) Paragraphs (b) and (c) of this clause shall become operative only with respect to any change or other modification made pursuant to one or more provisions of this contract which involves a price adjustment in excess of $100,000. The requirements of this clause shall be limited to such price adjustments.

(b) The Contractor shall require subcontractors hereunder to submit cost or pricing data under the following circumstances: (i) prior to award of any cost-reimbursement type, incentive or price redeterminable subcontract; (ii) prior to award of any subcontract, the price of which is expected to exceed $100,000; (iii) prior to the pricing of any subcontract change or other modification for which the price adjustment is expected to exceed $100,000; except, in the case of (ii) or (iii), where the price is based on adequate price competition, established catalog or market prices of commercial items sold in substantial quantities to the general public, or prices set by law or regulation.

(c) The Contractor shall require subcontractors to certify, in substantially the same form as that used in the certificate by the Prime Contractor to the Government, that to the best of their knowledge and belief the cost and pricing data submitted under (b) above is accurate, complete, and current as of the date of execution, which date shall be as close as possible to the date of agreement of the negotiated price of the contract modification,

(d) The Contractor shall insert the substance of this clause including this paragraph (d) in each subcontract hereunder which exceeds $100,000;

36. PRICE REDUCTION FOR DEFECTIVE COST OR PRICING DATA (SEP. 1964). (a) If the Contracting Officer determines that any price, including profit or fee, negotiated in connection with this contract was increased by any significant sums because the Architect-Engineer, or any subcontractor in connection with a subcontract covered by (c) below, furnished incomplete or inaccurate cost or pricing data or data not current as certified in the Architect-Engineer's Certificate of Current Cost or Pricing Data, then such price shall be reduced accordingly and the contract shall be modified in writing to reflect such adjustment.

(b) Failure to agree on a reduction shall be a dispute concerning a question of fact within the meaning of the "Disputes" clause of this contract.

(c) The Architect-Engineer agrees to insert the substance of paragraphs (a) and (c) of this clause in each of his cost-reimbursement type, price redeterminable, or incentive subcontracts hereunder, and in any other subcontract hereunder in excess of $100,000 unless the price is based on adequate price competition, established catalog or market prices of commercial items sold in substantial quantities to the general public, or prices set by law or regulation. In each such excepted subcontract hereunder which exceeds $100,000, the Architect-Engineer shall insert the substance of the following clause.

PRICE REDUCTION FOR DEFECTIVE COST OR PRICING DATA-PRICE AD-JUSTMENTS. (a) This clause shall become operative only with respect to any change or other modification made pursuant to one or more provisions of this contract which involves a price adjustment in excess of $100,000, except where the price is based on adequate price competition, established catalog or market prices of commercial items sold in substantial quantities to the general public, or prices set by law or regulation. The right to price reduction under this clause shall be limited to such price adjustments.

(b) If the Contractor determines that any price, including profit or fee, negotiated in connection with any price adjustment within the purview of paragraph (a) above was increased by any significant sum because the subcontractor or any of subcontractors in connection with a subcontract covered by paragraph (c) below, furnished incomplete or inaccurate cost or pricing data, or data not current as of the date of execution of the subcontractor's certificate of current cost or pricing data, then such price shall be reduced accordingly and the subcontract shall be modified in writing to reflect such adjustment.

(c) The subcontractor agrees to insert the substance of this clause in each subcontract hereunder which exceeds $100,000.

37. EQUAL OPPORTUNITY. (a) The "Equal Opportunity" clause in paragraph 24 is amended by deleting references to the President's Committee on Equal Employment Opportunity, Executive Order 10925 of March 6, 1961, as amended, and Section 303 of Executive Order No. 10925 of March 6, 1961, as amended, and substituting therfor the Secretary of Labor, Executive Order No. 11246 of September 24, 1965, and Section 204 of Executive Order 11246 of September 24, 1965, respectively.

(b) The Equal Opportunity representation in paragraph 24 is amended to insert after the reference to "Executive Order 10925" the following: "or the clause contained in Section 201 of Executive Order No. 11114."

(c) In accordance with regulations of the Secretary of Labor, the rules, regulations, orders, instructions, designations, and other directives issued by the President's Committee on Equal Employment Opportunity and those issued by the heads of various departments or agencies under or pursuant to any of the Executive Orders superseded by Executive Order 11246, shall, to the extent that they are not inconsistent with Executive Order 11246, remain in full force and effect unless and until revoked or superseded by appropriate authority. References in such directives to provisions of the superseded orders shall be deemed to be references to the comparable provisions of Executive Order 11246.

INDEX